Algorithms,
Architectures and
Information Systems
Security

Statistical Science and Interdisciplinary Research

Series Editor: Sankar K. Pal *(Indian Statistical Institute)*

Description:

In conjunction with the Platinum Jubilee celebrations of the Indian Statistical Institute, a series of books will be produced to cover various topics, such as Statistics and Mathematics, Computer Science, Machine Intelligence, Econometrics, other Physical Sciences, and Social and Natural Sciences. This series of edited volumes in the mentioned disciplines culminate mostly out of significant events — conferences, workshops and lectures — held at the ten branches and centers of ISI to commemorate the long history of the institute.

Vol. 1   Mathematical Programming and Game Theory for Decision Making
edited by S. K. Neogy, R. B. Bapat, A. K. Das & T. Parthasarathy
(Indian Statistical Institute, India)

Vol. 2   Advances in Intelligent Information Processing:
Tools and Applications
edited by B. Chandra & C. A. Murthy
(Indian Statistical Institute, India)

Vol. 3   Algorithms, Architectures and Information Systems Security
edited by Bhargab B. Bhattacharya, Susmita Sur-Kolay,
Subhas C. Nandy & Aditya Bagchi
(Indian Statistical Institute, India)

Platinum Jubilee Series

Statistical Science and
Interdisciplinary Research — Vol. 3

Algorithms,
Architectures and
Information Systems
Security

Editors

Bhargab B. Bhattacharya

Susmita Sur-Kolay

Subhas C. Nandy

Aditya Bagchi

Indian Statistical Institute, India

Series Editor: **Sankar K. Pal**

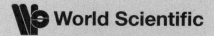

NEW JERSEY · LONDON · SINGAPORE · BEIJING · SHANGHAI · HONG KONG · TAIPEI · CHENNAI

Published by

World Scientific Publishing Co. Pte. Ltd.

5 Toh Tuck Link, Singapore 596224

USA office: 27 Warren Street, Suite 401-402, Hackensack, NJ 07601

UK office: 57 Shelton Street, Covent Garden, London WC2H 9HE

British Library Cataloguing-in-Publication Data
A catalogue record for this book is available from the British Library.

ALGORITHMS, ARCHITECTURES AND INFORMATION SYSTEMS SECURITY
Statistical Science and Interdisciplinary Research — Vol. 3

ISBN-13 978-981-283-623-6
ISBN-10 981-283-623-3

Printed in Singapore.

Foreword

The Indian Statistical Institute (ISI) was established on 17th December, 1931 by Prof. Prasanta Chandra Mahalanobis, a great visionary, to promote research in the theory and applications of statistics as a new scientific discipline in India. In 1959, Pandit Jawaharlal Nehru, the then Prime Minister of India introduced the ISI Act in the Parliament and designated it as an Institution of National Importance because of its remarkable achievements in statistical work as well as its contribution to economic planning for social welfare.

Today, the Indian Statistical Institute occupies a prestigious position in the academic firmament. It has been a haven for bright and talented academics working in a number of disciplines. Its research faculty has done India proud in the arenas of Statistics, Mathematics, Economics, Computer Science, among others. Over the last seventy five years, it has grown into a massive banyan tree, as epitomized in the emblem of the institute. The Institute now serves the nation as a unified and monolithic organization from different places, namely Kolkata, the Head Quarters, Delhi and Bangalore, two centers, a network of six SQC-OR Units located at Mumbai, Pune, Baroda, Hyderabad, Chennai and Coimbatore, and a branch (field station) at Giridih.

The platinum jubilee celebrations of ISI had been launched by Honorable Prime Minister Prof. Manmohan Singh on December 24, 2006, and the Govt. of India has declared 29th June as the "Statistics Day" to commemorate the birthday of Prof. Mahalanobis nationwide.

Prof. Mahalanobis was a great believer in interdisciplinary research, because he thought that this will promote the development of not only Statistics, but also the other natural and social sciences. To promote interdisciplinary research, major strides were made in the areas of computer science, statistical quality control, economics, biological and social sciences, physical and earth sciences.

The Institute's motto of 'unity in diversity' has been the guiding principle of all its activities since its inception. It highlights the unifying role of statistics in relation to various scientific activities.

v

In tune with this hallowed tradition, a comprehensive academic program, involving Nobel Laureates, Fellows of the Royal Society, and other dignitaries, has been implemented throughout the Platinum Jubilee year, highlighting the emerging areas of ongoing frontline research in its various scientific divisions, centres, and outlying units. It includes international and national-level seminars, symposia, conferences and workshops, as well as several special lectures. As an outcome of these events, the Institute is bringing out a series of comprehensive volumes in different subjects under the title Statistical Science and Interdisciplinary Research, published by the World Scientific Publishing, Singapore.

The present volume titled "Algorithms, Architectures, and Information Systems Security" is the third one in the series. It has sixteen chapters, written by eminent scientists from different parts of the world, dealing with three major topics of computer science. The first part of the book deals with computational geometric problems and related algorithms, which have several applications in areas like pattern recognition and computer vision, the second part addresses the issues of optimization in VLSI design and test architectures, and in wireless cellular networks, while the last part concerns with different problems, issues and methods of information systems security. I believe, the state-of-the art studies presented in this book will be very useful to the readers.

Thanks to the contributors for their excellent research articles and to volume editors Dr. B. B. Bhattacharya, Dr. S. Sur-Kolay, Dr. S. C. Nandy and Dr. A. Bagchi for their sincere effort in bringing out the volume nicely in time. Initial design of the cover by Mr. Indranil Dutta is acknowledged. Thanks are also due to World Scientific for their initiative in publishing the series and being a part of the Platinum Jubilee endeavor of the Institute. Sincere efforts by Prof. Dilip Saha and Dr. Barun Mukhopadhyay for editorial assistance are appreciated.

April 2008   Sankar K. Pal
Kolkata   Series Editor and Director

Preface

It is our great pleasure to compile the Platinum Jubilee Commemorative Monograph Series of the Indian Statistical Institute: Volume 3, titled Algorithms, Architectures, and Information Systems Security. This volume contains mostly a collection of invited papers from leading researchers. It also includes the extended versions of a few papers, which were presented at the Second International Conference on Information Systems Security (December 18–20, 2006), and in Track I of the International Conference on Computing: Theory and Applications (March 5–7, 2007), both held in Kolkata as part of the Platinum Jubilee celebration of the Institute (1931–2006).

There are sixteen chapters in this volume. The first five chapters (Chapters 1–5) address several challenging geometric problems and related algorithms. The next five chapters (Chapters 6–10) focus on various optimization issues in VLSI design and test architectures, and in wireless cellular networks. The last six chapters (Chapters 11–16) comprise scholarly articles on Information Systems Security.

Chapter 1 by Li and Klette presents two important rubberband algorithms for computing Euclidean shortest paths in a simple polygon, which have major applications in 2D pattern recognition, picture analysis, and in robotics. The second chapter by Cheng, Dey, and Levine contains the theoretical analysis of a Delaunay refinement algorithm for meshing various types of 3D domains such as polyhedra, smooth and piecewise smooth surfaces, volumes enclosed by them, and also non-manifold spaces. In Chapter 3, Pach and Tóth characterize the families of convex sets in a plane that are not representable by a point set of the same order type. Further, they establish the size of the largest subfamily representable by points and discuss related Ramsey-type geometric problems. The fourth chapter by Asano, Katoh, Mehlhorn, and Tokuyama describes efficient algorithms for some generalizations of least-squares method. These are useful in approximating a data set by a polyline with one joint that minimizes the total sum of squared vertical errors. A few other related geometric optimization problems have also been studied. Chapter 5 by Wei and Klette addresses the depth recovery problem from gradient

vii

vector fields. This has tremendous significance in 3D surface reconstruction and has several applications in computer vision. The authors present three schemes: a two-scan method, a Fourier-transform based method, and a wavelet-transform based method.

In Chapter 6, Börner, Leininger, and Gössel present a new design of a single-output convolutional compactor for guaranteed 6-bit error detection. In Electronic Design Automation, such detectors are of importance for compressing test and diagnostic data of large VLSI circuits. Bhattacharya, Seth, and Zhang address the problem of low-energy pattern generation for random testing VLSI chips in Chapter 7. The method suits well in scan-based systems, and reduces test application time significantly. Chapter 8 by Taghavi and Sarrafzadeh has a review of existing methodologies for estimation and reduction of routing congestion at the floorplanning and placement phases of VLSI design cycle, followed by a novel contribution on a more general and accurate approach. The ninth chapter by Sinha and Audhya deals with the channel assignment problem in a hexagonal cellular network with two-band buffering that supports multimedia services. New lower bounds on minimum bandwidth requirement are derived and algorithms for channel assignment are presented. Chapter 10 by Das, Das, and Nandy contains an extensive survey on range assignment problems in various types of wireless networks, and their computational geometric solutions.

Focusing on the emerging problems of privacy in the electronic society, Ardagna, Cremonini, Damiani, De Capitani di Vimercati, and Samarati have highlighted in Chapter 11, the issues related to the protection of personal data released in an open public network. This chapter considers the combination of different security policies and their enforcement against a laid down privacy policy or a possible privacy law. It also considers the protection of location information in location-based services. In Chapter 12, Chen and Atluri discuss a situational role-based access control and risk-based access control mechanism in a networked environment where personal data often kept with third parties, need stringent security measures to be relaxed only in case of an emergency. In Chapter 13, Jajodia and Noel propose a framework for Topological Vulnerability Analysis (TVA) of a network connecting individual components of a distributed system. It simulates the possible ways for incremental network penetration and builds complete maps of multi-step-attacks discovering all possible attack paths. TVA also computes network hardening options to protect critical resources against minimal network changes. Chapter 14 by Dash, Reddy, and Pujari presents a new malicious code detection technique using variable length n-grams based on the concept of episodes. The authors have pointed out that proper feature extraction and selection technique can help in efficiently detecting virus programs. The next chapter (Chapter 15) addresses an important area of research called digital image forensics, which stems from the need for creation, alteration and manipulation of digital images. Sencar and Memon provide an excellent survey of the recent developments covering image source identification, discrimination of synthetic images, and image forgery detection. The last chapter (Chapter 16) by Butler, Enck, Traynor, Plasterr, and McDaniel deals with privacy preserving web-based email. In spite of the privacy policies stipulated by the service providers of web-based applications, personal information of the users collected by them may have indefinite life and can later be used without restriction. The authors have proposed a method to create virtual channels over online services, through which messages and cryptographic keys are delivered for preserving privacy.

We take this opportunity to express our heartfelt gratitude to all the eminent contributors of this monograph on Algorithms, Architectures, and Information Systems Security. We are also grateful to Prof. Sankar K. Pal, Director of the Indian Statistical Institute, for his support and encouragement in preparing the volume. We earnestly hope that this collection of technical articles would be of archival value to the peer community. Finally, the help of Mr. Indranil Dutta to prepare the camera-ready version is gratefully acknowledged.

Bhargab B. Bhattacharya
Susmita Sur-Kolay
Subhas C. Nandy
Aditya Bagchi

Contents

Chapter 1

Euclidean Shortest Paths in a Simple Polygon

Fajie Li and Reinhard Klette

*Computer Science Department, The University of Auckland,
Auckland, New Zealand*

Let p and q be two points in a simple polygon Π. This chapter provides two rubberband algorithms for computing a shortest path between p and q that is contained in Π. The two algorithms use previously known results on triangular or trapezoidal decompositions of simple polygons, and have either $O(n)$ or $O(n \log n)$ time complexity (where the super-linear time complexity is only due to preprocessing, i.e. for the trapezoidal decomposition of the simple polygon Π).

Contents

1.1 Introduction

Algorithms for computing Euclidean shortest paths (ESPs) between two points p and q of a simple polygon Π, where the path is restricted to be fully contained in Π, have applications in two-dimensional (2D) pattern recognition, picture analysis, robotics, and so forth. They have been intensively studied.[1–4]

There is Chazelle's[5] linear-time algorithm for triangulating a simple polygon, or an easier to describe, but $O(n \log n)$ algorithm for partitioning a simple polygon into trapezoids.[6] The design of algorithms for calculating ESPs within a simple polygon may use one of both partitioning algorithms as a preprocess. This chapter shows how rubberband algorithms[7] may be used to calculate approximate or exact ESPs within simple polygons, using either decompositions into triangles or into trapezoids.

For a start we prove a basic property of exact ESPs for such cases; see also Ref. 8:

Proposition 1.1 *Each vertex* ($\neq p, q$) *of the shortest path is a vertex of* Π.

To see this, let $\rho = \langle p, p_1, p_2, \ldots, p_k, q \rangle$ be the shortest path from p to q completely contained in simple polygon Π. Assume that at least one $p_i \in \rho$ is not a vertex of Π. Also assume that each p_i is not *redundant*, which means that $p_{i-1} p_i p_{i+1}$ must be a triangle (i.e., three points p_{i-1}, p_i and p_{i+1} are not collinear), where $i = 1, 2, \ldots, k$ and $p_0 = p$, $p_{k+1} = q$.

Case 1: Non of the two edges $p_{i-1} p_i$ and $p_i p_{i+1}$ is on a tangent of Π (see Figure 1.1, left); then there exists a sufficiently small neighborhood of p_i, denoted by $U(p_i)$, such that for each point $p' \in U(p_i) \cap \triangle p_{i-1} p_i p_{i+1} \subset \Pi^{\bullet}$ (the topological closure of a simple polygon Π), both edges $p_{i-1} p_i$ and $p_i p_{i+1}$ are completely contained in Π. By elementary geometry, we have that $d_e(p_{i-1}, p') + d_e(p', p_{i+1}) < d_e(p_{i-1}, p_i) + d_e(p_i, p_{i+1})$, where d_e denotes Euclidean distance. Therefore we may obtain a shorter path from p to q by replacing p_i by p'. This is a contraction to the assumption that p_i is a vertex of the shortest path ρ.

Case 2: Both $p_{i-1} p_i$ and $p_i p_{i+1}$ are on tangents of Π (see Figure 1.1, middle); then we can also derive a contradiction. In fact, let p'_{i-1} and p'_{i+1} be the closest vertices of Π such that $p'_{i-1} p_i$ and $p_i p'_{i+1}$ are on tangents of Π. Analogous to the first case, there exists a point p' such that the polygonal path $p'_{i-1} p' p'_{i+1}$ is completely contained in $\Pi^{\bullet}$ and the length of $p'_{i-1} p' p'_{i+1}$ is shorter than $p'_{i-1} p_i p'_{i+1}$. This is a contradiction as well.

Case 3: Either $p_{i-1} p_i$ or $p_i p_{i+1}$ is a tangent of Π (see Figure 1.1, right); then we may arrive at the same result as in Case 2.

This chapter is organized as follows. At first we introduce into rubberband algorithms. Then we recall briefly decompositions of simple polygons and specify (as a preliminary result) two approximate rubberband algorithms; we provide examples of using them. These two algorithms are finally transformed into two exact rubberband algorithms; we analyze their correctness and time complexity.

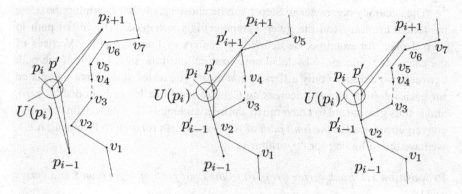

Fig. 1.1 Illustration that each vertex of a shortest path is a vertex of Π, where $v_1 v_2 v_3 v_4 v_5 \ldots$ is a polygonal part of the border of the simple polygon Π. Left, middle, right illustrate Cases 1, 2, 3 as discussed in the text, respectively.

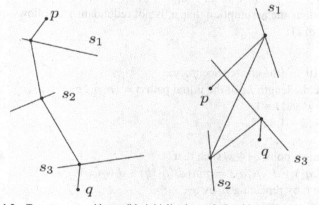

Fig. 1.2 Two step sets with possible initializations of Algorithm 1, both for $k = 3$.

1.2 Basics of Rubberband Algorithms

We explain basic ideas of a rubberband algorithm by using the following, very simple 2D example. In general, rubberband algorithms are for exact or approximate calculations of ESPs for 2D or 3D applications.[9]

Let Π be a plane. Assume that there are $k > 1$ line segments $s_i \subset \Pi$ (for $i = 1, 2, \ldots, k$) such that $s_i \cap s_j = \emptyset$, for $i \neq j$ and $i, j = 1, 2, \ldots, k$; see Figure 1.2. The following simple rubberband algorithm (see Figure 1.3) approximates a shortest path from p to q that intersects all the given segments s_i (at least once) in the given order.

The accuracy parameter in Step 1 can be chosen such that maximum possible numerical accuracy (on the given computer) is guaranteed. The initial path in Step 2 may, for example, be defined by centers of line segments. Vertices of the calculated path move by local optimization, until the total length of the path between two iterations only differs by ε at most. The series of lengths L calculated for each iteration forms a decreasing Cauchy sequence lower bounded by zero, and is thus guaranteed to converge to a minimum length. The path defined by this convergence is called *the limit path* of Algorithm 1. In relation to Proposition 1.1, we have the following for Algorithm 1:

Proposition 1.2 *Each vertex* ($\neq p, q$) *of the limit path of Algorithm 1 is a vertex of* Π.

Proof Let $\rho = \langle p, p_1, p_2, \ldots, p_k, q \rangle$ be the limit path from p to q of Algorithm 1. Let $i = 1, 2, \ldots,$ or k and $p_0 = p$, $p_{k+1} = q$. Assume that each $p_i \in \rho$ is not redundant. Then p_i must be an endpoint of s_i. (Otherwise, $p_i = p_{i-1}p_{i+1} \cap s_i$. This contradicts the assumption that p_i is not redundant.) It follows that p_i must be a vertex of Π. $\qquad\qquad\qquad\qquad\qquad\qquad\qquad\qquad\qquad\qquad\square$

1. Let $\varepsilon = 10^{-10}$ (the chosen accuracy).
2. Compute the length L_1 of the initial path $\rho = \langle p, p_1, p_2, \ldots, p_k, q \rangle$.
3. Let $q_1 = p$ and $i = 1$.
4. While $i < k - 1$ do:
4.1. Let $q_3 = p_{i+1}$.
4.2. Compute a point $q_2 \in s_i$ such that
$$d_e(q_1, q_2) + d_e(q_3, q_2) = \min\{d_e(q_1, q) + d_e(q_3, q) : q \in s_i\}.$$
4.3. Update ρ by replacing p_i by q_2.
4.4. Let $q_1 = p_i$ and $i = i + 1$.
5.1. Let $q_3 = q$.
5.2. Compute $q_2 \in s_k$ such that
$$d_e(q_1, q_2) + d_e(q_3, q_2) = \min\{d_e(q_1, q) + d_e(q_3, q) : q \in s_k\}.$$
5.3. Update ρ by replacing p_k by q_2.
6. Compute the length L_2 of the updated path $\rho = \langle p, p_1, p_2, \ldots, p_k, q \rangle$.
7. Let $\delta = L_1 - L_2$.
8. If $\delta > \varepsilon$, then let $L_1 = L_2$ and go to Step 3.
   Otherwise, stop.

Fig. 1.3   Algorithm 1: a simple rubberband algorithm for a given set of line segments.

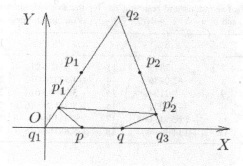

Fig. 1.4   Illustration of steps with joint endpoints.

The set $\{s_1, s_2, \ldots, s_k\}$ is a *step set* of a rubberband algorithm if its union contains all the vertices of the calculated path, and each s_i is a *step element* of the rubberband algorithm that contains at least one vertex of the calculated path, for $i = 1, 2, \ldots, k$.

In this chapter, step sets are sets of line segments, which may have joint endpoints, but cannot have further points in common. Furthermore, in this chapter, each step element contains exactly one vertex of the shortest path. For example, if the input for Algorithm 1 is as in Figure 1.4, with

$$s_1 = q_1 q_2, s_2 = q_2 q_3, q_1 = (0,0), q_2 = (2,4), q_3 = (3,0), p = (1,0), q = (2,0)$$

then we also have segments with joint endpoints. Assume a path initialization using p_1 and p_2, the centers of s_1 and s_2, respectively [i.e., $p_1 = (1,2)$, and $p_2 = (2.5,2)$]. We obtain that the length of the initialized polyline $\rho = \langle p, p_1, p_2, q \rangle$ is equal to 5.5616 (rounded to four digits). Algorithm 1 calculates an approximate shortest path $\rho = \langle p, p_1', p_2', q \rangle$ where $p_1' = (0.3646, 0.7291)$, $p_2' = (2.8636, 0.5455)$ and the length of it is equal to 4.4944 (see Table 1.1, which lists resulting δs for the number I of iterations). That means, Algorithm 1 is also able to deal with this input for the assumed initialization.

Table 1.1   Number I of iterations and resulting δs for the initialization illustrated by Figure 1.4 [i.e., with $p_1 = (1,2)$ and $p_2 = (2.5,2)$ as initial points on the path].

I	δ	I	δ	I	δ	I	δ
1	-0.8900	3	-0.0019	5	$-8.4435\mathrm{e}\text{-}008$	7	$-3.5740\mathrm{e}\text{-}012$
2	-0.1752	4	$-1.2935\mathrm{e}\text{-}005$	6	$-5.4930\mathrm{e}\text{-}010$		

Table 1.2   Number I of iterations and resulting δs, for the example shown in Figure 1.4, with $p_1 = (2-\delta', 2(2-\delta'))$ and $p_2 = (2+\delta', -4((2+\delta')-3))$ as initialization points and $\delta' = 2.221\text{e-}16$.

I	δ	I	δ	I	δ	I	δ
1	$-5.4831\text{e-}007$	7	-1.2313	13	$-7.0319\text{e-}010$	19	$8.8818\text{e-}016$
2	$-6.2779\text{e-}006$	8	-2.0286	14	$-4.5732\text{e-}012$	20	$8.8818\text{e-}016$
3	$-7.7817\text{e-}005$	9	-0.2104	15	$-3.0198\text{e-}014$	21	$-8.8818\text{e-}016$
4	$-9.6471\text{e-}004$	10	-0.0024	16	$-8.8818\text{e-}016$	22	$8.8818\text{e-}016$
5	-0.0119	11	$-1.6550\text{e-}005$	17	$8.8818\text{e-}016$	23	$-8.8818\text{e-}016$
6	-0.1430	12	$-1.0809\text{e-}007$	18	$-8.8818\text{e-}016$	24	0

However, if we assume a different initialization, such that $p_1 = p_2 = q_2$; in this case, Algorithm 1 will fail because the output of Step 4.2 in Algorithm 1 will be false: the calculated path equals $\rho = \langle p, p_1', p_2', q \rangle$, where $p_1' = q_2$ and $p_2' = q_2$, and its length equals 8.1231. (Referring to Lemma 16,[10] we see that $p_1 \neq p_0$ and p_2 in this example.)

We call a situation as in this initialization example a *degenerate path* within an application of a rubberband algorithm, and it may occur within initialization, or at a later iteration of the algorithm. In general, it is defined by the occurrence of at least two identical vertices of an initial or updated polygonal path. Such a degenerate case causes Step 4.2 in Algorithm 1 to fail.

A degenerate path can be dealt with approximately: we will not allow $p_2 = q_2$. To do so, we remove sufficiently small segments from both segments s_1 and s_2. The following shows how to handle such a degenerate case (for example) for the assumed data in Figure 1.4.

We modify the initial values of x_1 and x_2, and of y_1 and y_2 as follows:

$$\delta' = 2.221 \times 10^{-16} \text{ (for a reason, see below)}$$
$$x_1 = 2 - \delta' \quad \text{and} \quad y_1 = 2 \times x_1$$
$$x_2 = 2 + \delta' \quad \text{and} \quad y_2 = -4 \times (x_2 - 3)$$
$$p_1 = (x_1, y_1) \quad \text{and} \quad p_2 = (x_2, y_2)$$

Furthermore, let the accuracy be equals $\varepsilon = 1.0 \times 10^{-100}$. The length of the initialized polyline $\rho = \langle p, p_1, p_2, q \rangle$ is equal to 8.1231. Algorithm 1 will approximate a shortest path $\rho = \langle p, p_1', p_2', q \rangle$, where $p_1' = (0.3646, 0.7291)$ and $p_2' = (2.8636, 0.5455)$, and its length equals 4.4944 (see Table 1.2 for resulting δs in dependency of the number I of iterations).

Of course, if we leave the accuracy to be equals $\varepsilon = 1.0 \times 10^{-10}$ then the algorithm will stop sooner, after less iterations. – The algorithm was implemented on a Pentium 4 PC using Matlab 7.04. If we changed the value of δ' into $\delta' =$

2.22×10^{-16} then we obtained the same wrong result as that for identical points $p_1 = p_2 = q_2$. This is because the computer was not able to recognize a difference between x_1 and $x_1 \mp 2.22 \times 10^{-16}$. However, for practical applications, the value $\delta' = 2.221 \times 10^{-16}$ should be small or accurate enough in general (for this or a matching implementation environment).

With the example above we also illustrate that the approximate algorithm may be already *de facto an exact algorithm* if ε was chosen small enough (i.e., obtained result are accurate within the given numerical limits of the used implementation environment). But, later on, we even discuss (absolutely) exact algorithms.

1.3 Decompositions and Approximate ESPs

There are (at least) two ways of decomposing a simple polygon: into triangles[5] or trapezoids.[6] In the first case, Theorem 4.3[5] says that it is possible to compute a triangulation of a simple polygon in linear time (and the algorithm is "fairly complicated"). In the second case, Theorem 1[6] says that a given ("simple") algorithm for the decomposition into trapezoids has time complexity $O(n \log n)$, where n is the number of vertices of the original simple polygon Π.

Step sets can be defined by selecting edges of triangles or trapzoids of those decompositions.

1.3.1 *Triangulation*

Let Π be a simple polygon. Let $T_1 = \{\triangle_1, \triangle_2, \ldots, \triangle_m\}$ be such that $\Pi = \cup_{i=1}^m \triangle_i$ and $\triangle_i \cap \triangle_j = \emptyset$ or $= e_{ij}$, where e_{ij} is an edge of both triangles $\triangle_i$ and $\triangle_j$, $i \neq j$ and $i, j = 1, 2, \ldots, m$. We construct a corresponding simple graph $G = [V, E]$ where $V = \{v_1, v_2, \ldots, v_m\}$ and each edge $e \in E$ is defined as follows: If $\triangle_i \cap \triangle_j = e_{ij} \neq \emptyset$, then let $e = v_i v_j$ (where e_{ij} is an edge of both triangles $\triangle_i$ and $\triangle_j$); and if $\triangle_i \cap \triangle_j = \emptyset$, then there is not an edge between v_i and v_j, $i < j$ and $i, j = 1, 2, \ldots, m$. We say that G is a (*corresponding*) *graph* with respect to the triangulated simple polygon Π, denoted by G_Π.

Lemma 1.1 *For each triangulated simple polygon Π, its corresponding graph G_Π is a tree.*

Proof By contradiction. Suppose that G_Π is not a tree. Then there is a cycle $u_1 u_2 \cdots u_{m'} u_1$ in G_Π. Consequently, there are a sequence of triangles $\{\triangle'_1, \triangle'_2, \ldots, \triangle'_{m'}\} \subseteq T_1$ such that $\triangle'_i \cap \triangle'_j \neq \emptyset$, where $i \neq j$ and $i, j = 1, 2, \ldots, m'$. It follows that there is a polygonal curve $\rho = w_1 w_2 \cdots w_{m'} w_1 \subset \cup_{i=1}^{m'} \triangle'_i$. Since Π is a simple polygon, ρ can be contracted into a single point inside of Π. Note that

Input: the (original) tree T and two points $p', q' \in V(T)$.

Output: a unique path ρ from p' to q' in T.

1. Let $S_1 = \{v : d(v) = 1 \wedge v \in V(T)\} \setminus \{p', q'\}$.
2. If $S_1 = \emptyset$, stop (the current T is already a path from p' to q').
3. Otherwise, let $V_1 = \emptyset$.
4. Let the unique neighbor of $v \in S_1$ be n_v.
5. For each $v \in S_1$, do the following:
5.1. While $d(n_v) = 1$ do:
5.1.1. If $v = p'$ or q', then skip this while loop.
5.1.2. Otherwise, let $V_1 = V_1 \cup \{v\}$.
5.1.3. $v = n_v$
5.2. Update T by removing v from the set of neighbors of n_v.
5.3. Update T by removing V_1 from $V(T)$.
5.4. Let $V_1 = \emptyset$.
6. Goto Step 1.

Fig. 1.5   Procedure 1: step set calculation for a given triangulation.

we can find ρ such that there is a vertex of $\triangle'_1$, denoted by w, that is inside of the region enclosed by ρ. Therefore, w must be a redundant vertex. This contradicts to the fact that w is a vertex of Π. $\square$

Let T be a tree and $p \neq q$, $p, q \in V(T)$. The following procedure will compute a unique path from p to q in T. Although there exists a linear algorithm for computing the shortest path between two vertices in a positive integer weighted graph,[11] our procedure below is much simpler because here the graph is (just) a tree.

We apply Procedure 1 (see Figure 1.5) as follows: Let $T = G_\Pi$ and p', q' be the vertices of T corresponding to the triangle containing p, q, respectively. Let a sequence of triangles $\{\triangle'_1, \triangle'_2, \ldots, \triangle'_{m'}\}$ correspond to the vertices of the path calculated by Procedure 1. Let $\{e_1, e_2, \ldots, e_{m'-1}\}$ be a sequence of edges such that $e_i = \triangle_i \cap \triangle_{i+1}$, where $i = 1, 2, \ldots, m' - 1$. Let $\{e'_1, e'_2, \ldots, e'_{m'-1}\}$ be a sequence of edges such that e'_i is obtained by removing a sufficiently small segment (Assume that the length of the removed segment is δ'.) from both endpoints of e_i, where $i = 1, 2, \ldots, m' - 1$. Set $\{e'_1, e'_2, \ldots, e'_{m'-1}\}$ is the approximate step set we are looking for.

1. Apply Chazelle's algorithm to decompose Π into triangles.
2. Construct the corresponding graph with respect to the decomposed Π, denoted by G_Π.
3. Apply Procedure 1 to compute the unique path from p' to q', denoted by ρ.
4. Let $\delta' = \varepsilon$. Compute the step set from ρ, denoted by S, where removed segments have length δ'.
5. Let S, p and q as input, apply Algorithm 1 to compute the approximate ESP from p to q.

Fig. 1.6   Algorithm 2: approximate ESP after triangulation.

1.3.2 *Trapezoidal Decomposition*

Analogously to Section 1.3.1, let Π be a simple polygon, and let $T_2 = \{t_1, t_2, \ldots, t_m\}$ be such that $\Pi = \cup_{i=1}^m t_i$ and $t_i \cap t_j = \emptyset$ or e_{ij}, where e_{ij} is a part (a subset) of a joint edge of trapezoids t_i and t_j, $i \neq j$ and $i, j = 1, 2, \ldots, m$. We construct a corresponding simple graph $G = [V, E]$ where $V = \{v_1, v_2, \ldots, v_m\}$, and each edge $e \in E$ is defined as follows: If $t_i \cap t_j = e_{ij} \neq \emptyset$, then let $e = v_i v_j$ (where e_{ij} is a subset of a joint edge of trapezoids t_i and t_j); and if $t_i \cap t_j = \emptyset$, then there is not an edge between v_i and v_j, $i < j$ and $i, j = 1, 2, \ldots, m$. We say that G is a *(corresponding) graph* with respect to the trapezoidal decomposition of simple polygon Π, denoted by G_Π.

Analogously to Lemma 1.1, we also have the following

Lemma 1.2 *For each trapezoidal decomposition of a simple polygon Π, its corresponding graph G_Π is a tree.*

Following Section 1.3.1, we apply Procedure 1 as follows: Let $T = G_\Pi$ and p', q' be the vertices of T corresponding to the trapezoids containing p, q respectively. Let a sequence of trapezoids $\{t'_1, t'_2, \ldots, t'_{m'}\}$ correspond to the vertices of the path obtained by Procedure 1. Let $E' = \{e_1, e_2, \ldots, e_{m'-1}\}$ be a sequence of edges such that $e_i = t_i \cap t_{i+1}$, where $i = 1, 2, \ldots, m' - 1$. For each $i \in \{1, 2, \ldots, m' - 2\}$, if $e_i \cap e_{i+1} \neq \emptyset$, then update e_i and e_{i+1} in E' by removing sufficiently small segments from both sides of this intersection point. Then the updated set E' is the approximate step set.

1.3.3 *Two Approximate Algorithms*

Figures 1.6 and 1.7 show the main algorithms having decomposition, step set construction, and ESP approximation as their subprocedures. For Step 4, see the description following Lemma 1.1. For Step 5 note that the approximation is not due to Algorithm 1 but due to removing small segments of length δ'.

Modify Step 1 in Algorithm 2 as follows:
Apply a trapezoidal decomposition algorithm[6] to Π.

Fig. 1.7   Algorithm 3: approximate ESP after trapezoidal decomposition.

Table 1.3   Vertices of the simple polygon in Figure 1.8, where $p = (59,201)$ and $q = (707,382)$.

v_i	v_1	v_2	v_3	v_4	v_5	v_6	v_7	v_8	v_9	v_{10}	v_{11}	v_{12}	v_{13}	v_{14}
x_i	42	178	11	306	269	506	589	503	595	736	623	176	358	106
y_i	230	158	304	286	411	173	173	436	320	408	100	211	19	84

We illustrate Algorithms 2 and 3 by a few examples, using the simple polygon in Figure 1.8, with coordinates of vertices provided in Table 1.3.

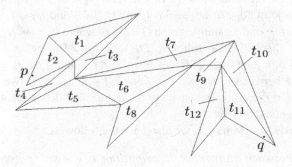

Fig. 1.8   A possible triangulation of a simple polygon.

After illustrating triangulation and Algorithm 2, we also illustrate decomposition into trapezoids and Algorithm 3.

Table 1.4   Vertices p_i calculated by Algorithm 2 for the simple polygon in Figure 1.8. The length of the path equals 1246.0330730004.

p_i	(x_i, y_i)	p_i	(x_i, y_i)
p_1	(177.9999999928, 157.9999999926)	p_6	(374.5899740372, 188.1320635957)
p_2	(178.000000018, 157.9999999861)	p_7	(506.0000000117, 172.9999999927)
p_3	(176.9605570407, 185.5452384224)	p_8	(589.0000000034, 172.9999999927)
p_4	(175.9999999835, 211.0000000093)	p_9	(589.0000000772, 173.0000001234)
p_5	(176.000000013, 211.0000000075)		

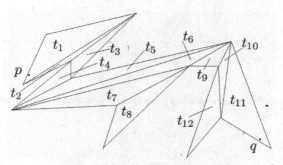

Fig. 1.9   Another possible triangulation of the simple polygon of Figure 1.8.

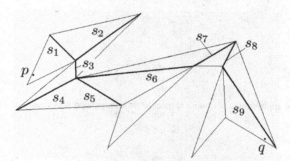

Fig. 1.10   The step set of the triangulation shown in Figure 1.8.

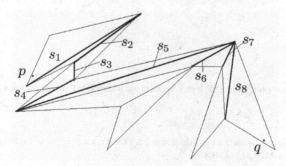

Fig. 1.11   The step set of the triangulation shown in Figure 1.9.

1.4 Improved and Exact Algorithms

We present an improved version of Algorithm 1.

For initialization, let p_i be the center of s_i; let a_i and b_i be the endpoints of s_i, l_i the line such that $s_i \subset l_i$, for $i = 1, 2, \ldots, k$. Now see Figure 1.19.

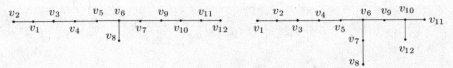

Fig. 1.12 Top (bottom): the corresponding graph (tree) with respect to the triangulated simple polygon in Figure 1.8 (Figure 1.9).

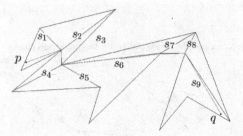

Fig. 1.13 The approximate ESP with respect to the triangulated simple polygon of Figure 1.8.

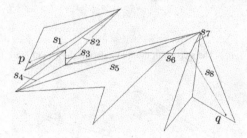

Fig. 1.14 The approximate ESP with respect to the triangulated simple polygon of Figure 1.9.

Table 1.5 Vertices p_i calculated by Algorithm 2 for the simple polygon in Figure 1.9. The length of the path equals 1323.510103408.

p_i	(x_i, y_i)	p_i	(x_i, y_i)
p_1	(123.3191615501, 175.7014459270)	p_5	(420.0869708340, 167.6376763887)
p_2	(178.000000018, 157.9999999861)	p_6	(510.0186257061, 170.4926523372)
p_3	(176.9605570407, 185.5452384224)	p_7	(589.0000000034, 172.9999999927)
p_4	(175.9999999835, 211.0000000093)	p_8	(609.1637118080, 208.7136929370)

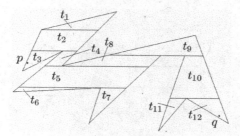

Fig. 1.15   A trapezoidal decomposition of the simple polygon of Figure 1.8.

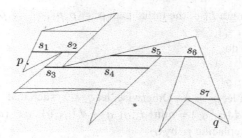

Fig. 1.16   The step set of those trapezoids in Figure 1.15.

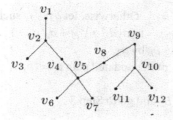

Fig. 1.17   Corresponding graph with respect to the trapezoidal decomposition in Figure 1.15.

Table 1.6   Vertices p_i calculated by Algorithm 3 for the simple polygon in Figure 1.15. The length of the path equals 1356.7016610946.

p_i	(x_i, y_i)	p_i	(x_i, y_i)
p_1	(170.9999999999, 149)	p_5	(504, 161)
p_2	(171.0000000001, 149)	p_6	(584, 161)
p_3	(171.9999999999, 202)	p_7	(669.1611374407582, 312)
p_4	(172.0000000001, 202)		

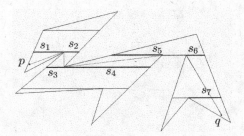

Fig. 1.18   The approximate ESP with respect to the trapezoidal decomposition in Figure 1.15.

1. Compute the length L_1 of the initial path $\rho = \langle p, p_1, p_2, \ldots, p_k, q \rangle$.
2. Let $q_1 = p$ and $i = 1$.
3. While $i < k - 1$ do:
3.1. Let $q_3 = p_{i+1}$.
3.2. Let $q'_2 = l_i \cap q_1 q_3$.
3.3. If $q'_2 \in s_i$ then let $q_2 = q'_2$. Otherwise, let $q_2 \in s_i$ such that
　　　$d_e(q_1, q_2) + d_e(q_3, q_2) = \min\{d_e(q_1, q) + d_e(q_3, q) : q \in \{a_i, b_i\} \cap V(\Pi)\}$.
3.4. Update ρ by replacing p_i by q_2.
3.5. Let $q_1 = p_i$ and $i = i + 1$.
4.1. Let $q_3 = q$.
4.2. Let $q'_2 = l_k \cap q_1 q_3$.
4.3. If $q'_2 \in s_k$ then let $q_2 = q'_2$. Otherwise, let $q_2 \in s_k$ such that
　　　$d_e(q_1, q_2) + d_e(q_3, q_2) = \min\{d_e(q_1, q) + d_e(q_3, q) : q \in \{a_k, b_k\} \cap V(\Pi)\}$.
4.4. Update ρ by replacing p_k by q_2.
5. Compute the length L_2 of the updated path $\rho = \langle p, p_1, p_2, \ldots, p_k, q \rangle$.
6. Let $\delta = L_1 - L_2$.
7. If $\delta > 0$, then let $L_1 = L_2$ and go to Step 3.
　　Otherwise, stop.

Fig. 1.19   Algorithm 1*: this allows to turn Algorithms 2 and 3 into exact algorithms.

We replace "Algorithm 1" by "Algorithm 1*" in Step 5 of Algorithm 2 and 3. Obviously, if Algorithm 1* provides an exact solution for any step set, then Algorithms 2 and 3 are provide exact ESPs.

1.4.1 *Proofs of Correctness*

In this subsection we present two versions of proofs to show that Algorithm 1 is correct for any sequence of disjointed segments. The first one is longer but leads to a stronger result: we not only prove that the algorithm is correct but also show that the ESP is unique. The second one is very short but without proving the uniqueness of the ESP.

We start with introducing a few definitions used in those proofs. Some of them are from mathematical analysis or multivariable calculus or from elementary topology textbook.

Definition 1.1 An *iteration* of Algorithm 1 is a complete pass through its loop. At the end of iteration $n \geq 1$ we obtain the nth *approximate ESP*, denoted by $AESP_n(S)$, for a given sequence of segments $S = \{s_1, s_2, \ldots, s_k\}$.

We assume that the sequence of the nth approximate ESPs is converging towards a polygonal path; let

$$AESP(S) = \lim_{n \to \infty} AESP_n(S)$$

be this polygonal path.

Let $p_i(t_{i_0})$ be the i-th vertex of the $AESP(S)$, for $i = 1, 2, \ldots, k$. Parameter $t_{i_0} \in [0,1]$ identifies the ith vertex of $AESP(S)$ that is on the segment s_i. Let $p_0 = p$, $p_{k+1} = q$, and

$$d_i = d_e(p_{i-1}, p_i) + d_e(p_i, p_{i+1})$$

for $i = 1, 2, \ldots,$ or k. Let

$$d(t_1, t_2, \ldots, t_k) = \sum_{i=1}^{k} d_i$$

Obviously, $d(t_1, t_2, \ldots, t_k)$ is an k-ary function on the domain $[0,1]^k$.

Let $p_i \in s_i$, for $i = 1, 2, \ldots, k$. We call the k tuple $(p_1, p_2, \ldots, p_k)$ a *point tuple* of S. We call it an *AESP critical point tuple* of S if it is the set of the vertices of the *AESP* of S.

Now let $P = (p_1, p_2, \ldots, p_k)$ be an *AESP* critical point tuple of S. Using P as an initial point set, defining $AESP_0(S)$, and n iterations of Algorithm 1, we get another critical point tuple of S, say $P' = (p'_1, p'_2, \ldots, p'_k)$, which defines (see above) the nth approximate polygonal path $AESP_n(S)$, or $AESP_n$ for short.

Definition 1.2 Let

$$\frac{\partial d(t_1, t_2, \ldots, t_k)}{\partial t_i}\bigg|_{t_i = t_{i0}} = 0$$

where $i = 1, 2, \ldots,$ or k. Then we say that $(t_{10}, t_{20}, \ldots, t_{k0})$ is a *critical point* of

$$d(t_1, t_2, \ldots, t_k)$$

Let $P = (p_1, p_2, \ldots, p_k)$ be a critical point tuple of S. Using P as an initial point set, n iterations of the Algorithm 1, we calculate an *n-rubberband transform* of P, denoted by $P \rightarrow_{rb_n} Q$, or $P \rightarrow Q$ for short, where Q is the resulting critical point tuple of S, and n is a positive integer.

Let $P = (p_1, p_2, \ldots, p_k)$ be a critical point tuple of S. For sufficiently small real $\varepsilon > 0$, the set

$$\{(p'_1, p'_2, \ldots, p'_k) : \ x'_i \in (x_i - \varepsilon, x_i + \varepsilon) \ \wedge \ y'_i \in (y_i - \varepsilon, y_i + \varepsilon)$$
$$\wedge \ p'_i = (x'_i, y'_i)$$
$$\wedge \ p_i = (x_i, y_i) \ \wedge \ i = 1, 2, \ldots, k\}$$

is the *ε-neighborhood* of P, denoted by $U_\varepsilon(P)$.

The ε-neighborhood of P is an open set in the Euclidean k-dimensional topological space $(\mathbb{R}^k, T)$; T is the topology, that means the family of all open sets in $\mathbb{R}^k$. We also use the following definition [see Definition 4.1 in Ref. 12]:

Definition 1.3 Let $Y \subset X$, where (X, T) is a topological space. Let T' be the family of sets defined as follows: A set W belongs to T' iff there is a set $U \in T$ such that $W = Y \cap U$. The family T' is called *the relativization of T to Y*, denoted by $T|_Y$.

1.4.2 *A Proof Without Using Convex Analysis*

We express a point

$$p_i(t_i) = (x_i + k_{x_i}t_i, y_i + k_{y_i}t_i)$$

on s_i in general form, with $t_i \in [0, 1]$, where $i = 1, 2, \ldots,$ or k. In the following, $p_i(t_i)$ will also be denoted by p_i for short, where $i = 1, 2, \ldots,$ or k.

The following is a multivariable version of Fermat's Theorem in mathematical analysis[13] (see Theorem 8.8.1). We will use it for proving Lemma 1.3; this lemma is then applied in the proofs of Lemmas 1.4 and Theorem 1.2.

Theorem 1.1 (Fermat's Theorem) *Let $f = f(t_1, t_2, \ldots, t_k)$ be a real-valued function defined on an open set U in $\mathbb{R}^k$. Let $C = (t_{10}, t_{20}, \ldots, t_{k0})$ be a point of U. Suppose that f is differentiable at C. If f has a local extremum at C, then*

$$\frac{\partial f}{\partial t_i} = 0$$

where $i = 1, 2, \ldots, k$.

Let $p_i(t_{i_0})$ be i-th vertex of an *AESP*, where $i = 1, 2, \ldots, k$. Then we have the following:

Lemma 1.3 $(t_{10}, t_{20}, \ldots, t_{k0})$ *is a critical point of* $d(t_1, t_2, \ldots, t_k)$.

Proof $d(t_1, t_2, \ldots, t_k)$ is differentiable at each point

$$(t_1, t_2, \ldots, t_k) \in [0, 1]^k$$

Because $AESP_n(S)$ is the nth polygonal path of S, where $n = 1, 2, \ldots$, and

$$AESP = \lim_{n \to \infty} AESP_n(S)$$

it follows that $d(t_{10}, t_{20}, \ldots, t_{k0})$ is a local minimum of $d(t_1, t_2, \ldots, t_k)$. By Theorem 1.1,

$$\frac{\partial d}{\partial t_i} = 0$$

where $i = 1, 2, \ldots, k$. By Definition 1.2, $(t_{10}, t_{20}, \ldots, t_{k0})$ is a critical point of $d(t_1, t_2, \ldots, t_k)$. $\qquad \square$

By Lemmas 1.2 and 1.3, we have the following:

Lemma 1.4 *Any sequence S of pairwise disjoint segments has only a finite number of AESP critical point tuples.*

This is our first important lemma in this subsection. In the rest of this subsection, based on Lemma 1.4, we show a much stronger result: S has actually only one (!) *AESP* critical point tuple.

Let $p_i = (p_{i_1}, p_{i_2})$ be on s_i, for $i = 1, 2, 3$. The proof of the following lemma specifies an explicit expression for the relation between parameter t and the optimum point p_2.

Lemma 1.5 *Optimum point $p_2 \in s_2$, defined by*

$$d_e(p_2, p_1) + d_e(p_2, p_3) = \min\{p'_2 : d_e(p'_2, p_1) + d_e(p'_2, p_3) \wedge p'_2 \in s_2\}$$

can be computed in $O(1)$ time.

Proof Let the two endpoints of s_2 be $a_2 = (a_{2_1}, a_{2_2})$ and $b_2 = (b_{2_1}, b_{2_2})$. Let $p_1 = (p_{1_1}, p_{1_2})$. Point p_2 can be written as

$$(a_{2_1} + (b_{2_1} - a_{2_1})t, a_{2_2} + (b_{2_2} - a_{2_2})t)$$

The formula

$$d_e(p_2, p_1) = \sqrt{\sum_{i=1}^{2} [(a_{2_i} - p_{1_i}) + (b_{2_i} - a_{2_i})t]^2}$$

can be simplified: We can rotate the coordinate system such that s_2 is parallel to one of the two coordinate axes. It follows that only one element of the set

$$\{b_{2_i} - a_{2_i} : i = 1, 2\}$$

is equal to a real number $\alpha \neq 0$, and the other is equal to 0. Without loss of generality we can assume that

$$d_e(p_2, p_1) = \sqrt{(\alpha t + A_1)^2 + B_1}$$

where A_1 and B_1 are functions of a_{2_i}, b_{2_i} and p_{1_i}, for $i = 1, 2$. – Analogously,

$$d_e(p_2, p_3) = \sqrt{(\alpha t + A_2)^2 + B_2}$$

where A_2 and B_2 are functions of a_{2_i}, b_{2_i} and p_{3_i}, for $i = 1, 2$. In order to find a point $p_2 \in s_2$ such that

$$d_e(p_2, p_1) + d_e(p_2, p_3) = \min\{p_2' : d_e(p_2', p_1) + d_e(p_2', p_3), p_2 \in s_2\}$$

we can solve the equation

$$\frac{\partial(d_e(p_2, p_1) + d_e(p_2, p_3))}{\partial t} = 0$$

The unique solution is

$$t = -1/\alpha \times (A_1 B_2 + A_2 B_1)/(B_2 + B_1)$$

This proves the lemma.   □

By the proof of Lemma 1.5, assuming the representation

$$p_i = (a_{i_1} + (b_{i_1} - a_{i_1})t_i, a_{i_2} + (b_{i_2} - a_{i_2})t_i)$$

we have defined a function f, $t_2 = f(t_1, t_3)$, for which we have the following:

Lemma 1.6 *The function $t_2 = f(t_1, t_3)$ is continuous at each tuple $(t_1, t_3) \in [0, 1]^2$.*

This is used to prove the following:

Lemma 1.7 *If $P \to_{rb_1} Q$, then for every sufficiently small real $\varepsilon > 0$, there is a sufficiently small real $\delta > 0$ such that $P' \in U_\delta(P)$ and $P' \to_{rb_1} Q'$ implies $Q' \in U_\varepsilon(Q)$.*

Proof By Lemma 1.5 and note that S has k segments; thus we use Lemma 1.6 repeatedly k times, and this proves this lemma.   □

By Lemma 1.7, we have the following:

Lemma 1.8 *If $P \to_{rb_n} Q$, then, for every sufficiently small real $\varepsilon > 0$, there is a sufficiently small real $\delta_\varepsilon > 0$ and a sufficiently large integer N_ε, such that $P' \in U_{\delta_\varepsilon}(P)$ and $P' \to_{rb_{n'}} Q'$ implies $Q' \in U_\varepsilon(Q)$, where n' is an integer and $n' > N_\varepsilon$.*

This lemma is used to prove Lemma 1.12; the latter one and the following three lemmas are then finally applied to prove the second important lemma (i.e., Lemma 1.13) in this section. Lemmas 1.13 and 1.3 imply then the main theorem (i.e., Theorem 1.2 below) of this section.

By Lemma 1.4, let $Q_1, Q_2, \ldots, Q_N$ with $N \geq 1$ be the set of all *AESP* critical point tuples of S. Let ε be a sufficiently small positive real such that

$$U_\varepsilon(Q_i) \cap U_\varepsilon(Q_j) = \emptyset$$

for $i, j = 1, 2, \ldots, N$ and $i \neq j$. Let

$$D_i = \{P : P \to Q' \wedge Q' \in U_\varepsilon(Q_i) \wedge P \in [0,1]^k\}$$

for $i = 1, 2, \ldots, N$.

The statements in the following two lemmas are obvious:

Lemma 1.9 *If $N > 1$ then $D_i \cap D_j = \emptyset$, for $i, j = 1, 2, \ldots, N$ and $i \neq j$.*

Lemma 1.10 $\bigcup_{i=1}^{N} D_i = [0,1]^k$

We consider the Euclidean topology T on $\mathbb{R}^k$, and its relativization $T = \mathbb{R}^k|_{[0,1]^k}$.

Lemma 1.11 *D_i is an open set of T, where $i = 1, 2, \ldots, N$ with $N \geq 1$.*

Proof By Lemma 1.8, for each $P \in D_i$, there is a sufficiently small real $\delta_P > 0$ such that

$$U_{\delta_P}(P) \subseteq D_i$$

So we have

$$\bigcup_{P \in D_i} U_{\delta_P}(P) \subseteq D_i$$

On the other hand, for $P \in U_{\delta_P}(P)$, we have

$$D_i = \cup P \subseteq \bigcup_{P \in D_i} U_{\delta_P}(P)$$

Note that $U_{\delta_P}(P)$ is an open set of T. Thus,

$$D_i = \bigcup_{P \in D_i} U_{\delta_P}(P)$$

is an open set of T. $\qquad\qquad\square$

The following basic lemma is characterizing open sets in general[14] (see Proposition 5.1.4).

Lemma 1.12 *Let $U \subset \mathbb{R}$ be an arbitrary open set. Then there are countably many pairwise disjoint open intervals U_n such that $U = \cup U_n$.*

Now we are prepared to approach the second important lemma in this subsection:

Lemma 1.13 *S has a unique AESP critical point tuple.*

Proof By contradiction. Suppose that $Q_1, Q_2, \ldots, Q_N$ with $N > 1$ are all the *AESP* critical point tuples of S. Then there exists $i \in \{1, 2, \ldots, N\}$ such that

$$D_i|_{s_j} \subset [0, 1]$$

where s_j is a segment in S, for $i, j = 1, 2, \ldots, N$. Otherwise we have

$$D_1 = D_2 = \cdots = D_N$$

This is a contradiction to Lemma 1.9.

Let

$$E = \{s_j : D_i|_{s_j} \subseteq [0, 1]\}$$

where s_j is any segment in S. We can select a critical point tuple of S as follows: go through each $s \in \{s_1, s_2, \ldots, s_k\}$. If $e \in E$, by Lemmas 1.11 and 1.12, select the minimum left endpoint of the open intervals whose union is $D_i|_s$. Otherwise select the midpoint of s. We denote the resulting critical point tuple as

$$P = (p_1, p_2, \ldots, p_k)$$

By the selection of P, we know that P is not in D_i. By Lemma 1.10, there is a $j \in \{1, 2, \ldots, N\} - \{i\}$ such that $P \in D_j$. Therefore, there is a sufficiently small real $\delta > 0$ such that $U_\delta(P) \subset D_j$. Again by the selection of P, there is a sufficiently small real $\delta' > 0$ such that $U'_\delta(P) \cap D_i \neq \emptyset$. Let $\delta'' = \min\{\delta, \delta'\}$. Then we have $U''_\delta(P) \subset D_j$ and $U''_\delta(P) \cap D_i \neq \emptyset$. This implies that $D_i \cap D_j \neq \emptyset$, and this is a contradiction to Lemma 1.9. $\qquad\square$

Let S be a sequence of pairwise disjoint segments. Let $AESP_n(S)$ be the nth approximate polygonal path of S, for $n = 1, 2, \ldots$. The subsection has shown that

$$AESP = \lim_{n \to \infty} AESP_n(S)$$

exists, and we can conclude the following main result of this section:

Theorem 1.2 *The AESP of S is the ESP of S, or, in short AESP = ESP.*

Proof By Lemma 1.13 and the proof of Lemma 1.3, $d(t_1, t_2, \ldots, t_k)$ has a unique local minimal value. This implies that the *AESP* of S is the ESP of S. $\square$

1.4.3 *A Shorter Proof by Using Convex Analysis*

This subsection gives a shorter proof of the correctness of Algorithm 1^* by applying some basic results from convex analysis (but without obtaining the uniqueness result for the ESP). We cite a few basic results of convex analysis:[15–17]

Proposition 1.3 (Ref. 15, page 27) *Each line segment is a convex set.*

Proposition 1.4 (Ref. 15, page 72) *Each norm on $\mathbb{R}^n$ is a convex function.*

Proposition 1.5 (Ref. 15, page 79) *A nonnegative weighted sum of convex functions is a convex function.*

Theorem 1.3 (Ref. 17, Theorem 3.5) *Let S_1 and S_2 be convex sets in $\mathbb{R}^m$ and $\mathbb{R}^n$, respectively. Then*

$$\{(x, y) : x \in S_1 \wedge y \in S_2\}$$

is a convex set in $\mathbb{R}^{m+n}$, where $m, n \in \mathbb{N}$.

Proposition 1.6 (Ref. 17, page 264) *Let f be a convex function. If x is a point where f has a finite local minimum, then x is a point where f has its global minimum.*

By Proposition 1.3, the interval $[0, 1]$ is a convex set. By Theorem 1.3, $[0, 1]^k$ is a convex set. For any $p, q \in \mathbb{R}^n$, $d_e(p, q)$ is a norm (see, for example Ref. 18, page 78). By Proposition 1.4 and 1.5, $d(t_1, t_2, \ldots, t_k)$ (see Section 1.4.2) is a convex function on $[0, 1]^k$. Since $d(t_1, t_2, \ldots, t_k)$ is continuous on $[0, 1]^k$, so its minimum is attained. It is clear that, for any sequence of pairwise disjoint segments S, Algorithm 1 will always produce an exact local minimum of the function $d(t_1, t_2, \ldots, t_k)$. By Proposition 1.6, each local minimum of $d(t_1, t_2, \ldots, t_k)$ is its global minimum. Therefore, we have proved Theorem 1.2 once again.

To recall, we proved that Algorithm 1* is correct for the family of sequences of pairwise disjoint segments, and that there is unique ESP for a given sequence of pairwise disjoint segments.

Although the proof in Subsection 1.4.2 is much more complicated than the one in Subsection 1.4.3, we proved a stronger result there, namely, that for each sequence of pairwise disjoint segments, Algorithm 1* will converge to a unique ESP.

See also Ref. 19 Lemma 1, Ref. 20 Lemma 3.3, and Ref. 21 Lemma 1, for proofs of the uniqueness of an ESP. Our proof is actually also completely suitable for the "curve case", where $p = q$.

1.4.4 *Computational Complexity*

Lemma 1.14 *Algorithm 1* can be computed in $\kappa(\varepsilon) \cdot O(k)$ time, where $\kappa(\varepsilon) = (L - L_0)/\varepsilon$, L be the true length of the ESP of S, L_0 that of an initial polygonal path, and k is the number of segments of the set S.*

Proof Let L_n be the true length of the polygonal path after n iterations. We slightly modify Algorithm 1* as follows:[a]

For each iteration, we update the vertices with odd indices first and then update those with even indices later (i.e., for each iteration, we update the following vertices p_1, p_3, p_5, ..., then the following vertices p_2, p_4, p_6, ...

Thus, $\{L_n\}_{n \to \infty}$ is a strict decreasing sequence with lower bound 0, since $L_0 - L$ can be written as $ak + b$ (i.e., it is a linear function of k), where a, b are constants such that $a \neq 0$. Because Algorithm 1* will not stop if $L_n - L_{n+1} > \varepsilon$ (see Step 8, Algorithm 1*), it follows that $L_n - L_{n+1}$ will also depend on k. Again, since $L_n - L_{n+1}$ can be written as $ck + d$, where c and d are constants such that $c \neq 0$. Then we have that

$$\lim_{k \to \infty} \frac{ak + b}{ck + d} = \frac{a}{c}$$

Therefore, Algorithm 1* will stop after at most $\lceil a/(c\varepsilon) \rceil$ iterations. (Note that, if we would not modify Algorithm 1, then it stops after at most

$$\lceil (L_0 - L)/\varepsilon \rceil$$

iterations.)

Thus, by Lemma 1.5, the time complexity of the original rubberband algorithm equals $\lceil (L_0 - L)/(\varepsilon) \rceil \cdot O(k) = \kappa(\varepsilon) \cdot O(k)$, where $\kappa(\varepsilon) = (L - L_0)/\varepsilon$, L be the true

---

[a]This is just for the purpose of time complexity analysis. By experience, Algorithm 1* runs faster without such a modification.

length of the ESP of S, L_0 that of an initial polygonal path, and k is the number of segments of the set S.   □

1.5 Conclusions

This chapter provided two exact algorithms for calculating ESPs in simple polygons. Depending on the used preprocessing step (triangular or trapezoidal decomposition), they are either linear time or $O(n \log n)$. But note that the trapezoidal decomposition algorithm[6] is substantially simpler than the triangulation algorithm.[5] The chapter illustrates that rubberband algorithms are of simple design, easy to implement, and can be used to solve ESP problems not only approximate but also in an exact way.

References

1. L. Guibas, J. Hershberger, D. Leven, M. Sharir, and R. E. Tarjan, Linear-time algorithms for visibility and shortest path problems inside triangulated simple polygons, *Algorithmica* **2** (1987) 209–233.
2. L. Guibas and J. Hershberger, Optimal shortest path queries in a simple polygon, *J. Computer System Sciences* **39** (1989) 126–152.
3. J. Hershberger, A new data structure for shortest path queries in a simple polygon, *Information Processing Letters* **38** (1991) 231–235.
4. J. S. B. Mitchell, Geometric shortest paths and network optimization. In J.-R. Sack, J. Urrutia, eds., *Handbook of Computational Geometry* (Elsevier Science Publishers, 2000), pp. 633–701.
5. B. Chazelle, Triangulating a simple polygon in linear time, *Discrete Computational Geometry* **6** (1991) 485–524.
6. F. Li and R. Klette, Decomposing a simple polygon into trapezoids, in *Proc. Computer Analysis Images Patterns* (Vienna, Springer, Berlin, 2007).
7. T. Bülow and R. Klette, Digital curves in 3D space and a linear-time length estimation algorithm, *IEEE Trans. Pattern Analysis Machine Intelligence* **24** (2002) 962–970.
8. D. T. Lee and F. P. Preparata, Euclidean shortest paths in the presence of rectilinear barriers, *Networks* **14** (1984) 393–410.
9. F. Li and R. Klette, Rubberband algorithms for solving various 2D or 3D shortest path problems, in *Proc. Computing: Theory and Applications*, Platinum Jubilee Conference of The Indian Statistical Institute (IEEE, 2007), pp. 9–18.
10. F. Li and R. Klette, Exact and approximate algorithms for the calculation of shortest paths. Report 2141 on www.ima.umn.edu/preprints/oct2006, 2006.
11. M. Thorup, Undirected single-source shortest paths with positive integer weights in linear time, *J. ACM* **3** (1999) 362–394.
12. T. O. Moore, *Elementary General Topology* (Prentice-Hall, Englewood Cliffs, N.J., 1964).
13. S. A. Douglass, *Introduction to Mathematical Analysis* (Addison-Wesley, 1996).

14. B. G. Wachsmuth, *Interactive Real Analysis*. See http://www.shu.edu/projects/reals/index.html, 2007.
15. S. Boyd and L. Vandenberghe, *Convex Optimization* (Cambridge University Press, Cambridge, UK, 2004).
16. A. W. Roberts and V. D. Varberg, *Convex Functions* (Academic Press, New York, 1973).
17. R. T. Rockafellar, *Convex Analysis* (Princeton University Press, Princeton, N.J., 1970).
18. R. Klette and A. Rosenfeld, *Digital Geometry: Geometric Methods for Digital Picture Analysis* (Morgan Kaufmann, San Francisco, 2004).
19. J. Choi, J. Sellen, and C.-K. Yap, Precision-sensitive Euclidean shortest path in 3-space, in *Proc. Annu. ACM Sympos. Computational Geometry* (1995), pp. 350–359.
20. M. Sharir and A. Schorr, On shortest paths in polyhedral spaces, *SIAM J. Comput.* **15** (1986) 193–215.
21. C.-K. Yap, Towards exact geometric computation, *Computational Geometry: Theory Applications* **7** (1997) 3–23.

Chapter 2

Theory of a Practical Delaunay Meshing Algorithm
for a Large Class of Domains

Siu-Wing Cheng[*], Tamal K. Dey[†] and Joshua Levine[†]

*Department of Computer Science and Engineering, HKUST, Hong Kong
†Department of Computer Science and Engineering, Ohio State University, USA

Recently a Delaunay refinement algorithm has been proposed that can mesh domains as general as piecewise smooth complexes. These domains include polyhedra, smooth and piecewise smooth surfaces, volumes enclosed by them, and above all non-manifold spaces. The algorithm is guaranteed to capture the input topology at the expense of four tests, some of which are computationally intensive and hard to implement. The goal of this paper is to present the theory that justifies a refinement algorithm with a single disk test in place of four tests of the previous algorithm.

The algorithm is supplied with a resolution parameter that controls the level of refinement. We prove that, when the resolution is fine enough (this level is reached very fast in practice), the output mesh becomes homeomorphic to the input while preserving all input features. Moreover, regardless of the refinement level, each k-manifold element in the input complex is meshed with a triangulated k-manifold. Boundary incidences among elements maintain the input structure. Implementation results reported in a companion paper corroborate our claims.

Contents

2.1 Introduction

Delaunay meshing of geometric domains is sought in a number of applications in science and engineering. Since its introduction by Chew,[1] the Delaunay refinement technique has been growing in its application to meshing geometric domains. Starting with polygonal and polyhedral complexes[2-6] where quality of elements were of prime concern, the technique has been extended to smooth domains where topology preservation is a major issue.[7-11] Non-smooth curved domains, the next in the order of difficulty, pose some fundamental obstacles to Delaunay refinement on which research results have started to appear.

There are two main challenges faced in extending Delaunay refinement to non-smooth domains. First, the sampling theory developed for smooth surfaces[12] is not applicable to non-smooth surfaces. Secondly, as in the polyhedral case,[13-16] small input angles possibly present at non-smooth regions pose problems for the termination of Delaunay refinement. Boissonnat and Oudot[17] successfully extended their algorithm for smooth surfaces to non-smooth ones but failed to admit small input angles. Cheng, Dey, and Ramos[18] removed this constraint on input angles enabling their algorithm to work on a large class of domains called piecewise smooth complex (PSC). This class includes polyhedral domains, smooth and piecewise smooth surfaces, volumes enclosed by them, and even non-manifolds. This algorithm protects non-smooth curves and vertices in the input complex with balls that are turned into weighted points during refinement stage. Staying away from non-smooth regions, the algorithm can afford to admit arbitrary small input angles. Notwithstanding its theoretical success, practical validity of the algorithm remains questionable since it employs costly computations during iterative refinement. The goal of this paper is to develop the theory further so that a simpler refinement strategy can be devised making it viable in practice.

The refinement procedure of Cheng, Dey and Ramos[18] performs four tests to guarantee topology preservation, namely (i) a Voronoi edge does not intersect the domain more than once, (ii) normals on the curves and surface patches do not vary beyond a threshold within Voronoi cells, (iii) no Delaunay edge in the restricted triangulation (defined later) connect vertices across different patches, and (iv) the restricted Delaunay triangles incident to points in a patch make a topological disk. After collecting restricted triangles, that is, triangles dual to the Voronoi edges intersecting the input domain, tests (i) and (iv) are only combinatorial. However, test (ii) is quite expensive and is a major obstacle in making the algorithm practical. Also, test (iii) requires recognizing intersections of Voronoi facets with the domain which is computationally harder than recognizing Voronoi edge-domain intersections.

We replace four tests with a single test that checks a topological disk condition similar to (iv). As long as the restricted triangles incident to each point do not form a topological disk, the refinement routine samples a new point from the domain. We argue that this procedure terminates. We prove the following guarantees. The algorithm is supplied with an input resolution parameter that determines how refined the output mesh should be. We prove that, regardless of the refinement level, a *k*-manifold in the input is meshed with a simplicial *k*-manifold. Boundary incidences among different manifold elements in the input complex are maintained in the output. Furthermore, if the refinement is sufficiently dense, homeomorphism between input and output is reached. More importantly, this refinement level is achieved very fast in practice giving us a provable practical algorithm for Delaunay meshing for a vast array of domains. We report the practical results in detail in a companion paper[19] (a sample is in Figure 2.1).

Fig. 2.1   Meshed PSCs, METABALL (Smooth), PART (Manifold PSC), and WEDGE (Non-manifold, PSC with small angles). Top row: surface mesh, bottom row: volume mesh.

2.2 Preliminaries

2.2.1 *Domain*

Throughout this paper, we assume a generic intersection property that a k-manifold $\sigma \subset \mathbb{R}^3$, $0 \leqslant k \leqslant 3$, and a j-manifold $\sigma' \subset \mathbb{R}^3$, $0 \leqslant j \leqslant 3$, intersect (if at all) in a $(k+j-3)$-manifold if $\sigma \not\subset \sigma'$ and $\sigma' \not\subset \sigma$. We will use both geometric and topological versions of closed balls. A *geometric* closed ball centered at point $x \in \mathbb{R}^3$ with radius $r > 0$, is denoted as $B(x,r)$. We use $\operatorname{int}\mathbb{X}$ and $\operatorname{bd}\mathbb{X}$ to denote the interior and boundary of a topological space $\mathbb{X}$, respectively.

The domain $\mathcal{D}$ is a piecewise smooth complex (PSC) where each element is a compact subset of a smooth (C^2) k-manifold, $0 \leqslant k \leqslant 3$. Each element is closed and hence contains its boundaries (possibly empty). We use $\mathcal{D}_k$ to denote the subset of all k-dimensional elements, the kth stratum. $\mathcal{D}_0$ is a set of *vertices*; $\mathcal{D}_1$ is a set of curves called *1-faces*; $\mathcal{D}_2$ is a set of surface patches called *2-faces*; $\mathcal{D}_3$ is a set of volumes called *3-faces*. For $1 \leqslant k \leqslant 2$, we use $\mathcal{D}_{\leqslant k}$ to denote $\mathcal{D}_0 \cup \ldots \cup \mathcal{D}_k$.

The domain $\mathcal{D}$ satisfies the usual proper requirements for being a complex: (i) interiors of the elements are pairwise disjoint and for any $\sigma \in \mathcal{D}$, $\operatorname{bd}\sigma \subset \mathcal{D}$; (ii) for any $\sigma, \sigma' \in \mathcal{D}$, either $\sigma \cap \sigma' = \emptyset$ or $\sigma \cap \sigma'$ is a union of cells in $\mathcal{D}$. We use $|\mathcal{D}|$ to denote the underlying space of $\mathcal{D}$. For $0 \leqslant k \leqslant 3$, we also use $|\mathcal{D}_k|$ to denote the underlying space of $\mathcal{D}_k$.

For any point x on a 2-face σ, we use $n_\sigma(x)$ to denote a unit outward normal to the surface of σ at x. For any point x on a 1-face σ, $n_\sigma(x)$ denotes a unit oriented tangent to the curve of σ at x. (We assume a consistent orientation of the tangents.)

2.2.2 *Complexes*

We will be dealing with weighted points and their Delaunay and Voronoi diagram. A weighted point p is represented as a ball $\hat{p} = B(p, w_p)$. The squared weighted distance of any point $x \in \mathbb{R}^3$ from $\hat{p}$ is given by $\|x - p\|^2 - w_p^2$. Under this distance metric, one can define weighted versions of Delaunay and Voronoi diagram. For a weighted point set $S \subset \mathbb{R}^3$, let $\operatorname{Vor}S$ and $\operatorname{Del}S$ denote the weighted Voronoi and Delaunay diagrams of S respectively. Each diagram is a cell complex where each k-face is a k-polytope in $\operatorname{Vor}S$ and is a k-simplex in $\operatorname{Del}S$. Each k-simplex ξ in $\operatorname{Del}S$ is dual to a $(3-k)$-face V_ξ in $\operatorname{Vor}S$ and vice versa.

Let S be a point set sampled from $|\mathcal{D}|$. For any sub-collection $\mathbb{X} \subset \mathcal{D}$ we define $\operatorname{Del}S|_{\mathbb{X}}$ to be the Delaunay subcomplex restricted to $\mathbb{X}$; i.e., each simplex $\xi \in \operatorname{Del}S|_{\mathbb{X}}$, called a restricted simplex, is the dual of a Voronoi face V_ξ where $V_\xi|_{\mathbb{X}} = V_\xi \cap |\mathbb{X}| \neq \emptyset$. By this definition, for any $\sigma \in \mathcal{D}$, $\operatorname{Del}S|_\sigma$ denote the Delaunay subcomplex restricted to σ and

$$\text{Del}S|_{\mathcal{D}_i} = \bigcup_{\sigma \in \mathcal{D}_i} \text{Del}S|_\sigma, \quad \text{Del}S|_{\mathcal{D}} = \bigcup_{\sigma \in \mathcal{D}} \text{Del}S|_\sigma.$$

An i-face $\sigma \in \mathcal{D}_i$ should be meshed with i-simplices. However, $\text{Del}S|_\sigma$ may have lower dimensional simplices not incident to any restricted i-simplex. Therefore, we compute special sub-complexes of restricted complexes. For $\sigma \in \mathcal{D}_i$, let $\text{Skl}^i S|_\sigma$ denote the following i-dimensional subcomplex of $\text{Del}S|_\sigma$:

$$\text{Skl}^i S|_\sigma = closure\{t \mid t \in \text{Del}S|_\sigma \text{ is an } i\text{-simplex}\}.$$

We extend the definition to strata:

$$\text{Skl}^i S|_{\mathcal{D}_i} = \bigcup_{\sigma \in \mathcal{D}_i} \text{Skl}^i S|_\sigma.$$

2.2.3 *Refinement*

Our strategy is to run Delaunay refinement with only the *disk condition* formulated as follows. See Figure 2.2 for more explanations.

Let p be a point on a 2-face σ. Let $\text{Umb}_{\mathcal{D}}(p)$ and $\text{Umb}_\sigma(p)$ be the set of triangles in $\text{Skl}^2 S|_{\mathcal{D}_2}$ and $\text{Skl}^2 S|_\sigma$ respectively which are incident to p. The following disk condition is used for refinement. Once the restricted Delaunay triangles are collected, this check is only combinatorial.

Disk_Condition(p) : (i) $\text{Umb}_{\mathcal{D}}(p) = \bigcup_{\sigma \ni p} \text{Umb}_\sigma(p)$, (ii) for each $\sigma \in \mathcal{D}_2$ containing p, underlying space of $\text{Umb}_\sigma(p)$ is a 2-disk which has all vertices in σ. Point p is in the interior of this 2-disk if and only if $p \in \text{int}\,\sigma$. Also, if p is in bd$\,\sigma$, it is not connected to any other point on $\mathcal{D}_1$ which is not adjacent to it.

2.3 Protection

The neighborhoods of the curves and vertices in $\mathcal{D}_{\leqslant 1}$ are regions of potential problems for Delaunay refinements. First, if the elements incident to these curves and vertices make small angles at the points of incidence, usual Delaunay refinement may not terminate. Second, these curves and vertices represent 'features' in the input which should be preserved in the output for many applications. Usual Delaunay refinement may destroy these features.[17,21] To overcome these problems Cheng et al.[18] protect elements in $\mathcal{D}_{\leqslant 1}$ with balls. We will not repeat the algorithm for computing these protecting balls here but mention only some of the key properties these balls satisfy.

PROTECTION PROPERTIES: Let $\omega \leqslant 0.076$ be a positive constant and $\mathcal{B}_p$ denote the protecting ball of a point p.

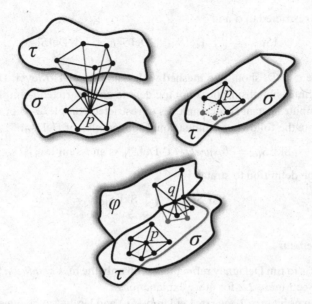

Fig. 2.2   Disk condition: (left):point $p \in \sigma$ has a disk in σ and another disk in $\tau \neq \sigma$ violating condition (i) (middle): point $p \in \sigma$ has a topological disk but some of its vertices (lightly shaded) belong to τ violating condition (ii), (right): Points p and q satisfy disk condition. Point p, an interior point in σ, lies in the interior of its disk in σ. Point q, a boundary point, has three disks for each of the three 2-faces.

(1) Any two adjacent balls on a 1-face overlap significantly without containing each other's centers.
(2) No three balls have a common intersection.
(3) Let $p \in \sigma$ be the center of a protecting ball. Further, let $B = B(p,R)$ be a ball with radius R and center p where $R \leqslant c\,\mathrm{radius}(\mathcal{B}_p)$ for some $c \leqslant 8$.
   (a) For $\tau = \sigma$ or any 2-face incident to σ, $\angle n_\tau(p), n_\tau(z) \leqslant 2\omega$ for any $z \in B \cap \tau$.
   (b) B intersects σ in a single open curve and any 2-face incident to σ in a topological disk.

In practice, balls satisfying (1) and (2) can be computed easily. If they are chosen small enough (3) is satisfied as well. After computing the protecting balls, each of them is turned into a weighted vertex. That is, for each protecting ball $\mathcal{B}_p$, we obtain the weighted point (p, w_p), where $w_p = \mathrm{radius}(\mathcal{B}_p)$. For technical reasons that will be clear later, we need to ensure that each 2-face is intersected by some Voronoi edge in the Voronoi diagram Vor S of the current point set. The weighted vertices ensure it for 2-faces that have boundaries. For 2-faces without boundary, initially we place three weighted points satisfying the protection properties.

After protection the meshing algorithm inserts points for further Delaunay refinement. These points are not weighted. Also, the refinement step never attempts to insert a point in the interior of any of the protecting balls. This is because no Voronoi point (equidistant from three or more points) can lie inside any protecting ball since no three of them have a common intersection. In essence, our algorithm maintains a point set S with the following two properties : (i) S contains all weighted points placed during protection phase, and (ii) other points in S are unweighted and they lie outside the protecting balls. We call such a point set *admissible*. The following Lemma proved in Cheng et al.[18] is an important consequence of the protection properties.

Lemma 2.1 *Let S be an admissible point set. Let p and q be adjacent weighted vertices on a 1-face σ. Let σ_{pq} denote the curve segment between p and q. V_{pq} is the only Voronoi facet in $\mathrm{Vor}\,S$ that intersects σ_{pq}, and V_{pq} intersects σ_{pq} exactly once.*

2.4 Algorithm

During refinement the size of the restricted triangles determines the location of the point to be inserted. For any triangle $t \in \mathrm{Skl}^2 S|_\sigma$, define $\mathrm{size}(t,\sigma)$ to be the maximum weighted distance between the vertices of t and points in $V_t|_\sigma$. Notice that if all vertices of t are unweighted, the maximum weighted distance is just the maximum Euclidean distance.

When we mesh volumes, we use the standard technique of inserting circumcenters of tetrahedra that have radius-edge ratio (denoted $\rho()$) greater than a threshold, $\rho_0 \geqslant 1$. If the insertion of the circumcenter threatens to delete any triangle in $\mathrm{Skl}^2 S|_{\mathcal{D}_2}$, the circumcenter is not inserted. In this case we say that the triangle is *encroached* by the circumcenter. Essentially, this strategy allows refining most of the tetrahedra except the ones near boundary. The following pseudo-code summarizes our algorithm.

DelPSC $(\mathcal{D}, \lambda, \rho_0)$

(1) PROTECTION. Protect elements in $\mathcal{D}_{\leqslant 1}$ with weighted points. Insert three weighted points in each element of $\mathcal{D}_2$ that has no boundary. Let S be the current admissible point set.
(2) Mesh2Complex.

    (a) Let (p,σ) be any tuple where $p \in \mathrm{Skl}^2 S|_\sigma$. If **Disk_Condition**$(p)$ is violated, find the triangle $t \in \mathrm{Umb}_{\mathcal{D}}(p)$ that maximizes $\mathrm{size}(t,\sigma)$ over all σ

containing p and insert $x \in V_t|_\sigma$ that realizes size(t,σ) into S. Go to step 2(c).

(b) If size$(t,\sigma) > \lambda$ for some tuple (t,σ), where $t \in \mathrm{Skl}^2 S|_\sigma$, insert $x \in V_t|_{\mathcal{D}}$ that realizes size(t,σ) into S.

(c) Update Del$S$ and Vor$S$.

(d) If S has grown in the last execution of step 2, repeat step 2.

(3) Mesh3Complex. For any tuple (t,σ) where t is a tetrahedron in $\mathrm{Skl}^3 S|_\sigma$

(a) If $\rho(t) > \rho_0$ insert the orthocenter of the Delaunay ball (orthoball) of t into S if it does not encroach any triangle in $\mathrm{Skl}^2 S|_{\mathcal{D}}$ or any ball $B(p,2r)$ where $B(p,r)$ is a protecting ball.

(b) Update Del$S$ and Vor$S$.

(c) If S has grown in the last execution of step 3, repeat step 3.

(4) Return $\bigcup_i \mathrm{Skl}^i S|_{\mathcal{D}}$.

2.4.1 *Guarantees*

The analysis of the algorithm establishes two main facts: (i) the algorithm terminates, (ii) at termination the output mesh satisfies properties T1-T3:

(T1) For each $\sigma \in \mathcal{D}_1$, $\mathrm{Skl}^1 S|_\sigma$ is homeomorphic to σ and two vertices are joined by an edge in $\mathrm{Skl}^1 S|_\sigma$ if and only if these two vertices are adjacent on σ.

(T2) For $0 \leqslant i \leqslant 2$ and $\sigma \in \mathcal{D}_i$, $\mathrm{Skl}^i S|_\sigma$ is a i-manifold with vertices only in σ. Further, bd$\mathrm{Skl}^i S|_\sigma = \mathrm{Skl}^{i-1} S|_{\mathrm{bd}\sigma}$. For $i = 3$, the statement is true only if the set $\mathrm{Skl}^i S|_\sigma$ is not empty at the end of Mesh2Complex.

(T3) There exists a $\lambda > 0$ so that the output mesh of DelPSC$(\mathcal{D},\lambda,\rho_0)$ is homeomorphic to $\mathcal{D}$. Further, this homeomorphism respects stratification with vertex restrictions, that is, for $0 \leqslant i \leqslant 3$, $\mathrm{Skl}^i S|_\sigma$ is homeomorphic to $\sigma \in \mathcal{D}_i$ where bd$\mathrm{Skl}^i S|_\sigma = \mathrm{Skl}^{i-1} S|_{\mathrm{bd}\sigma}$ and vertices of $\mathrm{Skl}^i S|_\sigma$ lie in σ.

2.5 Termination

We prove that during refinement DelPSC maintains a positive distance between each inserted point and all other existing points. Then the compactness of $\mathcal{D}$ allows the standard packing argument to claim termination. We need some results from sampling theory.[8,12,22] Let $\Sigma \subset \mathbb{R}^3$ be a smooth (C^2-smooth) closed surface. The local feature size $f(x)$ at a point $x \in \Sigma$ is its distance to the medial axis of Σ.

Lemma 2.2 ([8,12,22]) *Let $\varepsilon \in (0, 1/3)$ be some constant.*

(i) *For any two points x and y in Σ such that $\|x - y\| \leqslant \varepsilon f(x)$,*

   (a) *the angle between the surface normals at x and y is at most $\varepsilon/(1 - 3\varepsilon)$;*
   (b) *the angle between xy and the surface normal at x is at least $\arccos(\varepsilon/2)$.*

(ii) *Let pqr be a triangle with vertices on Σ and circumradius no more than $\varepsilon f(p)$. The angle between the normal of pqr and the surface normal at p is less than 7ε.*

We use Lemma 2.2 to prove the next result. This result says that if restricted triangles incident to a point in a 2-face σ have small sizes, V_p intersects σ nicely. Figure 2.3 explains more about its implications.

Lemma 2.3 *Let $p \in S$ be a point on a 2-face σ. Let σ_p be the connected component in $V_p|_\sigma$ containing p. There exists a constant $\lambda > 0$ so that following holds:*

If some edge of V_p intersects σ and $\text{size}(t, \sigma) < \lambda$ for each triangle $t \in \text{Skl}^2 S|_\sigma$ incident to p, then

   (i) *there is no 2-face τ where $p \notin \tau$ and τ intersects a Voronoi edge of V_p;*
   (ii) *$\sigma_p = V_p \cap B \cap \sigma$ where $B = B(p, 2\lambda)$ if p is unweighted and $B = B(p, 2\text{radius}(\mathcal{B}_p) + 2\lambda)$ otherwise;*
   (iii) *σ_p is a 2-disk;*
   (iv) *any edge of V_p intersects σ_p at most once;*
   (v) *any facet of V_p intersects σ_p in an empty set or an open curve.*

Proof Observe that, because of protection, p has a positive minimum distance to any 2-face τ not containing it. Therefore, any Voronoi edge of V_p intersecting τ violates the size condition when λ is sufficiently small. This proves (i).

Consider p is unweighted. For any facet F of V_p, we use H_F to denote the plane of F. Assume that λ is less than $1/32$ the local feature size of the surface of σ at p. Let $B = B(p, 2\lambda)$. It is known that $B \cap \sigma$ is a 2-disk.

First, we claim that for any facet F of V_p, if H_F intersects $B \cap \sigma$, then the angle between the normal to σ at p and the plane of F is at most $\arcsin(1/16)$, that is, $\angle n_\sigma(p), H_F \leqslant \arcsin(1/16)$ and both $H_F \cap B \cap \sigma$ and $F \cap B \cap \sigma$ contains no closed curve. The dual Delaunay edge pq of F has length at most 4λ, which is less than $1/8$ the local feature size. By Lemma 2.2(ib), $\angle n_\sigma(p), H_F = \pi/2 - \angle n_\sigma(p), pq \leqslant \arcsin(1/16)$. There is no closed curve in $H_F \cap B \cap \sigma$ because such a closed curve would bound a 2-disk in $B \cap \sigma$, which would contain a point x such that $\angle n_\sigma(x), H_F = \pi/2$. This is a contradiction because $\angle n_\sigma(x), H_F \leqslant \angle n_\sigma(x), n_\sigma(p) + \angle n_\sigma(p), H_F \leqslant 1/13 + \arcsin(1/16) < \pi/2$ by Lemma 2.2(ia).

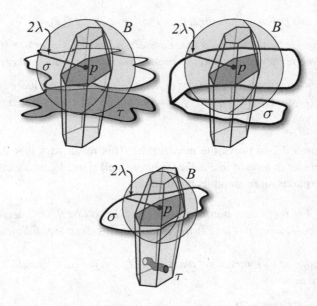

Fig. 2.3   (left): A 2-face τ where $p \notin \tau$ intersects some edge of V_p. This is not possible according to Lemma 2.3(i), (middle): also not possible since there is another component of σ within $B \cap V_p$ other than σ_p, (right): Within B, σ intersects V_p in a topological disk. It is possible that there is a different component (τ) which does not intersect any Voronoi edge and hence does not contribute any dual restricted triangle incident to p.

Since $H_F \cap B \cap \sigma$ contains no closed curve, neither does $F \cap B \cap \sigma$. This proves the claim.

Second, we claim that for any facet F of V_p, if H_F is within a distance of λ from p, $H_F \cap B \cap \sigma$ is a single open curve. Consider the disk $H_F \cap B$. Let $\vec{d}$ be the projection of $n_\sigma(p)$ onto H_F. Let $L \subset H_F$ be the line through the center of $H_F \cap B$ orthogonal to $\vec{d}$. Let x be any point in $H_F \cap B \cap \sigma$. The angle between px and the tangent plane at p is at most $\arcsin(1/32)$ by Lemma 2.2(ib). We already proved that $\angle n_\sigma(p), H_F \leqslant \arcsin(1/16)$. So the distance between x and L is less than $\|p - x\| \sin(2\arcsin(1/16)) \leqslant 2\lambda \sin(2\arcsin(1/16)) < 0.25\lambda$. Let $L^* \subset H_F$ be the strip of points at distance 0.25λ or less from L. Since radius of B is 2λ and H_F is at most λ distance from p, the radius of $H_F \cap B$ is at least $\sqrt{3}\lambda$. It follows that the boundary of $H_F \cap B$ intersects L^* in two disjoint circular arcs. We already proved that there is no closed curve in $H_F \cap B \cap \sigma$. It can be shown that if $H_F \cap B \cap \sigma$ contains two open curves, one of the curves, say C, must have both endpoints on the same arc in $H_F \cap B \cap L^*$. The radius of $H_F \cap B$ is at least $\sqrt{3}\lambda$. So some tangent to C must make an angle at most $\arcsin(0.25/\sqrt{3}) < 0.15$ with

$\bar{d}$. But this implies that the angle between the surface normal at some point on C and $n_\sigma(p)$ is at least $\pi/2 - 0.15 - \angle n_\sigma(p), H_F \geq \pi/2 - 0.15 - \arcsin(1/16) > 1$, contradicting Lemma 2.2(ia). This proves the claim.

Third, we claim that for any facet F of V_p, if F intersects $B \cap \sigma$, $F \cap B \cap \sigma$ is a single open curve with endpoints in bd$F$. We already proved that there is no closed curve in $F \cap B \cap \sigma$. Since F does not have any tangential contact with σ, $F \cap B \cap \sigma$ is a set of open curves and the endpoints of any open curve in $F \cap \sigma$ thus lie in bd$F$. Assume to the contrary that $F \cap B \cap \sigma$ contains two open curves, say ξ and ξ'. By our assumption, H_F is within a distance of λ from p. We have shown before that $H_F \cap B \cap \sigma$ is a single open curve. Follow $H_F \cap B \cap \sigma$ from ξ to ξ'. When we leave ξ, we must leave F at a Voronoi edge $e \subset$ bd$F$. Afterwards, we stay in the plane H_F and we must cross the support line of e again in order to reach ξ'. Therefore, some tangent to $H_F \cap B \cap \sigma$ is parallel to e. However, the angle between the surface normal at some point on $H_F \cap B \cap \sigma$ and $n_\sigma(p)$ would then be at least $\pi/2 - \angle n_\sigma(p), e \geq \pi/2 - 7/32$ by Lemma 2.2(ii). This contradicts Lemma 2.2(ia). This proves the claim.

Each facet of V_p that intersects $B \cap \sigma$ intersects it in a single open curve. Every curve endpoint is dual to some triangle incident to p. Thus, our assumption about sizes of restricted triangles incident to p ensures that every curve endpoint lies strictly inside B. This implies that the facets of V_p intersect $B \cap \sigma$ in a set of simple closed curves. We have analyzed this situation in Cheng et al.[9] and showed that exactly one face in this arrangement of closed curves lies inside V_p and it is a 2-disk. Of course, this face is σ_p. This proves that $\sigma_p = V_p \cap B \cap \sigma$ and it is a 2-disk. Take an edge e of V_p that intersects $B \cap \sigma$. By Lemma 2.2(ii), $\angle n_\sigma(p), e \leq 7/32$. Then, by Lemma 2.2(ia), $B \cap \sigma$ is monotone in the direction of e. Therefore, e intersects $B \cap \sigma$ exactly once. It follows that any edge of V_p intersects σ_p at most once. The correctness of (iii) follows from the third claim.

The above proves the lemma for the case that p is unweighted. The case of p being weighted can be handled similarly by setting B to be $B(p, 2\text{radius}(\mathcal{B}_p) + 2\lambda)$. $\qquad\square$

Now, we are ready to prove the termination of DelPSC.

Theorem 2.1 DelPSC *terminates.*

Proof   Consider a vertex p on a 2-face σ. Let λ be a constant that satisfies Lemma 2.3. By Lemma 2.3(i), no restricted triangle incident to p connects it to a vertex in τ where $p \notin \tau$. Notice that, because of protection, we can assume λ to be so small that no triangle incident to p connects two non-adjacent weighted vertices. This satisfies parts of disk condition. The only thing we need to show is that the restricted triangles incident to p form a topological disk if $\text{size}(t,\sigma) < \lambda$ for each such triangle t.

Observe that if no edge of the Voronoi cell V_p intersects σ, there is no restricted triangle incident to p and hence p does not exist in $\text{Skl}^2 S|_{\mathcal{D}_2}$ and can be ignored. So, assume that some edge of V_p intersects σ. Let σ_p be the connected component of $V_p|_\sigma$ containing p. By duality the conclusions in Lemma 2.3(ii)-(v) imply that the restricted triangles incident to p form a topological disk which contains p in the interior if and only if p lies in the interior of σ. In other words, disk condition is satisfied. Therefore, if disk condition is not satisfied for some point p, we can assume there is a triangle t incident to p for which $\text{size}(t,\sigma) > \lambda > 0$. In that case the new inserted point by Mes2Complex has a positive weighted distance λ from all other existing points.

Consider a tetrahedron t whose orthocenter is inserted by Mesh3Complex. Since $\rho_0 > 1$, closest point distance cannot decrease when t has all vertices unweighted. When t has a weighted vertex, the inserted orthocenter has a positive weighted distance from all other existing points. This is because the orthocenter is inserted only if it lies at least $2r$ distance away from any protecting ball of radius r. In sum, all points are inserted with a fixed lower bound on their distances to all existing points. A standard packing argument establishes termination of the refinement process. □

2.6 Topology Guarantees

Let M denote the output mesh of DelPSC. Property T1 follows immediately from Lemma 2.1.

Theorem 2.2 *M satisfies T1.*

Theorem 2.3 *M satisfies T2.*

Proof

Case(i): $\sigma \in \mathcal{D}_{\leqslant 1}$. It follows trivially from property T1.

Case(ii): $\sigma \in \mathcal{D}_2$. At the end of Mesh2Complex if $\text{Skl}^2 S|_\sigma$ is not empty, the **Disk_Condition** ensures that $\text{Skl}^2 S|_\sigma$ is a simplicial complex where each vertex v belongs to σ and has a 2-disk as its star. If σ has a non-empty boundary, the weighted points on $\text{bd}\,\sigma$ have some edge in their Voronoi cells which intersects σ (by Lemma 2.1). If σ has empty boundary, the same is guaranteed by the three weighted points which are initially placed on σ. In both cases $\text{Skl}^2 S|_\sigma$ remains non-empty. Insertions in Mesh3Complex does not disturb $\text{Skl}^2 S|_{\text{bd}\,\sigma'}$ for any $\sigma' \in \mathcal{D}_3$. Therefore, $\text{Skl}^2 S|_\sigma$ retains the disk property at each vertex even after

Mesh3Complex terminates. It follows from a result in PL topology that $\text{Skl}^2 S|_\sigma$ is a 2-manifold when DelPSC terminates.

Now we show that the boundary of $\text{Skl}^2 S|_\sigma$ is $\text{Skl}^1 S|_{\text{bd}\sigma}$. Each vertex $v \in \text{Skl}^2 S|_\sigma$ is an interior vertex if and only if it belongs to the interior of σ. Thus, the only vertices which are on the boundary of the manifold are the vertices on $\text{bd}\sigma$. These are the vertices of $\text{Skl}^1 S|_{\text{bd}\sigma}$ by T1. The edges of $\text{bd}\,\text{Skl}^2 S|_\sigma$ can then connect only vertices in $\text{Skl}^1 S|_{\text{bd}\sigma}$. Because of disk condition, these edges cannot connect non-adjacent vertices in $\text{bd}\sigma$. Therefore, they connect only adjacent vertices on $\text{bd}\sigma$ and are exactly the edges of $\text{Skl}^1 S|_{\text{bd}\sigma}$ by T1.

Case (iii): $\sigma \in \mathcal{D}_3$. Let $K = \text{Skl}^3 S|_\sigma$. First we argue that if K is not empty then it has vertices only in σ. Suppose not. Then there is a tetrahedron in K one of whose vertex, say p, is in $\sigma' \in \mathcal{D}$ where $\sigma' \not\subseteq \sigma$. The dual Voronoi vertex, say v, of this tetrahedron is in σ by definition. The Voronoi cell of p intersects both σ and σ'. Therefore there is a Voronoi edge of this Voronoi cell that intersects $\text{bd}\sigma$. But then $\text{Skl}^2 S|_\sigma$ has a triangle with vertex $p \in \sigma'$. This contradicts property T2.

Next we establish that $\text{Skl}^2 S|_{\text{bd}\sigma}$ is a 2-manifold which will help proving K is a 3-manifold. By definition $\text{Skl}^2 S|_{\text{bd}\sigma} = \cup \text{Skl}^2 S|_{\sigma_i}$ where each σ_i is a 2-face in $\text{bd}\sigma$. By property T2 any vertex $p \in \text{Skl}^2 S|_{\text{bd}\sigma}$ is in some $\sigma' \subseteq \text{bd}\sigma$. If p is in the interior of σ', $\text{Umb}_{\text{bd}\sigma}(p)$ is a 2-disk by property T2. If p is in $\text{bd}\sigma'$ we have two cases. If p is not a 0-face, it is incident to exactly two 2-faces, σ' and, say σ'', in $\text{bd}\sigma$. The two 2-disks $\text{Umb}_{\sigma'}(p)$ and $\text{Umb}_{\sigma''}(p)$ meet along two edges to form a 2-disk with p in its interior. If p is a 0-face, this construction generalizes to all 2-faces incident to p on $\text{bd}\sigma$. All 2-disks around p meet along common edges to form a 2-disk with p in its interior. Therefore, all vertices in $\text{Skl}^2 S|_{\text{bd}\sigma}$ have a star which is a closed 2-disk with the vertex in its interior. It implies from PL topology that $\text{Skl}^2 S|_{\text{bd}\sigma}$ is a closed 2-manifold.

Now we show that $\text{bd}\,K = \text{Skl}^2 S|_{\text{bd}\sigma}$ as required. Let T be the set of triangles in K that are incident to only one tetrahedron in K. We claim that $T = \text{Skl}^2 S|_{\text{bd}\sigma}$. The dual Voronoi edge of any triangle $t \in T$ intersects $\text{bd}\sigma$. Otherwise, its two endpoints lie in σ whose two dual tetrahedra being incident to t contradict that t is incident only to one tetrahedron. Therefore, $t \in \text{Skl}^2 S|_{\text{bd}\sigma}$ by definition. It follows that $T \subseteq \text{Skl}^2 S|_{\text{bd}\sigma}$. Being on the boundary of a union of tetrahedra, each edge of the triangles in T is incident to positive and even number of triangles in T. If T is a strict subset of $\text{Skl}^2 S|_{\text{bd}\sigma}$ some edge in T will have only a single triangle incident to it since $\text{Skl}^2 S|_{\text{bd}\sigma}$ is a connected closed manifold. Therefore T is exactly equal to $\text{Skl}^2 S|_{\text{bd}\sigma}$. $\qquad\square$

To prove T3 we need a result of Edelsbrunner and Shah[20] about the topological ball property (TBP). For smooth surface meshing, the TBP played a key role in

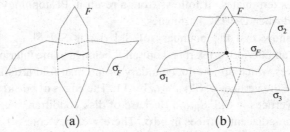

Fig. 2.4 F is a Voronoi facet ($k = 2$). In (a), F intersects a 2-face ($j = 2$) in a closed topological interval (1-ball). Voronoi edges ($k = 1$) intersect σ in a single point, a 0-ball. In (b), F intersects the 1-face (σ_F) in a single point, and for $1 \leqslant i \leqslant 3$, $F \cap \sigma_i$ are closed topological 1-balls. F intersects only 2-faces that are incident to σ_F.

proving topological guarantees.[7,9] It turns out that an extended version of the TBP (also given in Edelsbrunner and Shah[20]) is needed to prove topological guarantees for PSCs. It can be shown that the following two properties P1 and P2 imply the extended TBP.[18] Therefore, according to Edelsbrunner-Shah[20] result, $\mathrm{Del}\,S|_{\mathcal{D}}$ has underlying space homeomorphic to the $|\mathcal{D}|$ if P1 and P2 hold (see Figure 2.4).

Let F be a k-face of $\mathrm{Vor}\,S$ where S is the output vertex set.

(P1) If F intersects an element $\sigma \in \mathcal{D}_j \subseteq \mathcal{D}$, the intersection is a closed $(k + j - 3)$-ball.

(P2) There is a unique lowest dimensional element $\sigma_F \in \mathcal{D}$ so that F intersects σ_F and only elements that are incident to σ_F.

Lemma 2.1 and Lemma 2.3 together almost provide condition P1 except that the precondition of intersecting Voronoi edges of Lemma 2.3 needs to be proved and the case of a surface patch avoiding all Voronoi edges as in Figure 2.3(right) needs to be avoided. These two facts are proved in Lemma 2.4 and Lemma 2.5 respectively. Notice that we require stronger condition on triangle sizes for proving these results. Both Lemma 2.4 and Lemma 2.5 require that all restricted triangles have small sizes in contrast to restricted triangles incident to a particular point as in Lemma 2.3. It is interesting to note that the termination guarantee does not require this stronger condition.

Lemma 2.4 *There exists a constant $\lambda > 0$ for which following holds.   If* $\mathrm{size}(t, \sigma') < \lambda$ *for each tuple* (t, σ'), $t \in \mathrm{Skl}^2\,S|_{\sigma'}$, *then for each point $p \in S$ and each 2-face σ where $p \in \sigma$, an edge of V_p intersects σ.*

Proof Suppose that no edge of V_p intersects σ. Let $q \in \sigma$ be a weighted point. By the initialization step each 2-face has at least one such point. Because of Lemma 2.1 an edge of V_q has to intersect σ. Consider walking on a path in σ from

p to q. Let $p = p_0, p_1, .., p_k = q$ be sequence of vertices whose Voronoi cells are encountered along this walk. Since no edge of V_p intersects σ and some edge of V_q intersects σ, there exists two consecutive vertices p_i and p_{i+1} in this sequence so that no edge of V_{p_i} intersects σ whereas some edge of $V_{p_{i+1}}$ does intersect σ. By Lemma 2.3 we can claim that $\sigma_{p_{i+1}}$ is a disk. A boundary cycle of σ_{p_i} overlaps with the boundary of $\sigma_{p_{i+1}}$. This is impossible as the curves on the boundary of $\sigma_{p_{i+1}}$ intersect Voronoi edges whereas those on the boundary of σ_{p_i} do not. $\qquad\square$

Lemma 2.5 *There exists a constant* $\lambda > 0$ *such that if* $\mathrm{size}(t, \sigma') < \lambda$ *for each tuple* (t, σ'), $t \in \mathrm{Skl}^2 S|_{\sigma'}$, *then for each point* p *and 2-face* σ *where* $p \in \sigma$

(i) $V_p|_\sigma$ *is a 2-disk;*

(ii) *any edge of* V_p *intersects* $V_p|_\sigma$ *at most once;*

(iii) *any facet of* V_p *intersects* $V_p|_\sigma$ *in an empty set or an open curve.*

(iv) V_p *does not intersect any* σ' *where* $p \notin \sigma'$.

Proof Recall that σ_p denotes the connected component in $V_p|_\sigma$ containing p. Because of Lemma 2.4 we can apply Lemma 2.3 to claim that σ_p satisfies properties (i)-(iii) for each σ containing p. Therefore, it suffices to prove that $V_p|_\mathcal{D} = \bigcup_{\sigma \ni p} \sigma_p$ to claim (i)-(iv).

Assume to the contrary that $\bigcup \sigma_p \subset V_p|_\mathcal{D}$. By Lemma 2.3, $\sigma_p = V_p \cap B \cap \sigma$ and no edge of V_p intersects $\mathcal{D}$ outside B. This implies that for any connected component $C \in V_p|_\mathcal{D} \setminus \bigcup \sigma_p$, $\mathrm{bd}\,C$ lies inside facets of V_p. Also, V_p can intersect a 1-face only if it contains p (by Lemma 2.1). Therefore, $\mathrm{bd}\,C$ cannot have endpoints implying that $\mathrm{bd}\,C$ is a set of closed curves not intersecting any Voronoi edge.

Let $C \subset \sigma'$. Voronoi cells partition σ'. A path on σ' from $\mathrm{bd}\,C$ to a sample point $q \in \sigma'$ passes through the connected components of this partition. We must encounter two adjacent components along this path, say C' and C'', where $C' \neq \sigma'_s$ for any s and $C'' = \sigma'_r$ for some $r \in \sigma'$. This is because the first and last components satisfy this property. Then, we reach a contradiction since $\mathrm{bd}\,C'' = \mathrm{bd}\,\sigma'_r$ intersects Voronoi edges by Lemma 2.3 whereas $\mathrm{bd}\,C'$ does not. $\qquad\square$

Lemma 2.1 and Lemma 2.5 establish two facts for $\sigma \in \mathcal{D}_{\leqslant 2}$ before Mesh3Complex begins : (i) σ intersects V_p if and only if $p \in \sigma$, and (ii) property P1 holds for σ. Since Mesh3Complex does not insert any point which encroaches a triangle in $\mathrm{Skl}^2 S|_\sigma$, (i) and (ii) still hold for any $\sigma \in \mathcal{D}_{\leqslant 2}$. This means (i) also holds for any 3-face σ at termination. Let p be in the interior of $\sigma \in \mathcal{D}_3$. Then, V_p cannot intersect $\mathrm{bd}\,\sigma$ due to (i). In this case σ satisfies P1 with V_p trivially. If p is in $\mathrm{bd}\,\sigma$, the intersection $V_p \cap \sigma$ should satisfy P1 again since $\mathrm{bd}\,\sigma \cap V_p$ does so. This establishes P1. Next lemma establishes Property P2.

Lemma 2.6 *There exists a $\lambda > 0$ so that if* $\text{size}(t, \sigma) < \lambda$ *for each tuple* (t, σ), $t \in \text{Skl}^2 S|_\sigma$, *then the following holds. Let F be a k-face in VorS. There is an element $\sigma_F \in \mathcal{D}$ so that F intersects σ_F and only elements in $\mathcal{D}$ that have σ_F on their boundary.*

Proof Case 1: F is a Voronoi cell V_p. Let $\sigma_F \in \mathcal{D}$ be the lowest dimensional element containing p. We claim all elements in $\mathcal{D}$ intersecting F have σ_F in their boundaries and thus σ_F is unique. If not, let there be another $\sigma' \in \mathcal{D}$ where $\sigma_F \not\subset \text{bd}\,\sigma'$. Notice that $p \notin \sigma_F \cap \sigma'$ since otherwise $\sigma_F \cap \sigma'$ is either an element of $\mathcal{D}$ whose dimension is lower than σ_F or $\sigma_F \cap \sigma' = \sigma_F$ both of which are impossible. It follows that $p \notin \sigma'$. But we already argued above that V_p intersects only elements in $\mathcal{D}$ that contain p.

Case 2: F is a Voronoi facet V_{pq}. Let σ_F be a lowest dimensional element that F intersects. Assume there is another σ' intersecting F where $\sigma_F \not\subset \text{bd}\,\sigma'$. We go over different dimensions of σ' each time reaching a contradiction. If $\sigma' \in \mathcal{D}_3$, $\text{bd}\,\sigma'$ must intersect F. Otherwise, F intersects only σ' contradicting that it also intersects $\sigma_F \neq \sigma'$. Renaming $\text{bd}\,\sigma'$ as σ' we can use the contradiction reached for the case below.

Assume $\sigma' \in \mathcal{D}_{\leqslant 2}$; σ' intersects F and does not contain σ_F on its boundary. Two cases can arise. Either (i) σ_F and σ' are disjoint within V_p or V_q, or (ii) σ_F and σ' have a common boundary in V_p and V_q. Case (i) cannot happen due to the claim in Case 1. For (ii) to happen both p and q have to be on the common boundaries of σ_F and σ'. This means that p and q have to be on some element in $\mathcal{D}_{\leqslant 1}$. Observe that p and q are non-adjacent since otherwise V_{pq} has to intersect the common boundary of σ_F and σ' whose dimension is lower than that of σ_F. But this would contradict the disk condition that no two non-adjacent vertices in $\mathcal{D}_1$ is connected by a restricted edge.

The above argument implies that all elements intersecting F have σ_F as a subset.

Case 3: F is a Voronoi edge. Certainly F cannot intersect a 2-face σ_F more than once due to Lemma 2.5. The other possibility is that $F = V_t$, $t \in \text{Skl}^2 S|_{\sigma_F}$, and F intersects $\sigma' \neq \sigma_F$. But then a Voronoi cell adjacent to F would intersect two 2-faces σ_F and σ' and t is in both $\text{Skl}^2 S|_{\sigma_F}$ and $\text{Skl}^2 S|_{\sigma'}$ violating the disk condition. $\qquad \square$

Theorem 2.4 *M satisfies T3.*

Proof For sufficiently small $\lambda > 0$, the triangles in $\text{Skl}^2 S|_\mathcal{D}$ satisfy the conditions for Lemma 2.5 and Lemma 2.6. This means that properties P1 and P2 are satisfied when λ is sufficiently small. Also when P1 and P2 are satisfied

$\bigcup_i \text{Skl}^i S|_{\mathcal{D}} = \text{Del} S|_{\mathcal{D}}$. It follows that Edelsbrunner-Shah conditions are satisfied for the output M of DelPSC. Thus, M has an underlying space homeomorphic to $|\mathcal{D}|$. The homeomorphism constructed by Edelsbrunner and Shah actually respects the stratification, that is, for each $\sigma \in \mathcal{D}_i$, $\text{Skl}^i S|_{\sigma}$ is homeomorphic to σ. Furthermore, property T2 holds for any output of DelPSC. This means, for $0 \leqslant i \leqslant 2$, $\text{Skl}^i S|_{\sigma} = \text{Skl}^{i-1} S|_{\text{bd}\sigma}$ and $\text{Skl}^i S|_{\sigma}$ has vertices only in σ. We can claim the same condition even for $i = 3$ since $\text{Skl}^3 S|_{\sigma}$, $\sigma \in \mathcal{D}_3$, is not empty at the end of Mesh2Complex when λ is sufficiently small. A Voronoi edge intersecting bdσ cannot intersect it more than once as we argued in Case 3 of Lemma 2.6. $\square$

2.7 Conclusions

We have presented a practical algorithm to mesh a wide variety of geometric domains with Delaunay refinement technique. The output mesh maintains a manifold property and with increasing level of refinement captures the topology of the input. An interesting aspect of the algorithm is that the input 'non-smooth features' are preserved in the output.

A number of experimental results that validate our claims. This is the first practical algorithm to produce Delaunay meshes for a large class of three dimensional geometric domains with theoretical guarantees. It can handle arbitrarily small input angles and preserve input features. When applied to volumes, the algorithm guarantees bounded radius-edge ratio for most of the tetrahedra except near boundary. It can be easily extended to guarantee bounded aspect ratio for most triangles and bounded radius-edge ratio for most tetrahedra except the ones near non-smooth elements.

An obvious open question is to analyze the time and space complexity of the algorithm. Is it possible to claim non-trivial or even optimal bounds for these complexities. Very few results exist for optimal Delaunay refinement.[23] Finally, can the algorithm be improved even more by incorporating easier primitives?

Acknowledgment

Research supported by NSF, USA (CCF-0430735 and CCF-0635008) and RGC, Hong Kong, China (HKUST 6181/04E)

References

1. L. P. Chew, Guaranteed-quality triangular meshes, Report TR-98-983, Comput. Sci. Dept., Cornell Univ., Ithaca, New York, 1989.
2. T. K. Dey, C. Bajaj and K. Sugihara, On good triangulations in three dimensions, *Internat. J. Comput. Geom.* **2** (1992) 75–95.

3. X.-Y. Li and S.-H. Teng, Generating well-shaped Delaunay meshes in 3D, *Proc. 12th. Ann. ACM-SIAM Sympos. Discrete Algorithm* (2001), pp. 28–37.
4. G. L. Miller, D. Talmor, S.-H. Teng and N. Walkington, A Delaunay based numerical method for three dimensions: generation, formulation, and partition, *Proc. 27th Ann. ACM Sympos. Theory Comput.* (1995), pp. 683–692.
5. J. Ruppert, A Delaunay refinement algorithm for quality 2-dimensional mesh generation, *J. Algorithms* **18** (1995) 548–585.
6. J. R. Shewchuk, Tetrahedral mesh generation by Delaunay refinement, *Proc. 14th Ann. Sympos. Comput. Geom.* (1998), pp. 86–95.
7. J.-D. Boissonnat and S. Oudot, Provably good surface sampling and meshing of surfaces, *Graphical Models* **67** (2005) 405–451.
8. H.-L. Cheng, T. K. Dey, H. Edelsbrunner, and J. Sullivan, Dynamic skin triangulation, *Discrete Comput. Geom.* **25** (2001) 525–568.
9. S.-W. Cheng, T. K. Dey, E. A. Ramos and T. Ray, Sampling and meshing a surface with guaranteed topology and geometry, *Proc. 20th Ann. Sympos. Comput. Geom.* (2004) 280–289.
10. L. P. Chew, Guaranteed-quality mesh generation for curved surfaces, *Proc. 9th Ann. Sympos. Comput. Geom.* (1993) 274–280.
11. S. Oudot, L. Rineau, and M. Yvinec, Meshing volumes bounded by smooth surfaces, *Proc. 14th Internat. Meshing Roundtable* (2005), pp. 203–219.
12. N. Amenta and M. Bern, Surface reconstruction by Voronoi filtering, *Discr. Comput. Geom.* **22** (1999) 481–504.
13. S.-W. Cheng and S.-H. Poon, Three-dimensional Delaunay mesh generation, *Discrete Comput. Geom.* **36** (2006) 419–456.
14. S.-W. Cheng, T. K. Dey, E. A. Ramos and T. Ray, Quality meshing for polyhedra with small angles, *Internat. J. Comput. Geom. Appl.* **15** (2005) 421–461.
15. S. Pav and N. Walkington, Robust three dimensional Delaunay refinement, *13th Internat. Meshing Roundtable*, 2004.
16. J. R. Shewchuk, Mesh generation for domains with small angles, *Proc. 16th Ann. Sympos. Comput. Geom.* (2000), pp. 1–10.
17. J.-D. Boissonnat and S. Oudot, Provably good sampling and meshing of Lipschitz surfaces, *Proc. 22nd Ann. Sympos. Comput. Geom.* (2006), pp. 337–346.
18. S.-W. Cheng, T. K. Dey, and E. A. Ramos, Delaunay refinement for piecewise smooth complexes, *Proc. 18th Ann. ACM-SIAM Sympos. Discrete Algorithms* (2007), pp. 1096–1105.
19. S.-W. Cheng, T. K. Dey, and J. A. Levine, A practical Delaunay meshing algorithm for a large class of domains, *Proc. 16th Internat. Meshing Roundtable* (2007), to appear.
20. H. Edelsbrunner and N. Shah, Triangulating topological spaces, *Internat. J. Comput. Geom. Appl.* **7** (1997) 365–378.
21. T. K. Dey, G. Li, and T. Ray, Polygonal surface remeshing with Delaunay refinement, *Proc. 14th Internat. Meshing Roundtable* (2005), pp. 343–361.
22. N. Amenta, S. Choi, T. K. Dey and N. Leekha, A simple algorithm for homeomorphic surface reconstruction, *Internat. J. Comput. Geom. Applications* **12** (2002) 125–141.
23. S. Har-Peled and A. Üngör, A time optimal Delaunay refinement algorithm in two dimensions, *Proc. Ann. Sympos. Comput. Geom.* (2005), pp. 228–236.

Chapter 3

Families of Convex Sets not Representable by Points

János Pach* and Géza Tóth†

City College of New York, CUNY and
Rényi Institute, Hungarian Academy of Sciences, Budapest, Hungary

Let (A,B,C) be a triple of disjoint closed convex sets in the plane such that each of them contributes at least one point to the boundary ∂ of the convex hull of their union. If there are three points $a \in A, b \in B, c \in C$ that belong to ∂ and follow each other in clockwise (counterclockwise) order, we say that the *orientation* of the triple (A,B,C) is clockwise (counterclockwise). We construct families of disjoint closed convex sets $\{C_1,\ldots,C_n\}$ in the plane whose every triple has a unique orientation, but there are no points $p_1,\ldots,p_n$ in general position in the plane whose triples have the same orientations. In other words, these families cannot be represented by a point set of the same *order type*. This answers a question of A. Hubard and L. Montejano. We also show the size of the largest subfamily representable by points, which can be found in any family of n disjoint closed convex sets in general position in the plane, is $O(n^{\log 8/\log 9})$. Some related Ramsey-type geometric problems are also discussed.

Contents

3.1 Introduction

Let $\mathcal{C}$ be a family of disjoint closed convex sets in the plane in *general position*, that is, assume that

*Supported by NSF grant CCF-05-14079 and by grants from NSA, PSC-CUNY, BSF, and OTKA-K-60427.
†Supported by OTKA-K-60427.

(1) no *three* of them have a common tangent line, and
(2) the convex hull of the union of two members $A, B \in \mathcal{C}$ never contains a third member $C \in \mathcal{C}$, that is, $\mathrm{conv}(A \cup B \cup C) \neq \mathrm{conv}(A \cup B)$ holds for every triple of distinct members $A, B, C \in \mathcal{C}$.

Analogously, we say that a set of points P in the plane is in general position if no *three* elements of P are collinear.

The type $\mathrm{tp}(A, B, C)$ of an ordered triple of members of $\mathcal{C}$ is defined as $+1$ (or -1) if there are three points $a \in A, b \in B, c \in C$ belonging to the boundary $\mathrm{Bd\,conv}(A \cup B \cup C)$ that follow each other in clockwise (counterclockwise) order along this boundary. Notice that the same triple (A, B, C) may have *two* types at the same time if one of the sets contributes two arcs to $\mathrm{Bd\,conv}(A \cup B \cup C)$.

A point set P in general position in the plane is said to *represent* the order type of $\mathcal{C}$ if there is a one-to-one correspondence $f : \mathcal{C} \to P$ such that

$$\mathrm{tp}(f(A), f(B), f(C)) = \mathrm{tp}(A, B, C) \quad \text{for all } A, B, C \in \mathcal{C} \text{ with a unique type.}$$

As far as we know, order types of set families were first studied by Bisztriczky and G. Fejes Tóth.[1,2] They generalized a famous conjecture of Erdős and Szekeres[3,4] to families of convex sets in the plane as follows: Any family $\mathcal{C}$ of at least $2^{n-2} + 1$ disjoint closed convex sets in general position has n members in convex position. Recently, A. Hubard and L. Montejano suggested that this stronger conjecture may actually be equivalent to the original one. More precisely, they suspected that the order type of every family $\mathcal{C}$ with the above property can be represented by points. In the present note, we show that this is not the case.

Theorem 3.1 *There exists a family of* nine *pairwise disjoint segments in general position in the plane, whose order type cannot be represented by points.*

Let $r = r(n)$ denote the largest integer such that every family $\mathcal{C}$ of n disjoint closed convex sets in general position in the plane has r members whose order type can be represented by points. By definition, the order type of any subfamily of $\mathcal{C}$ in convex position can be represented by points. According to Pach and Tóth,[5] every family $\mathcal{C}$ with the above property has at least $\log_{16} n$ members in convex position. Therefore, we have $r(n) \geqslant \log_{16} n$. Iterating the construction in Theorem 3.1, we obtain

Theorem 3.2 *For every n, there exists a family of n pairwise disjoint segments in general position in the plane which has no subfamily of size $\lfloor n^{\log 8 / \log 9} \rfloor$ whose order type is representable by points.*

A collection of two-way infinite (unbounded) non-selfintersecting curves in the plane is called a family of *pseudolines* if any two curves have precisely one point in common, at which they properly cross. It is said to be *simple* if no three pseudolines pass through the same point. A family of pseudolines $\mathcal{P}$ is *stretchable* if there exists a family of lines $\mathcal{L}$ such that the cell decompositions induced by $\mathcal{P}$ and $\mathcal{L}$ are topologically isomorphic. It was known already to Hilbert[6] and Levi[7] that there are nonstretchable families of pseudolines. The first example of a non-stretchable *simple* arrangement was given by Ringel.[8] It was shown by Mnev[9,10] that it is a computationally hard (*NP*-hard) problem to decide whether an arrangement of pseudolines is stretchable (see also Shor.[11])

In complete analogy to the above problem, we can try to determine the size of the largest stretchable subfamily contained in every simple family of n pseudolines.

Theorem 3.3 *Let $s = s(n)$ denote the largest integer such that every simple family $\mathcal{P}$ of n pseudolines has a stretchable subfamily of size s. For every n, we have*

$$\log_4 n \leqslant s(n) \leqslant \lfloor n^{\log 8 / \log 9} \rfloor.$$

Theorem 3.1 is established in Section 3.2, Theorems 3.2 and 3.3 are proved in Section 3.3. In the last section, we discuss some related problems.

3.2 A Nonrepresentable Order Type of Segments

The aim of this section is to establish Theorem 3.1. The proof is based on Ringel's[8] construction of a nonstretchable arrangement of *nine* pseudolines, that can be obtained by modifying the *Pappus configuration*; see Figure 3.1. It is known that every arrangement of fewer than nine pseudolines is stretchable.[12,13]

Let $\mathcal{S} = \{S_1, \ldots, S_n\}$ be a family of disjoint segments in general position in the plane. We say that $\mathcal{S}$ can be *flattened* if for any $\varepsilon > 0$ there are two disks of radius ε at unit distance, D_1 and D_2, and another family of disjoint segments $\mathcal{S}' = \{S_1', \ldots, S_n'\}$ with the same order type such that each $S_i' \in \mathcal{S}'$ has one endpoint in D_1 and one in D_2.

Lemma 3.1 *There exists a family $\mathcal{S}$ of nine segments in general position in the plane*

(i) *which can be flattened, and*
(ii) *the order type of which cannot be represented by points.*

Proof Start with the Pappus configuration (see Figure 3.1), and slightly perturb its points so that its originally collinear triples receive the following orientations:

$$\mathrm{tp}(p_1^1, p_2^1, p_3^1) = +1, \quad \mathrm{tp}(p_1^2, p_2^2, p_3^2) = +1, \quad \mathrm{tp}(p_1^3, p_2^3, p_3^3) = +1,$$

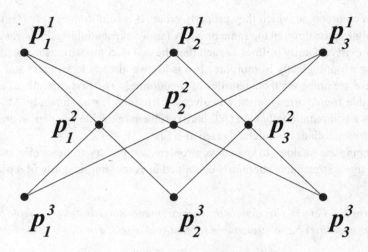

Fig. 3.1   The Pappus configuration.

$$\text{tp}(p_1^1, p_1^2, p_2^3) = +1, \quad \text{tp}(p_1^1, p_2^2, p_3^3) = -1,$$

$$\text{tp}(p_2^1, p_1^2, p_1^3) = -1, \quad \text{tp}(p_2^1, p_3^2, p_3^3) = +1,$$

$$\text{tp}(p_3^1, p_3^2, p_2^3) = -1, \quad \text{tp}(p_3^1, p_2^2, p_1^3) = -1.$$

This can be achieved, for example, by taking

$$p_1^1 = (-2, 1), \quad p_2^1 = (0, 1 + 110\delta), \quad p_3^1 = (2, 1),$$

$$p_1^2 = (-1, -2\delta), \quad p_2^2 = (0, -\delta), \quad p_3^2 = (1, -2\delta),$$

$$p_1^3 = (-2, -1 - 50\delta), \quad p_2^3 = (0, -1 - 20\delta), \quad p_3^3 = (2, -1).$$

It follows from Ringel's result,[8] by duality, that there are no points $\overline{p}_j^i$ ($1 \leqslant i, j \leqslant 3$), for which the above *nine* triples have the same types (orientations) as for the points p_j^i ($1 \leqslant i, j \leqslant 3$), except that the orientation of the last triple is opposite, that is, $\text{tp}(\overline{p}_3^1, \overline{p}_2^2, \overline{p}_1^3) = +1$. In other words, this modified order type τ is not representable by points (see Figure 3.2).

Next we show that there is a family of segments whose order type is τ.

Let L be a fixed very large number. For any $1 \leqslant i, j \leqslant 3$, let S_j^i be a segment of length L with slope $1/2$, whose right endpoint is p_j^i. Observe that the set of segments $\{S_j^i \mid 1 \leqslant i, j \leqslant 3\}$ has the same order type as the point set $\{p_j^i \mid 1 \leqslant i, j \leqslant 3\}$. Now slightly rotate S_3^1 about p_3^1 in the clockwise direction to a position in

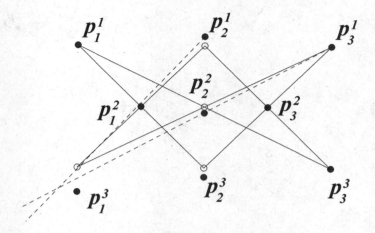

Fig. 3.2   The set $\{p_j^i \mid 1 \leqslant i, j \leqslant 3\}$.

which p_2^2, p_1^3, and hence the entire segment S_1^3, lie below it. Next, applying a small counterclockwise rotation about p_2^2, bring the segment S_2^2 into a position where it lies below S_1^1. See Figure 3.3. With a slight abuse of notation, the new segments are also denoted by S_j^i. Notice that during the above transformation no triple of segments S_j^i switched orientations, except one: the orientation of (S_3^1, S_2^2, S_1^3) has become clockwise. That is, we have

$$\mathrm{tp}(S_1^1, S_2^1, S_3^1) = +1, \quad \mathrm{tp}(S_1^2, S_2^2, S_3^2) = +1, \quad \mathrm{tp}(S_1^3, S_2^3, S_3^3) = +1,$$

$$\mathrm{tp}(S_1^1, S_1^2, S_2^3) = +1, \quad \mathrm{tp}(S_1^1, S_2^2, S_3^3) = -1,$$

$$\mathrm{tp}(S_2^1, S_1^2, S_1^3) = -1, \quad \mathrm{tp}(S_2^1, S_3^2, S_3^3) = +1,$$

$$\mathrm{tp}(S_3^1, S_3^2, S_2^3) = -1, \quad \mathrm{tp}(S_3^1, S_2^2, S_1^3) = +1.$$

Consequently, we found a family of segments whose order type τ is not representable by points.

It remains to argue that the family $\mathcal{S} = \{S_j^i \mid 1 \leqslant i, j \leqslant 3\}$ can be flattened. To see this, notice that for any $\varepsilon > 0$, one can choose a sufficiently small δ and a sufficiently large L so that, after appropriate scaling, all left endpoints and all right endpoints of the segments lie in two disks of radius ε at unit distance from each other. $\qquad \square$

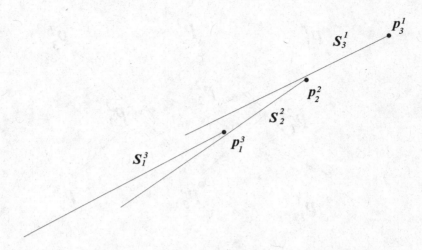

Fig. 3.3   The triple (S_3^1, S_2^2, S_1^3) switched its orientation.

3.3 The Iteration Step

Theorem 3.2 is an immediate corollary of Lemma 3.1 and the following statement.

Lemma 3.2 *Suppose there exists a family $\mathcal{S}$ of k disjoint segments in general position in the plane, which can be flattened and which has an order type that cannot be represented by points. Then, for every $i = 1, 2, \ldots,$ there is a family $\mathcal{S}^i$ of k^i segments in general position in the plane, which can be flattened and which does not have any subfamily of size larger than $(k-1)^i$ whose order type can be represented by points.*

Proof   For any $\delta > 0$, let $\mathcal{S}^1(\delta)$ be a family of k unit segments whose order type cannot be represented by points and their left endpoints and right endpoints lie in two disks of radius δ. Let $i \geqslant 1$ and suppose, recursively, that we have already constructed a family $\mathcal{S}^i = \{S_1, \ldots, S_{k^i}\}$ of unit segments such that their left and right endpoints lie in two disks of radius $\varepsilon/2$ and $\mathcal{S}^i$ has no subfamily of size larger than $(k-1)^i$, whose order type can be represented by points.

Let δ be a small positive number to be specified later. Replace each segment $S_j \in \mathcal{S}^i$ by a congruent copy $\mathcal{S}_j^1(\delta)$ of $\mathcal{S}^1(\delta)$, with the same orientation, in such a way that S_j coincides with a member of $\mathcal{S}_j^1(\delta)$. Let $\mathcal{S}^{i+1}$ be the union of these copies. Obviously, we have $|\mathcal{S}^{i+1}| = k^{i+1}$. Furthermore, if $\delta > 0$ is sufficiently small, then

(1) the members of $\mathcal{S}^{i+1}$ are disjoint and are in general position;

(2) the left endpoints and the right endpoints of the segments in $\mathcal{S}^{i+1}$ lie in two disks of radius at most $\varepsilon/2 + \delta \leqslant \varepsilon$;

(3) for any three distinct indices j, j', j'', the orientation of any triple of segments belonging to the subfamilies replacing $S_j, S_{j'}, S_{j''}$, respectively, is the same as the orientation of the triple $(S_j, S_{j'}, S_{j''})$.

It is easy to show that one cannot select more than $(k-1)^{i+1}$ segments from $\mathcal{S}^{i+1}$ such that their order type can be represented by points. Indeed, by property 3 above and by the induction hypothesis, such a set cannot contain segments belonging to more than $(k-1)^i$ subfamilies $\mathcal{S}_j^1(\delta)$ replacing distinct elements $S_j \in \mathcal{S}^i$. On the other hand, from each subfamily $\mathcal{S}_j^1(\delta)$, we can select at most $k-1$ segments. $\qquad\square$

In the same way, as Theorem 3.2 can be deduced from Theorem 3.1 using Lemma 3.2, one can establish the upper bound in Theorem 3.3 by iterating Ringel's construction of a nonstretchable arrangement of nine pseudolines.[8] By the result of Goodman[14] every arrangement of pseudolines can be represented such that all pseudolines are x-monotone curves, in particular, it follows that Ringel's construction can be "flattened" in the following sense: An arrangement of pseudolines $\mathcal{P} = \{\pi_i \mid i \in I\}$ can be *flattened* if for any $\varepsilon > 0$, there exist real functions $f_i : \mathbf{R} \to \mathbf{R}$, $i \in I$ satisfying two conditions.

(1) The graphs of the functions f_i form an arrangement of pseudolines such that the cell decomposition of the plane induced by them is isomorphic to the cell decomposition induced by $\mathcal{P}$.

(2) For every $i \in I$ and $x \in \mathbf{R}$, we have $|f_i(x)| < \varepsilon$.

To prove the upper bound in Theorem 3.3, instead of Lemma 3.2 we have to use the following statement (the straightforward recursive proof of which is left to the reader).

Lemma 3.3 *Suppose there exists a simple nonstretchable arrangement $\mathcal{P}$ of k pseudolines in the plane, which can be flattened. Then, for every $i = 1, 2, \ldots$, there is a simple nonstretchable arrangement $\mathcal{P}^i$ of k^i pseudolines which can be flattened and which does not have any stretchable subarrangement of size larger than $(k-1)^i$.*

It was first pointed out by Goodman and Pollack[15] that every finite arrangement of pseudolines is isomorphic to an arrangement of x-*monotone* pseudolines (see also Goodman and Pollack[13] and Goodman et al.[16] for a much stronger statement). Therefore, to prove the lower bound in Theorem 3.3, it is enough to restrict our attention to families of x-monotone pseudolines.

Let $\mathcal{P} = \{\ell_1, \ell_2, \ldots, \ell_n\}$ be a simple arrangement of x-monotone pseudolines. We say that $\ell_1, \ell_2, \ldots, \ell_n$ form a *cap* (a *cup*) if

(1) in the cell decomposition determined by them there is an unbounded cell whose boundary contains a piece of each member of $\mathcal{P}$, in this clockwise (counterclockwise) order, and

(2) this cell lies below (above) every pseudoline $\ell_i \in \mathcal{P}$.

It is clear that all caps and cups $\mathcal{P}$ are stretchable, since the cell decomposition of the plane induced by them is isomorphic to the cell decomposition induced by $|\mathcal{P}|$ distinct tangent lines of an open semicircle.

Thus, the lower bound in Theorem 3.3 follows from

Lemma 3.4 *Any simple arrangement $\mathcal{P}$ of more than $\binom{k+m-4}{m-2}$ x-monotone pseudolines contains a cap of size at least k or a cup of size at least m.*

Proof The statement can be established by dualizing and adapting the original proof of the Erdős-Szekeres theorem.[3] The lemma holds if $k \leqslant 2$ or $m \leqslant 2$. Suppose that $k, m > 2$ are fixed and that we have already proved the statement for all pairs (k', m') with $k' < k$ or $m' < m$.

Let $\mathcal{P}$ be a simple arrangement of $\binom{k+m-4}{m-2} + 1$ x-monotone pseudolines. Using the induction hypothesis and the fact that $\binom{k+m-4}{m-2} + 1 > \binom{k+m-5}{m-2} + 1$, we obtain that $\mathcal{P}$ contains a cap of size $(k-1)$ or a cup of size m. In the latter case we are done. So we may assume that $\mathcal{P}$ contains a cap of size $(k-1)$. Delete the first member of such a cap from the arrangement. We still have more than $\binom{k+m-5}{m-2} + 1$ pseudolines, so $\mathcal{P}$ must have another cap of size $(k-1)$. Again, delete its first member, and repeat this procedure as long as there are more than $\binom{k+m-5}{m-2}$ pseudolines left. Then stop. We have deleted altogether $\binom{k+m-4}{m-2} + 1 - \binom{k+m-5}{m-2} = \binom{k+m-5}{m-3} + 1$ pseudolines. Therefore, by the induction hypothesis, the set of deleted pseudolines must contain a cap of size k or a cup of size $(m-1)$. In the first case, we are done. In the second case, there exists a cup $\mathcal{C}_1$ of size $(m-1)$ such that its first member is also the first member of a cap $\mathcal{C}_2$ of size $(k-1)$. It is easy to verify (see Figure 3.4) that

(1) either one can extend $\mathcal{C}_1$ by the second member of $\mathcal{C}_2$ to a cup of size m,

(2) or one can extend $\mathcal{C}_2$ by the second member of $\mathcal{C}_1$ to a cap of size k.

This completes the proof. $\square$

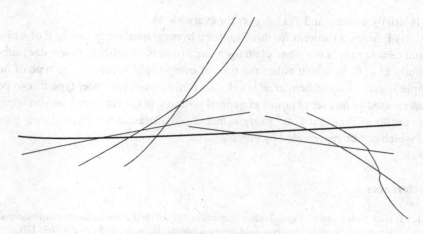

Fig. 3.4   A cap and a cup of size four, having the same first member.

3.4 Concluding Remarks

The *lower* bounds for the functions $r(n)$ and $s(n)$ (in Theorem 3.3) follow from Erdős-Szekeres type results: from the existence of large *convex* subconfigurations. To further improve these bounds, we need to find larger classes Γ of "unavoidable" configurations, representable by points, such that for any k, every sufficiently large system contains a subconfiguration of size k belonging to Γ. To improve the *upper* bounds, on the other hand, we have to define more complicated operations for building large nonrepresentable order types or nonstretchable arrangements of pseudolines, using smaller examples.

One can extend the definition of *order type* to arbitrary families of disjoint convex sets, no three of which have a common tangent line, as follows. Suppose that a member of $\mathcal{C}$ is allowed to lie "between" two other members, that is, it can be contained in the convex hull of the union of two others. If, for example, $B \subset \text{conv}(A \cup C)$ for some $A, B, C \in \mathcal{C}$, let $\text{tp}(A, B, C)$ and the type of every permutation of these three members be *zero*. In all other cases, let us define $\text{tp}(A, B, C)$ as before.

We can now say that a point set P in general position in the plane represents the order type of $\mathcal{C}$ if there is a one-to-one correspondence $f : \mathcal{C} \to P$ such that

$$\text{tp}(f(A), f(B), f(C)) = \text{tp}(A, B, C) \quad \text{for all } A, B, C \in \mathcal{C} \text{ with a unique nonzero type.}$$

Note that, according to this definition, any set of n points in general position represents the order type of a family of disjoint convex bodies lying between and touching the graphs of two functions $f, g : \mathbf{R} \to \mathbf{R}$, where f is strictly concave,

g is strictly convex, and $f(x) \leqslant g(x)$ for every $x \in \mathbf{R}$.

By Ramsey's theorem for three-uniform hypergraphs, every family $\mathcal{C}$ of n disjoint convex sets in the plane contains a large (that is, roughly $\log \log n$ size) subfamily $\mathcal{C}' \subseteq \mathcal{C}$, in which either the type of every triple is zero or the type of no triple is zero (see Graham *et al.*[17]). In the former case, the order type $\mathcal{C}'$ can be represented by any set of points in general position, in the latter one, we can apply the results of Pach and Tóth[5] to argue that $\mathcal{C}'$ has a large subfamily, the order type of which can be represented by points.

References

1. T. Bisztriczky and G. Fejes Tóth, A generalization of the Erdős-Szekeres convex n-gon theorem, *Journal für die reine und angewandte Mathematik* **395** (1989) 167–170.
2. T. Bisztriczky and G. Fejes Tóth, Convexly independent sets, *Combinatorica* **10** (1990) 195–202.
3. P. Erdős and G. Szekeres, A combinatorial problem in geometry, *Compositio Mathematica* **2** (1935) 463–470.
4. P. Erdős and G. Szekeres, On some extremum problems in elementary geometry, *Ann. Universitatis Scientiarum Budapestinensis, Eötvös, Sectio Mathematica* **III–IV** (1960–61) 53–62.
5. J. Pach and G. Tóth, A generalization of the Erdős-Szekeres theorem to disjoint convex sets, *Discrete and Computational Geometry* **19** (1998) 437–445.
6. D. Hilbert, *The Foundations of Geometry*, 2nd edn. (Open Court, Chicago, 1910).
7. F. Levi, Die Teilung der projektiven Ebene durch Gerade oder Pseudogerade, *Ber. Math.-Phys. Kl. sächs. Akad. Wiss. Leipzig* **78** (1926) 256–267.
8. G. Ringel, Teilungen der projectiven Ebene durch Geraden oder topologische Geraden, *Math. Z.* **64** (1956) 79–102.
9. N. E. Mnev, Varieties of combinatorial types of projective configurations and convex polyhedra (in Russian), *Dokl. Akad. Nauk SSSR* **283** (1985) 1312–1314.
10. N. E. Mnev, The universality theorems on the classification problem of configuration varieties and convex polytopes varieties, in *Topology and geometry — Rohlin Seminar, Lecture Notes in Mathematics* **1346** (Springer, Berlin, 1988), pp. 527–543.
11. P. W. Shor, Stretchability of pseudolines is NP-hard, in: *Applied Geometry and Discrete Mathematics, DIMACS Ser. Discrete Math. Theoret. Comput. Sci.* **4** (Amer. Math. Soc., Providence, RI, 1991), pp. 531–554.
12. J. E. Goodman and R. Pollack, Proof of Grünbaum's conjecture on the stretchability of certain arrangements of pseudolines, *J. Comb. Theory, Ser. A* **29** (1980) 385–390.
13. J. E. Goodman and R. Pollack, Allowable sequences and order types in discrete and computational geometry, in: *New Trends in Discrete and Computational Geometry* (J. Pach, ed.), *Algorithms and Combinatorics* **10** (Springer, Berlin, 1993), pp. 103–134.
14. J. E. Goodman, Proof of a conjecture of Burr, Grünbaum, and Sloane, *Discrete Mathematics* **32** (1980) 27–35.

15. J. E. Goodman and R. Pollack, Semispaces of configurations, cell complexes of arrangements, *J. Comb. Theory, Ser. A* **37** (1984) 257–293.
16. J. E. Goodman, R. Pollack, R. Wenger, and T. Zamfirescu, Arrangements and topological planes, *Amer. Math. Monthly* **101** (1994) 866–878.
17. R. L. Graham, B. L. Rothschild, and J. H. Spencer, *Ramsey Theory*, 2nd edn. (John Wiley, New York, 1990).
18. G. Fejes Tóth, Recent progress on packing and covering, in: *Advances in Discrete and Computational Geometry* (B. Chazelle *et al.*, eds.), *Contemporary Mathematics*, Vol. **223** (AMS, Providence, 1999).
19. J. Pach and G. Tóth, Erdős–Szekeres-type theorems for segments and non-crossing convex sets, *Geometriae Dedicata* **81** (2000) 1–12.

Chapter 4

Some Generalizations of Least-Squares Algorithms

Tetsuo Asano*, Naoki Katoh†, Kurt Mehlhorn‡ and Takeshi Tokuyama§

*Japan Advanced Institute of Science and Technology, Nomi, 923-1292, Japan
†Kyoto University, Kyoto, 606-8501, Japan
‡Max-Planck-Institut für Informatik, D-66123, Saarbrücken, Germany
§Tohoku University, Sendai, Aobaku, 980-8579, Japan

Least-squares method is often used for solving optimization problems such as line fitting of a point sequence obtained by experiments. This paper describes some extensions of the method and presents an application to some completely different geometric optimization problem. Given a sorted sequence of points $(x_1, y_1), (x_2, y_2), \ldots, (x_n, y_n)$, a line $y = ax + b$ that optimally approximates the sequence can be computed by determining the constants a and b that minimizes the sum of squared distances to the line, which is given by $\sum_{i=1}^{n} (ax_i + b - y_i)^2$. It suffices to solve a system of linear equations derived by differentiating the sum by a and b. In this paper we extend the problem of approximating a point sequence by a 1-joint polyline. Another problem we consider is a geometric optimization problem. Suppose we are given a set of points in the plane. We want to insert a new point so that distances to existing points are as close as those distances specified as input data. If the criterion is to minimize the sum of squared errors, an application of the same idea as above combined with a notion of arrangement of lines leads to an efficient algorithm.

Keywords: Algorithm; computational geometry; least-squares method; combinatorial optimization; polyline approximation of point sequence.

Contents

4.1   Introduction

The least square method for approximating a set of data by a line that minimizes the total sum of squared vertical errors is applied in various situations. It is a natural idea to extend this idea so as to approximate a set of data points by a piece-wise linear function, i.e., a polyline with bounded number of joints instead of a single line. We consider the problem of approximating a set of points by a polyline with one joint to minimize the sum of squared errors. We show that this is also done in linear time.

In addition to the least-squares criterion, we could consider several different optimization criteria. The second criterion is to minimize the sum of absolute error, and the third one to minimize the sum of the maximum error associated with each edge of the resulting polyline. We show that these three problems can be handled within a similar framework. An important observation is that our objective functions are all convex when a given point set is partitioned into subsets each associated with an edge of a polyline. The computational complexities of the polyline approximation may be different by the criteria. For one-joint case the problem for the least sum of squared errors can be solved in linear time if points are sorted in advance. On the other hand, it seems that no linear time algorithm exists for other criteria. However, a good news is that a general k-joints case can be solved in polynomial time both in n and k for the criterion of minimizing the sum of the maximum errors.[1]

There are few related works. One important related work is Imai, Katoh and Yamamoto[2] which presents a linear-time algorithm for finding an optimal line approximating a set of points to minimize the sum of absolute error instead of squared errors.

Another problem we consider is a geometric optimization problem. Suppose we are given a set of points in the plane. We want to insert a new point so that distances to existing points are as close as those distances specified as input data. If the criterion is to minimize the sum of squared errors, an application of the same idea as above combined with a notion of arrangement of lines lead to an efficient algorithm.

4.2 Polyline Approximation of a Point Sequence

Experimental data are often accompanied by noise. A process of quantitatively estimating the underlying function known as regression or curve fitting is required. Most common is the least squares method. It is well known that given a number of 2-dimensional data the best line approximating the data is obtained in linear time. A number of variations have been considered. One of the generalizations is the following.[a]

- Approximate a set of points by a polyline with a bounded number of joints, and
- Minimize the error defined under three different criteria: sum of least square errors, sum of absolute errors, and sum of maximum errors, where error for each point is measured by the vertical difference from the approximating polyline.

More formally, the problem is described as follows:

[Problem] Given a set of points $S = \{(x_1, y_1), (x_2, y_2), \ldots, (x_n, y_n)\}$ such that $x_1 < x_2 < \cdots < x_n$ in the plane and an integer k, find an x-monotone polyline $P = ((u_0, v_0), (a_1, b_1), (u_1, v_1), (a_2, b_2), (u_2, v_2), \ldots, (u_{k+1}, v_{k+1}), (a_{k+1}, b_{k+1}), (u_{k+2}, v_{k+2}))$ where $-\infty = u_0 < u_1 < \ldots < u_{k+1} < u_{k+2} = +\infty$ and $v_i = a_i u_i + b_i = a_{i+1} u_i + b_{i+1}$, $i = 1, 2, \ldots, k+1$ for each of the following three optimization conditions:

$$(A) \quad \sum_{i=0}^{k+1} \sum_{u_i \leqslant x_j < u_{i+1}} (a_i x_j + b_i - y_j)^2 \to \min,$$

$$(B) \quad \sum_{i=0}^{k+1} \sum_{u_i \leqslant x_j < u_{i+1}} |a_i x_j + b_i - y_j| \to \min,$$

$$(C) \quad \sum_{i=0}^{k+1} \max_{u_i \leqslant x_j < u_{i+1}} (a_i x_j + b_i - y_j) \to \min,$$

subject to $a_i x_j + b_i - y_j \geqslant 0$ for $u_i \leqslant x_j < u_{i+1}$.

Figure 4.1 gives an example of a polyline approximating a set of points in the plane.

The author developed a general theory to deal with those problems.[1] In this paper we focus on a special case of 1-joint polyline, which is most interesting among them.

---

[a]The result described here is partially included in the paper.[1]

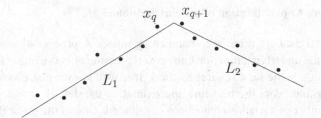

Fig. 4.1   A 1-joint polyline consisting of two half lines approximating a set of points.

4.3 Known Results

4.3.1 *Least Square Error*

An algorithm for finding a line minimizing the sum of squared errors is well known as a least squares method. Once we compute $\sum_{i=1}^{n} x_i^2, \sum_{i=1}^{n} x_i, \sum_{i=1}^{n} y_i$, and $\sum_{i=1}^{n} x_i y_i$ in linear time, we have an optimal approximating line $y = ax + b$ by solving the system of equations defined by those values.

4.3.2 *Minimum Absolute Error*

A linear-time algorithm for finding an optimal line that minimizes the sum of absolute errors is known. An idea behind the efficient implementation is prune-and-search. See Imai, Katoh and Yamamoto[2] for more detail.

4.3.3 *Sum of Max Vertical Errors*

The problem in this case is to find a pair of parallel lines of minimum vertical width that contains all the given points between them. Such a pair is characterized by a minimum vertical gap between upper and lower envelopes of a set of lines dual to the given points. Thus, if points are sorted, the gap can be computed in linear time.

4.4 Approximation by a 1-joint Polyline

In this section we consider the problem of approximating a set of points by a 1-joint polyline to minimize the objective functions defined above. For the time being we fix a partition of a given point set into subsets: $S_1 = \{(x_1,y_1),(x_2,y_2),\ldots,(x_q,y_q)\}$ and $S_2 = \{(x_{q+1},y_{q+1}),\ldots,(x_n,y_n)\}$, where each point p_i is given by (x_i,y_i) throughout the paper. Our objective polyline consists of two half lines. The left half line $y = a_1x + b_1$ approximates points of S_1 and the right one $y = a_2x + b_2$ those of S_2. We call a tuple (a_1, b_1, a_2, b_2) *feasible* if the two lines $y = a_1x + b_1$

and $y = a_2x + b_2$ meet in the interval $[x_q, x_{q+1}]$. Our goal here is to find a tuple (a_1, b_1, a_2, b_2) for the following three problems:

$R_1(q)$: minimize $D_1(a_1, b_1, a_2, b_2) =$

$$\sum_{i=1}^{q}(a_1x_i + b_1 - y_i)^2 + \sum_{j=q+1}^{n}(a_2x_j + b_2 - y_j)^2 \qquad (4.1)$$

$R_2(q)$: minimize $D_2(a_1, b_1, a_2, b_2) =$

$$\sum_{i=1}^{q}|a_1x_i + b_1 - y_i| + \sum_{j=q+1}^{n}|a_2x_j + b_2 - y_j| \qquad (4.2)$$

$R_3(q)$: minimize $D_3(a_1, b_1, a_2, b_2) =$

$$\max_{1 \leqslant i \leqslant q}(a_1x_i + b_1 - y_i) + \max_{q+1 \leqslant j \leqslant n}(a_2x_j + b_2 - y_j) \qquad (4.3)$$

subject to $\quad a_1x_i + b_1 - y_i \geqslant 0, \ 1 \leqslant i \leqslant q,$

and $\quad a_2x_j + b_2 - y_j \geqslant 0, \ q+1 \leqslant j \leqslant n.$

A crucial observation here is the convexity of the objective functions.

Lemma 4.1 *Fixed a partition of a point set S into subsets S_1 and S_2, the problem of finding (a_1, b_1, a_2, b_2) that minimizes three objective functions above are all decomposed into two convex programming problems.*

Proof It is an easy observation that each term (1)-(3) above is a convex function in the (a_1, b_1, a_2, b_2) space. Thus, each function is the sum of two convex functions and hence it is also convex in the space. Also, the constraint that the two lines $y = a_1x + b_1$ and $y = a_2x + b_2$ meet in the interval $[x_q, x_{q+1}]$ can be expressed as linear constraints depending on whether $a_1 \leqslant a_2$ holds or not. If $a_1 \leqslant a_2$ holds, the constraint is

$$x_q(a_2 - a_1) \leqslant b_1 - b_2 \leqslant x_{q+1}(a_2 - a_1). \qquad (4.4)$$

Otherwise, it is

$$x_q(a_1 - a_2) \leqslant b_1 - b_2 \leqslant x_{q+1}(a_1 - a_2). \qquad (4.5)$$

Thus, the problem $R_i(q), q = 1, 2, 3$ under the constraint of (4.4) or (4.5) is a convex program. In particular, for $i = 1$, it is a convex quadratic program while for $i = 2, 3$ it is a linear program. □

From this lemma, all three problems $R_i(q)$ are polynomially solvable, and thus 1-joint polyline approximation can be solved in polynomial time.

In order to solve a 1-joint problem more efficiently, let us consider the following 0-joint (i.e. linear approximation) problems for S_1 and S_2.

$$Q_1^1 : \text{minimize} \sum_{i=1}^{q} (a_1 x_i + b_1 - y_i)^2 \tag{4.6}$$

$$Q_1^2 : \text{minimize} \sum_{i=q+1}^{n} (a_1 x_i + b_1 - y_i)^2 \tag{4.7}$$

$$Q_2^1 : \text{minimize} \sum_{i=1}^{q} |a_1 x_i + b_1 - y_i| \tag{4.8}$$

$$Q_2^2 : \text{minimize} \sum_{i=q+1}^{n} |a_1 x_i + b_1 - y_i| \tag{4.9}$$

$$Q_3^1 : \text{minimize} \left\{ \max_{1 \leqslant i \leqslant q} (a_1 x_i + b_1 - y_i) \mid a_1 x_i + b_1 - y_i \geqslant 0,\ 1 \leqslant i \leqslant q \right\} \tag{4.10}$$

$$Q_3^2 : \text{minimize} \left\{ \max_{q+1 \leqslant i \leqslant n} (a_1 x_i + b_1 - y_i) \mid a_1 x_i + b_1 - y_i \geqslant 0,\ q+1 \leqslant i \leqslant n \right\} \tag{4.11}$$

Lemma 4.2 *Let (a_1^*, b_1^*) and (a_2^*, b_2^*) be optimal solutions of Q_i^1 and Q_i^2, respectively. (i) If the two lines $y = a_1^* x + b_1^*$ and $y = a_2^* x + b_2^*$ do not meet in the open interval (x_q, x_{q+1}), then an optimal 1-joint polyline for $R_i(q)$ has its joint on $x = x_q$ or $x = x_{q+1}$. (ii) Otherwise, $(a_1^*, b_1^*, a_2^*, b_2^*)$ is optimal to $R_i(q)$.*

Proof Let $(\hat{a}_1, \hat{b}_1, \hat{a}_2, \hat{b}_2)$ be an optimal solution of $R_i(q)$, and let us assume $a_1 \leqslant a_2$ without loss of generality. Suppose an optimal 1-joint polyline has its joint in the open interval (x_q, x_{q+1}). Then the constraint (4.4) is not tight. This means that an optimal solution of $R_i(q)$ can be obtained without imposing the constraint (4.4). Therefore, $(\hat{a}_1, \hat{b}_1) = (a_1^*, b_1^*)$ and $(\hat{a}_2, \hat{b}_2) = (a_2^*, b_2^*)$. This is a contradiction.   □

4.4.1 *Least Square Error*

From Lemma 4.2, we have a simple algorithm:

Algorithm for finding an optimal 1-joint polyline

(input) $((x_1, y_1), (x_2, y_2), \ldots, (x_n, y_n)), x_1 < x_2 < \cdots < x_n.$
(Step 0) Compute the followings:

$$\sum_{i=1}^{n} x_i^2,\ \sum_{i=1}^{n} y_i^2,\ \sum_{i=1}^{n} x_i,\ \sum_{i=1}^{n} y_i,\ \sum_{i=1}^{n} x_i y_j.$$

(Step 1) for q=1 to n do{

(Step 2)   update the summations: for example, add x_i^2 to $\sum_{i=1}^{q-1} x_i^2$ to get $\sum_{i=1}^{q} x_i^2$.

(Step 3)   Solve the left problem defined by $S_1 = \{(x_1,y_1),\ldots,(x_q,y_q)\}$ and
the right one by $S_2 = \{(x_{q+1},y_{q+1}),\ldots,(x_n,y_n)\}$ independently.

(Step 4)   If the resulting two lines $y = a_1x + b_1$ and $y = a_2x + b_2$ meet
between x_q and x_{q+1}, then continue to the next q.

(Step 5)   Solve the problem with additional constraint that the two lines
meet at $x = x_q$.

}

(Step 6) return the best polyline found so far.

The problem with the additional constraint can be solved using the Kuhn-Tucker condition.

Define

$$f(a_1,b_1,a_2,b_2) = \sum_{i=1}^{q}(a_1x_i+b_1-y_i)^2 + \sum_{j=q+1}^{n}(a_2x_j+b_2-y_j)^2,$$

$$g(a_1,b_1,a_2,b_2) = a_1x+b_1-a_2x-b_2,$$

$$L(a_1,b_1,a_2,b_2) = f(a_1,b_1,a_2,b_2) - \lambda g(a_1,b_1,a_2,b_2),$$

where

$$x = x_q.$$

Then, an optimal solution minimizing $f(a_1,b_1,a_2,b_2)$ also minimizes $L(a_1,b_1,a_2,b_2)$. So, by the Kuhn-Tucker condition an optimal solution $(a_1^*,b_1^*,a_2^*,b_2^*)$ for a fixed partition of S into S_1 and S_2 should satisfy

$$\frac{\partial L}{\partial a_1}(a_1^*,b_1^*,a_2^*,b_2^*) = \frac{\partial L}{\partial b_1}(a_1^*,b_1^*,a_2^*,b_2^*) = 0,$$

$$\frac{\partial L}{\partial a_2}(a_1^*,b_1^*,a_2^*,b_2^*) = \frac{\partial L}{\partial b_2}(a_1^*,b_1^*,a_2^*,b_2^*) = 0,$$

and

$$g(a_1^*,b_1^*,a_2^*,b_2^*) = a_1^*x+b_1^*-a_2^*x-b_2^* = 0.$$

We can compute optimal parameter values a_1^*, b_1^*, a_2^* and b_2^* and further the value of our objective function in constant time if the summations

$$\sum_{i=1}^{q}x_i^2, \ \sum_{i=1}^{q}y_i^2, \ \sum_{i=1}^{q}x_i, \ \sum_{i=1}^{q}y_i, \ \sum_{i=1}^{q}x_iy_i,$$

$$\sum_{j=q+1}^{n}x_j^2, \ \sum_{j=q+1}^{n}y_j^2, \ \sum_{j=q+1}^{n}x_j, \ \sum_{j=q+1}^{n}y_j, \ \sum_{j=q+1}^{n}x_jy_j,$$

are all available. It is also obvious that we can update those summations in constant time at each iteration of the loop in the algorithm.

Theorem 4.1 *The algorithm described above finds an optimal 1-joint polyline in linear time.*

4.4.2 Minimum Absolute Error

Let us now turn to the next criterion, minimization of the sum of absolute errors. Given a set S of points $\{(x_1,y_1),(x_2,y_2),\ldots,(x_n,y_n)\}$ such that $x_1 < x_2 < \cdots < x_n$ in the plane, find an x-monotone 1-joint polyline, $P = ((a_1,b_1),(u_1,v_1),(a_2,b_2))$, such that

(1) each of two intervals $[-\infty,u_1]$ and $[u_1,\infty]$ contains at least one point from S, where u_1 gives the intersection of the two lines $y = a_1x + b_1$ and $y = a_2x + b_2$,
(2) $\sum_{x_i \in [-\infty,u_1]} |a_1x_i + b_1 - y_i| + \sum_{x_j \in [u_1,\infty]} |a_2x_j + b_2 - y_j|$ is smallest among those satisfying the above conditions.

An optimal approximating polyline can be found in polynomial time using linear programming. First of all we sort points in the increasing order of their x-coordinate. So, we assume $x_1 < x_2 < \ldots < x_n$. Then, there are $n-1$ cases depending on the value of u_1: case 1: $x_1 \leqslant u_1 < x_2$, case 2: $x_2 \leqslant u_1 < x_3$, $\ldots$, case $n-1$: $x_{n-1} \leqslant u_1 < x_n$. In each case we can find an optimal approximating polyline by solving the following problem:

Problem for Case q

$$\min \sum_{i=1}^{q} |a_1x_i + b_1 - y_i| + \sum_{j=q+1}^{n} |a_2x_j + b_2 - y_j|$$

$$\text{subject to:} \quad x_q \leqslant \frac{b_2 - b_1}{a_1 - a_2} < x_{q+1}.$$

This problem can be converted to the following two linear programs depending on the sign of $a_1 - a_2$. Required optimal solution is obtained by taking a better one among the two solutions.

Linear Program $P^+(q)$ for Case q

$$\min \sum_{i=1}^{n} z_i$$

subject to: $\quad z_i \geqslant a_1 x_i + b_1 - y_i, i = 1, 2, \ldots, q,$

$$z_i \geqslant -a_1 x_i - b_1 + y_i, i = 1, 2, \ldots, q,$$

$$z_j \geqslant a_2 x_j + b_2 - y_j, j = q + 1, \ldots, n,$$

$$z_j \geqslant -a_2 x_j - b_2 + y_j, j = q + 1, \ldots, n,$$

$$x_q(a_1 - a_2) \leqslant b_2 - b_1 < x_{q+1}(a_1 - a_2).$$

Linear Program $P^-(q)$ for Case q

$$\min \sum_{i=1}^{n} z_i$$

subject to: $\quad z_i \geqslant a_1 x_i + b_1 - y_i, i = 1, 2, \ldots, q,$

$$z_i \geqslant -a_1 x_i - b_1 + y_i, i = 1, 2, \ldots, q,$$

$$z_j \geqslant a_2 x_j + b_2 - y_j, j = q + 1, \ldots, n,$$

$$z_j \geqslant -a_2 x_j - b_2 + y_j, j = q + 1, \ldots, n,$$

$$x_q(a_1 - a_2) \geqslant b_2 - b_1 > x_{q+1}(a_1 - a_2).$$

Each of the programs has $2n + 2$ constraints and $n + 4$ variables $z_1, \ldots, z_n, a_1, b_1,$ a_2, b_2. Thus, it is indeed solved in polynomial time in n. However, we have a more efficient algorithm based on some useful observations.

Again, our objective is to find a 1-joint polyline that optimally approximates a given set of points. Consider Case q, that is, the case where the joint lies between x_q and x_{q+1}. If $q = 1$ or $q = n - 1$ then one of the parts contains exactly one point. Thus, if we calculate an optimal approximating line for the other part, we can easily obtain an optimal approximating monotone polyline by connecting the line with the vertical line passing through the remaining one point. So, we assume that each part contains two or more points. Now, by S_1 we denote the set of the first q points in the sorted order, and by S_2 the set of remaining points, that is,

$$S_1 = \{(x_1, y_1), (x_2, y_2), \ldots, (x_q, y_q)\},$$
$$S_2 = \{(x_{q+1}, y_{q+1}), \ldots, (x_n, y_n)\}.$$

Is there a more efficient algorithm? A framework similar to that for the least square error also applies in this case. That is, when we partition a given point set into left and right parts at x_q, an optimal solution is obtained either by a combination of two independent optimal solutions in the both sides or by solving the problem with an additional constraint that the two lines must meet at the partition point x_q. For the first part we can apply a linear-time algorithm by Imai et al.[2] Unfortunately, the second problem cannot be solved in a fashion similar to that for the least square errors since it is hard to apply the Kuhn-Tucker condition to our objective function defined by the absolute function. So, we need a different scheme.

We present an $O(\min\{n^2 \log_2 B, n^2 \log^3 n\})$-time algorithm where $\log_2 B$ is a problem size. A key lemma to the algorithm is the following: Let $R_2^+(q)$ and $R_2^-(q)$ be the problem $R_2(q)$ with the additional constraint (4.4) and the one with (4.5), respectively.

Lemma 4.3 *Let (a_1^*, b_1^*) and (a_2^*, b_2^*) be optimal solutions to Q_2^1 and Q_2^2, respectively. If $a_1^* < a_2^*$ then an optimal 1-joint polyline for the problem $R_2^-(q)$ is a straight line without a joint, that is, $a_1 = a_2$. Otherwise, an optimal 1-joint polyline for the problem $R_2^+(q)$ is a straight line without a joint.*

Proof We only prove the first part. The optimal solution to $R_2^+(q)$ is $(a_1^*, b_1^*, a_2^*, b_2^*)$. Let (a_1', b_1', a_2', b_2') be an optimal solution to $R_2^-(q)$. Suppose $a_1' > a_2'$. The constraint $a_1 \geqslant a_2$ in $R_2^-(q)$ is not tight in the optimal solution. So, (a_1', b_1', a_2', b_2') remains optimal after removing the constraint $a_1 \geqslant a_2$. However, it contradicts to the optimality of $(a_1^*, b_1^*, a_2^*, b_2^*)$. Hence, the optimal solution to $R_2^-(q)$ should satisfy $a_1' = a_2'$. If $a_1' = a_2'$ holds, the constraint $(a_1 - a_2)x_q \geqslant b_2 - b_1 \geqslant (a_1 - a_2)x_{q+1}$ is meaningless since it is meaningful only if $a_1 \neq a_2$. Therefore, an optimal solution to $R_2^-(q)$ also satisfies $a_1' = a_2'$. $\square$

Hereafter, we present an algorithm for solving $R_2^+(q)$ under the assumption that $a_1^* \leqslant a_2^*$. If the two lines $y = a_1^* x + b_1^*$ and $y = a_2^* x + b_2^*$ meet in the interval $[x_q, x_{q+1}]$, then we have an optimal solution. So, hereafter we assume that their intersection $x = u^*$ lies outside the interval.
Case 1: If $u^* > x_{q+1}$ then the joint of an optimal solution to $P^+(q)$ is x_{q+1} because if the constraint $b_2 - b_1 \leqslant x_{q+1}'(a_1 - a_2)$ is not tight in the optimal solution to $R_2^+(q)$, then we have a contradiction as we had before.
Case 2: If $u^* < x_q$ then the joint of an optimal solution to $R_2^+(q)$ is x_k.

In the following we describe an algorithm for solving $R_2^+(q)$ under the condition that the joint is at x_q.

Let b be the y-coordinate of the joint. Then, two lines become $y = a_1(x - x_q) + b$ and $y = a_2(x - x_q) + b$. For simplicity we assume $x_q = 0$. Note that this problem is again a linear program.

Let D_1 and D_2 be the dual planes for S_1 and S_2, respectively. If we fix b, an optimal a_1 value corresponds to the median among those intersections between the horizontal line through b and the dual lines in D_1. This value can be computed in $O(n)$ time. An optimal a_2 value is also obtained in $O(n)$ time. So, if $a_1 \leqslant a_2$ then we are done. Problem occurs when $a_1 > a_2$. However, an optimal solution to $R_2^-(q)$ under the condition $a_1 \geqslant a_2$ is a straight line ($a_1 = a_2$) by Lemma 4.3, and hence if $a_1 > a_2$ then it cannot be optimal to $R_2(q)$. Thus, we do not need to care the condition $a_1 \leqslant a_2$.

Let $f(b)$ be an optimal value when b is fixed. Then, $f(b)$ is a convex function since it is a linear program. Thus, $R_2(q)$ can be computed in $O(n \log B)$ time using binary search on b, where $\log_2 B$ is a problem size. Alternatively, we can apply the parametric search which gives us $O(n \log^3 n)$ time algorithm.

Theorem 4.2 *We can find an optimal 1-joint polyline with minimum absolute error in $O(n^2 \log^3 n)$ time.*

4.4.3 *Sum of Maximum Vertical Errors*

Given a set S of points $\{(x_1, y_1), (x_2, y_2), \ldots, (x_n, y_n)\}$ such that $x_1 < x_2 < \cdots < x_n$ in the plane, find an x-monotone 1-joint polyline, $P = ((a_1, b_1), (u_1, v_1), (a_2, b_2))$, such that

(1) each of two intervals $[-\infty, u_1]$ and $[u_1, \infty]$ contains at least one point from S, where u_1 gives the intersection of the two lines $y = a_1 x + b_1$ and $y = a_2 x + b_2$, that is, $a_1 u_1 + b_1 = a_2 u_1 + b_2$,
(2) $\sum_{x_i \leqslant u_1} (a_1 x_i + b_1 - y_i) + \sum_{x_j > u_1} (a_2 x_j + b_2 - y_j) \to \min$,
    subject to $a_1 x_i + b_1 - y_i \geqslant 0$ for each $x_i \leqslant u_1$ and $a_2 x_j + b_2 - y_j \geqslant 0$ for each $x_j > u_1$ among those satisfying the above conditions.

Given a partition of a point set into left and right parts at $x = x_q$, we construct arrangements of dual lines for the two parts and find upper and lower envelopes for both of them. Then, an optimal solution to the constrained problem is characterized as a pair of two vertical segments satisfying the following condition:

(1) one segment spans the upper and lower envelopes of the arrangement for the left point set, and the other segment spans them in the right one,
(2) the two segments have the same length,
(3) the slope of the line passing through the upper (resp. lower) endpoints of the segments is between x_q and x_{q+1},
(4) the pair of segments has a shortest possible length among all those satsifying the above conditions.

See Figure 4.2 for pictorial illustration.

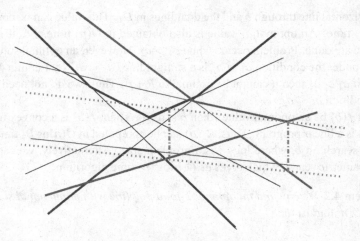

Fig. 4.2   Two arrangements of dual lines and two vertical segments to span the upper and lower envelopes in both sides.

We want to optimize the sum (not the max) of the two vertical widths. The feasible region is convex (namely the region above the upper envelope) and for each point in the feasible region the objective value is the distance to the lower envelope of the dual lines.

The algorithm: The algorithm follows the general outline. For each q, we determine the optimal solution of both subproblems (the unconstrained optima) and also the optimal solution under the assumption that the lines intersect on $x = x_q$ (the constrained optima).

We observe first that envelopes can be obtained in amortized constant time. Assume we add the point (x_i, y_i). The dual line has larger slope than all preceding lines. We can therefore update the upper envelope by walking along it from the right and the lower envelope by walking along it from the left.

If we maintain the envelopes in a binary tree structure, we can determine the closest points in time $O(\log n)$ by binary search.

Lemma 4.4 *The unconstrained optima can be determined in total time $O(n \log n)$.*

We turn to the constrained optima. The lines (a_1, b_1) and (a_2, b_2) have to meet at $x = x_q$.

Lemma 4.5 *The vertical width of the left solution is a convex function of the y-coordinate of its intersection with $x = x_q$.*

Proof A line (a_1, b_1) intersects $x = x_q$ at $y = a_1 x_q + b_1$. A fixed value of y, say $y = y_0$, therefore, restricts consideration to the pairs (a, b) with $y_0 = a_1 x_q + b_1$ or $b_1 = y_0 - a_1 x_q$. This line intersecting the feasible region is a sloped line segment. The slope is $-x_q$. The lower envelope minus this sloped line is a convex function. The minimum distance between the sloped line is either at an interior point (if there is a vertex on the lower envelope within the range of the line segment where on of the incident edges has slope larger than $-x_q$ and one has slope smaller than $-x_q$) or at one of the endpoints. Rest of the proof is easier. . □

The Lemma above paves the way for finding the optimal height of the intersection with $x = x_q$. We use binary search. For a fixed height, the lemma above tells us the optimal slope, the vertical width and also the derivative of the vertical width. Adding the derivatives for both sides tells us in which direction to proceed. We have thus shown.

Theorem 4.3 *The 1-joint sum of vertical widths problem can be solved in time* $O(n \log n)$.

Proof For each partition of a point set we can determine the optimal constrained optimum in time $O(\log n)$. □

4.5 Inserting a Point with Designated Distances to Existing Points

In this section we show a similar idea applies to a completely different geometric problem. Suppose that we are given a set S of n points in the plane, $S = \{p_1, \ldots, p_n\}$ and we are requested to insert a new point p. When a target distance δ_i from the new point p to each existing point p_i is given as input, we want to insert p at the distance as close as the given distance δ_i, as shown in Figure 4.3. More exactly, our objective function is given by

$$f(p) = \sum_{i=1}^{n} |d(p, p_i)^2 - \delta_i^2|, \tag{4.12}$$

where $d(p, p_i)$ is the Euclidean distance between the two points p and p_i. We could define an objective function without squares, that is,

$$g(p) = \sum_{i=1}^{n} |d(p, p_i) - \delta_i|. \tag{4.13}$$

In this case the problem is harder since a special case where all target distances are 0 is a hard problem known as the Fermat-Weber problem.[3]

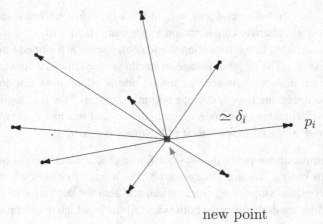

$\simeq \delta_i$ p_i

new point

Fig. 4.3   Inserting a point so that the distances to existing points are roughly preserved.

Let (x,y) be the coordinates of the new point p and (x_i,y_i) be those of the existing point p_i. Then, the equation (4.12) becomes

$$f(x,y) = \sum_{i=1}^{n} |(x - x_i)^2 + (y - y_i)^2 - \delta_i^2|. \qquad (4.14)$$

A serious difference from the previous sections is the absolute symbols in the right terms. Without them we can differentiate the objective function by x and y. In other words, we can use the least-squares method if we can remove the absolute symbols.

We use a standard technique in computational geometry. That is, we draw a circle C_i of radius δ_i centered at each point p_i in the plane. Then, the plane is partitioned into $O(n^2)$ cells (connected regions bounded by those n circles). Each region (cell, hereafter) R is characterized by two sets: one of all circles that contain R in it and the other of all other circles.

Let $p(x,y)$ be any point in a cell R. To characterize the cell R we define a variable σ_i by

$$\sigma_i = \begin{cases} 1, & \text{if the point } p \text{ (or the cell } R) \text{ is outside the circle } C_i, \\ -1, & \text{otherwise.} \end{cases} \qquad (4.15)$$

Then, the objective function value $f_R(x,y)$ at the point $p(x,y)$ in cell R can be calculated by

$$f_R(x,y) = \sum_{i=1}^{n} \sigma_i[(x - x_i)^2 + (y - y_i)^2 - \delta_i^2]. \qquad (4.16)$$

The resulting expression has no absolute symbol, and thus we can find an optimal solution within the cell R by solving a system of equations obtained by differentiating the objective function by x and y.

$$\frac{\partial f_R}{\partial x} = 2\left(\sum_{i=1}^{n} \sigma_i\right)x - 2\left(\sum_{i=1}^{n} \sigma_i x_i\right) = 0,$$

$$\frac{\partial f_R}{\partial y} = 2\left(\sum_{i=1}^{n} \sigma_i\right)y - 2\left(\sum_{i=1}^{n} \sigma_i y_i\right) = 0,$$

that is,

$$p^*(x,y) = \left(\frac{\sum_{i=1}^{n} \sigma_i x_i}{\sum_{i=1}^{n} \sigma_i}, \frac{\sum_{i=1}^{n} \sigma_i y_i}{\sum_{i=1}^{n} \sigma_i}\right). \tag{4.17}$$

If the point p^* giving the solution lies in the cell, it is the solution for the cell. Otherwise, there are two cases depending on the sign of the objective function $f_R(x,y)$. Recall that

$$f_R(x,y) = \left(\sum_{i=1}^{n} \sigma_i\right)(x^2 + y^2) - 2\left(\sum_{i=1}^{n} \sigma_i x_i\right)x - 2\left(\sum_{i=1}^{n} \sigma_i y_i\right)y$$

$$+ \sum_{i=1}^{n}(\sigma_i x_i^2 + \sigma_i y_i^2 - \sigma_i \delta_i^2). \tag{4.18}$$

If the coefficient of the leading term, $\sum_{i=1}^{n} \sigma_i$, is positive, the function is convex, and otherwise it is concave. If it is convex, a point on the cell boundary that is closest to the point p^* minimizes the objective function. On the other hand, if it is concave, a point that is farthest from p^* is optimal. See Figure 4.4. The coefficient $\sum_{i=1}^{n} \sigma_i$ may happen to be 0. In Figure 4.4 the cell painted dark is contained in exactly two circles and outside exactly two circles, and thus we have $\sum_{i=1}^{n} \sigma_i = 0$ for the cell. In the case the objective function is a linear function in x and y. Indeed, it is

$$f_R(x,y) = -2\left(\sum_{i=1}^{n} \sigma_i x_i\right)x - 2\left(\sum_{i=1}^{n} \sigma_i y_i\right)y + \sum_{i=1}^{n}(\sigma_i x_i^2 + \sigma_i y_i^2 - \sigma_i \delta_i^2). \tag{4.19}$$

Thus, a point in the cell that minimizes the objective function is either a vertex on the cell boundary or a tangent point of a line of the form $(\sum_{i=1}^{n} \sigma_i x_i)x + (\sum_{i=1}^{n} \sigma_i y_i)y = c$ for some constant c with a circular arc forming the cell boundary. Thus, if we compute an optimal point for each cell, then we get a global optimum.

Implementing the algorithm above in a naive manner requires $O(n^2)$ space. It is also very slow since it takes $O(n^3)$ time in total although it is not so hard to

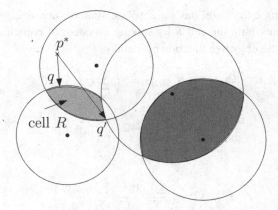

Fig. 4.4   An arrangement of four circles. The point p^* is a point that minimizes the objective function defined for the cell R. The points q and q' are those closest to and farthest from p^*, respectively, on the cell R. The cell painted dark is contained by exactly two circles and outside exactly two circles.

improve the time complexity to $O(n^2 \log n)$. To improve it further to $O(n^2)$ is a challeging open problem.

We can save the space complexity by using a plane sweep technique which is a standard technique to reduce the space complexity and also the running time. It moves a vertical sweep line from left to right while maintaining a set of circles intersecting the sweep line and also information about the current cell. The system of equations derived by differentiating the objective function by x and y can be updated in constant time at each event point where the sweep line starts to hit some circle, leaves some circle and arrives at intersection of two circles. Since the status of the sweep line (insertion and deletion) can be maintained in $O(\log n)$ time at each event point, the total time we need is $O(n^2 \log n)$.

It is still open whether we can improve the time complexity to $O(n^2)$. Arrangement of straight lines instead of that of circles can be searched in linear time using a technique called Topological Sweep[4] or Topological Walk.[5] It is not known whether a similar algorithm can be applied to an arrangement of circles.

4.6 Conclusion

In this paper we have revisited the least-squares method to extend it to approximate a point sequence by two half lines and also to solve a completely different geometric problem. Extension to a more general case is still open. The algorithm for the geometric problem given in this paper is just a draft and more detailed description and analysis will be needed.

Acknowledgments

The first part of this paper is based on an unpublished paper which is a joint work with Boris Aronov, Kurt Mehlhorn, and Naoki Katoh. The author would like to express his sincere thanks to those coauthors. The part of this research was partially supported by the Ministry of Education, Science, Sports and Culture, Grant-in-Aid for Scientific Research on Priority Areas and Scientific Research (B).

References

1. B. Aronov, T. Asano, N. Katoh, K. Mehlhorn, and T. Tokuyama, Polyline fitting of planar points under min-sum criterion, *International Journal on Computational Geometry and Applications*, **16** (2006), 97-116.
2. H. Imai, K. Kato, and P. Yamamoto, A linear-time algorithm for linear L_1 approximation of points, *Algorithmica* **4** (1989) 77–96.
3. P. Bose, A. Maheshwari, P. Morin, Fast approximations for sums of distances, clustering and the Fermat-Weber problem, *Computational Geometry: Theory and Applications* **24** (2003) 135–146.
4. H. Edelsbrunner and L. J. Guibas, Topologically sweeping an arrangement, *J. Comp. and Sys. Sci.* **38** (1989) 165–194.
5. T. Asano, L. J. Guibas and T. Tokuyama, Walking in an Arrangement Topologically, *Int. J. Computational Geometry and Applications* **4** (1994) 123–151.
6. P. K. Agarwal and K. R. Varadarajan, Efficient algorithms for approximating polygonal chains, *Discrete Comput. Geom.* **23** (2000) 273–291.
7. M. Goodrich, Efficient piecewise-linear function approximation using the uniform metric, *Discrete Comput. Geom.* **14** (1995) 445–462.
8. S. Hakimi and E. Schmeichel, Fitting polygonal functions to a set of points in the plane, *Graphical Models and Image Processing* **53** (1991) 132–136.
9. H. Imai and M. Iri, Polygonal approximations of a curve — Formulations and algorithms, *Computational Morphology*, Elsevier Science Publishers B.V. (North Holland, 1988), pp. 71–86.
10. Y. Kurozumi and W. A. Davis, Polygonal approximation by the minimax method, *Computer Graphics and Image Processing* **19** (1982) 248–264.
11. A. Melkman and J. O'Rourke, On polygonal chain approximation, *Computational Morphology*, Elsevier Science Publishers B.V. (North Holland, 1988), pp. 87–95.
12. J. O'Rourke and G. Toussaint, Pattern recognition, Chapter 43 of *Handbook of Discrete and Computational Geometry*, eds. J. Goodman and J. O'Rourke (CRC Press, 1997).
13. D. P. Wang, N. F. Huang, H. S. Chao, and R. C. T. Lee, Plane sweep algorithms for polygonal approximation problems with applications, in *Proc. 4th Int. Symp. Algorithms Comput. (ISAAC 2003)*, LNCS 762 (1993), pp. 515–522.
14. Robert Weibel, Generalization of spatial data: principles and selected algorithms, *Algorithmic Foundations of Geographic Information Systems*, LNCS 1340 (1997), pp. 99–152.

Appendix

Denote

$$D_1(a,b) = \sum_{i=1}^{n}(ax_i + b - y_i)^2 \tag{4.20}$$

$$D_2(a,b) = \sum_{i=1}^{n}|ax_i + b - y_i| \tag{4.21}$$

$$D_3(a,b) = \max_{1 \leqslant i \leqslant n}(ax_i + b - y_i) \mid ax_i + b - y_i \geqslant 0, i = 1,2,\ldots,n. \tag{4.22}$$

Lemma 4.6 *The three functions are all convex.*

Proof It suffices to show

$$\theta D_j(a_1,b_1) + (1-\theta)D_j(a_2,b_2) \geqslant D_j(\theta a_1 + (1-\theta)a_2, \theta b_1 + (1-\theta)b_2)) \tag{4.23}$$

holds for $j = 1,2,3$ for any $\theta, 0 \leqslant \theta \leqslant 1$ and for any (a_1,b_1,a_2,b_2).

$$\theta D_1(a_1,b_1) + (1-\theta)D_1(a_2,b_2) - D_1(\theta a_1 + (1-\theta)a_2, \theta b_1 + (1-\theta)b_2))$$

$$= \sum_{i=1}^{n}[\theta(a_1x_i + b_1 - y_i)^2 + (1-\theta)(a_2x_i + b_2 - y_i)^2$$

$$- ((\theta a_1 + (1-\theta)a_2)x_i + \theta b_1 + (1-\theta)b_2 - y_i)^2$$

$$\geqslant \sum_{i=1}^{n}[\theta(a_1x_i + b_1 - y_i) + (1-\theta)(a_2x_i + b_2 - y_i)]^2$$

$$- ((\theta a_1 + (1-\theta)a_2)x_i + \theta b_1 + (1-\theta)b_2 - y_i)^2$$

$$\text{since } (\theta A^2 + (1-\theta)B^2 \geqslant (\theta A + (1-\theta)B)^2$$

$$= \sum_{i=1}^{n}[2\theta(a_1x_i + b_1 - y_i) + 2(1-\theta)(a_2x_i + b_2 - y_i)] \times 0$$

$$= 0.$$

$$\theta D_2(a_1,b_1) + (1-\theta)D_2(a_2,b_2) - D_2(\theta a_1 + (1-\theta)a_2, \theta b_1 + (1-\theta)b_2))$$

$$= \sum_{i=1}^{n}[\theta|a_1x_i + b_1 - y_i| + (1-\theta)|a_2x_i + b_2 - y_i|$$

$$- |(\theta a_1 + (1-\theta)a_2)x_i + \theta b_1 + (1-\theta)b_2 - y_i|$$

$$\geqslant \sum_{i=1}^{n}|\theta(a_1x_i + b_1 - y_i) + (1-\theta)(a_2x_i + b_2 - y_i)|$$

$$- |(\theta a_1 + (1-\theta)a_2)x_i + \theta b_1 + (1-\theta)b_2 - y_i|$$

$$\text{since } (|A| + |B| \geqslant |A + B|)$$

$$= 0.$$

$$\theta D_3(a_1,b_1)+(1-\theta)D_3(a_2,b_2)-D_3(\theta a_1+(1-\theta)a_2,\theta b_1+(1-\theta)b_2))$$

$$= \max_{1\leqslant i\leqslant n}\theta(a_1x_i+b_1-y_i)+\max_{1\leqslant i\leqslant n}(1-\theta)(a_2x_i+b_2-y_i)$$

$$-\max_{1\leqslant i\leqslant n}(\theta a_1+(1-\theta)a_2)x_i+\theta b_1+(1-\theta)b_2-y_i)$$

$$\geqslant \max_{1\leqslant i\leqslant n}[\theta(a_1x_i+b_1-y_i)+(1-\theta)(a_2x_i+b_2-y_i)]$$

$$-\max_{1\leqslant i\leqslant n}(\theta a_1+(1-\theta)a_2)x_i+\theta b_1+(1-\theta)b_2-y_i)$$

(since two maxs are independent)

$$= 0.$$

$\square$

Define

$$D_1^+(a_1,b_1,a_2,b_2) = \sum_{i=1}^{q}(a_1x_i+b_1-y_i)^2 + \sum_{j=q+1}^{n}(a_2x_j+b_2-y_j)^2, \qquad (4.24)$$

$$D_2^+(a_1,b_1,a_2,b_2) = \sum_{i=1}^{q}|a_1x_i+b_1-y_i| + \sum_{j=q+1}^{n}|a_2x_j+b_2-y_j|, \qquad (4.25)$$

$$D_3^+(a_1,b_1,a_2,b_2) = \max_{1\leqslant i\leqslant q}(a_1x_i+b_1-y_i) + \max_{q+1\leqslant j\leqslant n}(a_2x_j+b_2-y_j). \qquad (4.26)$$

For $j = 1,2,3$, $D_j^+(a_1,b_1,a_2,b_2)$ is defined only if the intersection of the two lines $y = a_1x+b_1$ and $y = a_2x+b_2$ lies in the interval $[x_q,x_{q+1}]$ and $a_1 \geqslant a_2$. Moreover, $D_3^+(a_1,b_1,a_2,b_2)$ is defined only if $a_1x_i+b_1-y_i \geqslant 0, i = 1,2,\ldots,q$ and $a_2x_j+b_2-y_j \geqslant 0, j = q+1,\ldots,n$.

Similarly we define $D_1^-(a_1,b_1,a_2,b_2), D_2^-(a_1,b_1,a_2,b_2)$, and $D_3^-(a_1,b_1, a_2,b_2)$ by the same conditions except $a_1 \leqslant a_2$ instead of $a_1 \geqslant a_2$.

Lemma 4.7 *The six functions are all convex.*

Proof Let $\Gamma^+(q)$ is a set of all tuples (a_1,b_1,a_2,b_2) such that their associated intersections lie in the interval $[x_q,x_{q+1}]$ and $a_1 \geqslant a_2$. $\Gamma^-(q)$ is defined similarly but with $a_1 \leqslant a_2$.

We can rewrite $D_1^+(a_1,b_1,a_2,b_2)$ as

$$D_1^+(a_1,b_1,a_2,b_2) = D_{11}(a_1,b_1)+D_{12}(a_2,b_2), \qquad (4.27)$$

where

$$D_{11}(a_1,b_1) = \sum_{i=1}^{q}(a_1x_i+b_1-y_i)^2, \qquad (4.28)$$

$$D_{12}(a_2,b_2) = \sum_{j=q+1}^{n}(a_2x_j+b_2-y_j)^2. \qquad (4.29)$$

Then, $D_{11}(a_1, b_1)$ and $D_{12}(a_2, b_2)$ are both convex. Here note that the distinction between D_1^+ and D_1^- arises in distinct sets of feasible solutions, but the formula for computation are just the same.

For any $\theta, 0 \leqslant \theta \leqslant 1$, if $(a_1, b_1, a_2, b_2) \in \Gamma^+(q)$ and $(a_1', b_1', a_2', b_2') \in \Gamma^+(q)$, then so is $(\theta a_1 + (1 - \theta)a_1', \theta b_1 + (1 - \theta)b_1', \theta a_2 + (1 - \theta)a_2', \theta b_2 + (1 - \theta)b_2')$. Thus, for any $\theta, 0 \leqslant \theta \leqslant 1$ and for any such (a_1, b_1, a_2, b_2) and (a_1', b_1', a_2', b_2'), we have

$$
\begin{aligned}
&\theta D_1(a_1, b_1, a_2, b_2) + (1 - \theta)D_1(a_1', b_1', a_2', b_2') \\
&\quad - D_1(\theta a_1 + (1 - \theta)a_1', \theta b_1 + (1 - \theta)b_1', \theta a_2 + (1 - \theta)a_2', \theta b_2 + (1 - \theta)b_2') \\
&= \theta(D_{11}(a_1, b_1) + D_{12}(a_2, b_2)) + (1 - \theta)(D_{11}(a_1', b_1') + D_{12}(a_2', b_2')) \\
&\quad - (D_{11}(\theta a_1 + (1 - \theta)a_1', \theta b_1 + (1 - \theta)b_1') \\
&\quad + D_{12}(\theta a_2 + (1 - \theta)a_2', \theta b_2 + (1 - \theta)b_2')) \\
&\geqslant 0.
\end{aligned}
$$

Just the same argument applies to D_2^+ and D_3^+ and others.   □

Note that if $(a_1, b_1, a_2, b_2) \in \Gamma_1$ and $(a_1', b_1', a_2', b_2') \in \Gamma_2$ and $\Gamma_1 \neq \Gamma_2$, that is, $a_1 \neq a_2$ and $a_1' \neq a_2'$ and there exists some x_q between $(b_2 - b_1)/(a_1 - a_2)$ and $(b_2' - b_1')/(a_1' - a_2')$ then we may not have $\theta D_1(a_1, b_1, a_2, b_2) + (1 - \theta)D_1(a_1', b_1', a_2', b_2') - D_1(\theta a_1 + (1 - \theta)a_1', \theta b_1 + (1 - \theta)b_1', \theta a_2 + (1 - \theta)a_2', \theta b_2 + (1 - \theta)b_2') \geqslant 0$.

Chapter 5

On Depth Recovery from Gradient Vector Fields

Tiangong Wei and Reinhard Klette

CITR, Department of Computer Science, The University of Auckland,
Tamaki Campus, Auckland, New Zealand

Depth recovery from gradient vector fields is required when reconstructing a surface (in three-dimensional space) from its gradients. Such a reconstruction task results, for example, for techniques in computer vision aiming at calculating surface normals (such as shape from shading, photometric stereo, shape from texture, shape from contours and so on). Surprisingly, discrete integration has not been studied very intensively so far. This chapter presents three classes of methods for solving problems of depth recovery from gradient vector fields: a two-scan method, a Fourier-transform based method, and a wavelet-transform based method. These methods extend previously known techniques, and related proofs are given in a short but concise form.

The two-scan method consists of two different scans through a given gradient vector field. The final surface height values can be determined by averaging these two scans. Fourier-transform based methods are noniterative so that boundary conditions are not needed, and their robustness to noisy gradient estimates can be improved by choosing associated weighting parameters. The wavelet-transform based method overcomes the disadvantage of the Fourier-transform based method, which implicitly require that a surface height function is periodic. Experimental results using synthetic and real images are also presented.

Contents

5.1 Introduction

Discrete integration maps a dense but discrete gradient vector field into a surface representation, normally identified as "height" or "depth". The authors studied discrete integration in the context of computer vision. Here, this way of surface recovery may be part of techniques such as shape from shading (SFS), photometric stereo, shape from texture, or shape from contours. The SFS problem is to reconstruct the 3D shape of an object from a single 2D image of the object using shading and or lighting models for surface normal calculation. Algorithms for solving the SFS problem (see, for example,[1-5]) consist typically of two steps: the first step is to obtain the estimates of surface gradients or surface normals for a discrete set of visible points on the object surface (i.e., discrete gradient vector fields), and the second step is to recover the surface height from the estimated surface orientation. This second step results in the problem of depth recovery from gradient vector fields.

·   The problem of depth recovery from gradients also arises when applying the photometric stereo method (PSM).[6] PSM is to recover the 3D shape of an object from more than one image taken at the same attitude but for varying illumination. PSM allows an approximate solution for surface normals.[6-8] A subsequent integration step (i.e., depth recovery from gradients) is again required to convert estimated surface normals into an estimate of the surface shape.

Shape from texture is another area that leads to the depth recovery from gradients problem. Smith et al.[9,10] proposed a technique for the recovery of a surface texture relief. The recovered texture relief has an useful potential for both visualization and numerical assessments of surface roughness, or for obtaining other parameters such as peak height or peak count. The technique utilizes three or more images to determine a dense gradient field at first. This gradient field is then integrated to obtain the surface texture relief. On the other hand, texture gradient is related to surface shape parameters (orientation, curvature). Shape recovery is made possible by measuring the texture gradient in the image.

The problem of depth recovery from gradients also results when inferring surface shape from a Gauss map or surface shape from the Hessian matrix,[11] where the task is the estimation of an unknown surface height from a set of measurements of the gradients of some surface function. Therefore, it turns out that the entire shape reconstruction process can often be decomposed into two independent steps: gradient computation and gradient integration.

As mentioned above, several active fields of research related to computer vision produce gradient values for a discrete set of visible points on object surfaces. In order to achieve the relative height or depth values of the surface, these surface gradients have to be integrated by using gradient integration techniques.

Only a few numerical methods for the depth recovery from gradients problem have been developed. These methods have been classified traditionally into two categories: *local integration methods*[12–16] and *global integration methods*.[17–19]

So far, the problem of depth recovery from gradients has not been studied often. Just for illustration, recent papers[20,21] still apply algorithms for discrete integration as proposed about 20 years ago. In this chapter, we present theory, algorithms, and experiments for three classes of methods (different o those two categories mentioned before) for depth recovery from gradients.

The structure of the chapter is as follows. Section 5.2 analyzes the integrability of vector fields, and gives a brief review of related numerical methods for depth recovery from gradients. Section 5.3 discusses a two-scan method. Section 5.4 deals with the Fourier-transform based method for depth recovery from gradients. Section 5.5 presents a wavelet-transform based method. Section 5.6 presents experimental results using synthetic and real images. Section 5.7 concludes the chapter.

5.2 Depth Recovery from Gradient Vector Fields

It is a nontrivial problem of computing a surface height function $Z(x,y)$ from an estimated surface gradient field $(p(x,y), q(x,y))$. First, given a vector field (p,q), it may not correspond to a gradient field of any surface height function $Z(x,y)$ at all. Second, different surface height functions will have the same gradients (for example, looking orthogonally onto a stair case may produce the same gradient field as looking onto a plane). Therefore, the problem of depth recovery from gradients is ill-posed. It is only reasonable to determine the surface height function up to an additive constant, (i.e., the *relative surface height*). Notice that the relative surface height map is sometimes sufficient to recognize or inspect an object (e.g., for surface planarity tests in industrial surface inspection); and the relative surface height map may be transformed into absolute values if height values are available for proper scaling.

5.2.1 *Integrability of Vector Fields*

Generally, a given gradient field $p(x,y), q(x,y)$ may not correspond to any surface height function at all. In order of $p(x,y), q(x,y)$ to be the gradients of a surface height function, the given gradient field must be integrable. A vector field $(p(x,y), q(x,y))$ over a simply connected domain Ω is *integrable* if there exists some surface height function $Z(x,y) \in C^1(\Omega)$ such that it satisfies the *weak integrability condition*

$$Z_x(x,y) = p(x,y) \tag{5.1}$$
$$Z_y(x,y) = q(x,y) \tag{5.2}$$

for all $(x,y) \in \Omega$, where the subscripts denote partial derivatives. In other words, a gradient vector field is integrable if it is the gradient field of some surface height function.

Given a vector field $(p(x,y), q(x,y))$ over a simply connected domain Ω. Integrability can be characterized for two slightly different cases:

(i) Assume that components $p(x,y)$ and $q(x,y)$ are continuously differentiable; then the vector field $(p(x,y), q(x,y))$ is integrable if and only if

$$p_y(x,y) = q_x(x,y) \tag{5.3}$$

for all points in Ω; in other words, the surface height function $Z(x,y) \in C^2(\Omega)$ satisfies the *strong integrability condition*

$$Z_{xy}(x,y) = Z_{yx}(x,y) \tag{5.4}$$

(ii) Only assume that the components $p(x,y)$ and $q(x,y)$ are continuous; then the vector field is integrable if and only if

$$\oint_\gamma p(x,y)dx + q(x,y)dy = 0$$

for any closed curve γ in Ω; in other words, the surface height function $Z(x,y) \in C^1(\Omega)$ satisfies the *partial integrability condition*

$$\oint_\gamma Z_x(x,y)dx + Z_y(x,y)dy = 0$$

If the components of a vector field are continuously differentiable, then it is easy to determine whether or not it is integrable or nonintegrable by using the strong integrability condition.

5.2.2 *Local and Global Integration Methods*

In local integration methods,[12-15] an arbitrary initial height value Z_0 is preset for a starting point (x_0, y_0) somewhere in the image of the surface. Then, the relative heights at every point (x, y), which are consistent with this arbitrarily given height value Z_0, will be calculated according to a local approximation rule. Therefore, local integration approaches strongly depend on data accuracy (to avoid error propagation).

To compute surface height $Z(x, y)$ from the estimated surface gradient fields $(p(x, y), q(x, y))$, global integration techniques[1,17] are based on minimizing the quadratic error functional (cost functional) between ideal and given gradient values:

$$W = \iint_{\Omega} [|Z_x - p|^2 + |Z_y - q|^2] dx dy \qquad (5.5)$$

The above functional is invariant when a constant value is added to the surface height $Z(x, y)$. This expresses the fact that depth recovery from gradients can only reconstruct a surface height up to a constant.

Generally speaking, there are three possible methods to solve this optimization problem: variational approaches, direct discretization methods, and expansion methods. The variational approach[1] results in an Euler-Lagrange equation as the necessary condition for a minimum. Then there is a need to solve this Poisson equation. In order to solve the minimization problem (5.5) numerically, the continuous functional W is converted into a discrete problem directly by a discretization method. The disadvantage of variational approaches or discretization methods is the requirement of boundary conditions. But boundary information is not easy to obtain when dealing with depth recovery from gradients.

The surface height is expressed as a linear combination of a set of basis functions in expansion methods such as proposed and studied by Frankot and Chellappa[17] (and used by Klitt, Koschan and Schlüns[8] for formulating a *Frankot-Chellappa algorithm*). Nevertheless, the errors of this algorithm are high for imperfect estimates of surface gradients, or noisy gradient vector fields.[15] Also, the algorithm is very sensitive to abrupt changes in orientation. In this chapter, we will focus on expanding the surface height function using the Fourier basis functions and third-order Daubechies' scaling basis functions.

5.3 Two-Scan Method

This section presents a local method for depth recovery from gradients. Suppose that the surface normals at four grid points $\{(i, j), (i+1, j), (i, j+1), (i+1, j+1)\}$

are represented by the following surface gradients:

$$\mathbf{n}_{i,j} = (p_{i,j}, q_{i,j}, -1)^T$$
$$\mathbf{n}_{i+1,j} = (p_{i+1,j}, q_{i+1,j}, -1)^T$$
$$\mathbf{n}_{i,j+1} = (p_{i,j+1}, q_{i,j+1}, -1)^T$$
$$\mathbf{n}_{i+1,j+1} = (p_{i+1,j+1}, q_{i+1,j+1}, -1)^T$$

Consider grid points $(i, j+1)$ and $(i+1, j+1)$; since the line connecting points $(i, j+1, Z_{i,j+1})$ and $(i+1, j+1, Z_{i+1,j+1})$ is approximately perpendicular to the average normal between these two points, the dot product of the slope of this line and the average normal is equal to zero. This gives

$$Z_{i+1,j+1} = Z_{i,j+1} + \frac{1}{2}(p_{i,j+1} + p_{i+1,j+1})$$

Similarly, we obtain the following regressive relation for grid points $(i+1, j)$ and $(i+1, j+1)$:

$$Z_{i+1,j+1} = Z_{i+1,j} + \frac{1}{2}(q_{i+1,j} + q_{i+1,j+1})$$

Adding above two recursions together, and dividing the result by 2 gives

$$Z_{i+1,j+1} = \frac{1}{2}(Z_{i,j+1} + Z_{i+1,j})$$
$$+ \frac{1}{4}(p_{i,j+1} + p_{i+1,j+1} + q_{i+1,j} + q_{i+1,j+1}) \tag{5.6}$$

Suppose further that the total number of points on the object surface be $N \times N$. If two arbitrary initial height values are preset at grid points $(1, 1)$ and (N, N), then the two-scan algorithm consists of two stages; the first stage starts at the left-most, bottom-most corner of the given gradient field, and determines the height values along x-axis and y-axis by discretizing (5.1) in terms of the forward differences

$$Z_{i,1} = Z_{i-1,1} + p_{i-1,1} \tag{5.7}$$
$$Z_{1,j} = Z_{1,j-1} + q_{1,j-1} \tag{5.8}$$

where $i = 2, ..., N, j = 2, ..., N$. Then scan the image vertically using (5.6). The second stage starts at the right-top corner of the given gradient field and sets the height values by

$$Z_{i-1,N} = Z_{i,N} - p_{i,N} \tag{5.9}$$
$$Z_{N,j-1} = Z_{N,j} - q_{N,j} \tag{5.10}$$

Then scan the image horizontally using the following recursive equation

$$Z_{i-1,j-1} = \frac{1}{2}(Z_{i-1,j} + Z_{i,j-1}) - \frac{1}{4}(p_{i-1,j} + p_{i,j} + q_{i,j-1} + q_{i,j})$$

Since the estimated height values may be affected by the choice of the initial height value, we take an average of the two scan values for the surface height.

5.4 Fourier-Transform Based Methods

Frankot and Chellappa[17] suggested a solution for the SFS problem to enforce the weak integrability condition (5.1) by using the theory of projections onto convex sets. Their method is to project the given (possibly, non-integrable gradient field) onto the nearest integrable gradient field in the least-square sense. In order to improve the accuracy and robustness, and to strengthen the relation between the surface height function and the given gradient field, we introduce two new constraints as follows:

$$Z_{xx}(x,y) = p_x(x,y)$$
$$Z_{yy}(x,y) = q_y(x,y)$$

The two new constraints model the behavior of a change rate in second-order derivatives between the variables. Therefore, the changes of surface height will be more regular. Having the new constraints, we consider the following energy functional

$$W = \iint_{\Omega} \left[|Z_x - p|^2 + |Z_y - q|^2 \right] dxdy$$
$$+ \lambda \iint_{\Omega} \left[|Z_{xx} - p_x|^2 + |Z_{yy} - q_y|^2 \right] dxdy$$
$$+ \mu_1 \iint_{\Omega} \left(|Z_x|^2 + |Z_y|^2 \right) dxdy$$
$$+ \mu_2 \iint_{\Omega} \left(|Z_{xx}|^2 + 2|Z_{xy}|^2 + |Z_{yy}|^2 \right) dxdy \qquad (5.11)$$

where non-negative parameters λ, μ_1, and μ_2 establish a trade-off between those constraints (i.e., they are used to adjust the weighting between the constraints). This cost function reflects the relations among $Z(x,y)$, $p(x,y)$, and $q(x,y)$ more effectively, and makes the best use of the information provided by the given gradient field because it not only constraints the tangent line of the surface, but also constraints its concavity and convexity. The following objective is to solve for the unknown $Z(x,y)$ subject to an optimization process which minimizes the cost function W.

To solve this minimization problem (5.11), Fourier-transform techniques can be applied. The two-dimensional Fourier transform of the surface function $Z(x,y)$

is defined by

$$Z_F(u,v) = \iint_\Omega Z(x,y)e^{-j(ux+vy)}\,dxdy \qquad (5.12)$$

and the inverse Fourier transform is defined by

$$Z(x,y) = \frac{1}{2\pi}\iint_\Omega Z_F(u,v)e^{j(ux+vy)}\,dudv \qquad (5.13)$$

where $j = \sqrt{-1}$ is the imaginary unit, and u and v represent the two-dimensional frequencies in the Fourier domain. The following differentiation properties can be obtained easily:

$$Z_x(x,y) \leftrightarrow juZ_F(u,v)$$
$$Z_y(x,y) \leftrightarrow jvZ_F(u,v)$$
$$Z_{xx}(x,y) \leftrightarrow -u^2Z_F(u,v)$$
$$Z_{yy}(x,y) \leftrightarrow -v^2Z_F(u,v)$$
$$Z_{xy}(x,y) \leftrightarrow -uvZ_F(u,v)$$

where the sign $\leftrightarrow$ means that the Fourier transform of the function on the left-hand side is equal to the one on the right-hand side.

Let $P(u,v)$ and $Q(u,v)$ be the Fourier transforms of the given gradients $p(x,y)$ and $q(x,y)$, respectively. Taking the Fourier transform in the functional (5.11), and using the differentiation properties of the Fourier transform and the following Parseval's formula

$$\iint_\Omega |Z(x,y)|^2\,dxdy = \frac{1}{2\pi}\iint_\Omega |Z_F(u,v)|^2\,dudv \qquad (5.14)$$

we obtain that

$$\frac{1}{2\pi}\iint_\Omega \left[|juZ_F - P|^2 + |jvZ_F - Q|^2\right]dudv$$

$$+ \frac{\lambda}{2\pi}\iint_\Omega \left[|-u^2Z_F - juP|^2 + |-v^2Z_F - jvQ|^2\right]dudv$$

$$+ \frac{\mu_1}{2\pi}\iint_\Omega \left[|juZ_F|^2 + |jvZ_F|^2\right]dudv$$

$$+ \frac{\mu_2}{2\pi}\iint_\Omega \left[|-u^2Z_F|^2 + 2|-uvZ_F|^2 + |-v^2Z_F|^2\right]dudv$$

$$\to minimum$$

where $Z_F = Z_F(u,v)$, $P = P(u,v)$, and $Q = Q(u,v)$. The left-hand side of the above expression can be expanded into

$$\frac{1}{2\pi} \iint_\Omega [u^2 Z_F Z_F^* - ju Z_F P^* + ju Z_F^* P + PP^*$$

$$+ v^2 Z_F Z_F^* - jv Z_F Q^* + jv Z_F^* Q + QQ^*] \, dudv$$

$$+ \frac{\lambda}{2\pi} \iint_\Omega [u^4 Z_F Z_F^* - ju^3 Z_F P^* + ju^3 Z_F^* P + u^2 PP^*$$

$$+ v^4 Z_F Z_F^* - jv^3 Z_F Q^* + jv^3 Z_F^* Q + v^2 QQ^*] \, dudv$$

$$+ \frac{\mu_1}{2\pi} \iint_\Omega (u^2 + v^2) Z_F Z_F^* \, dudv$$

$$+ \frac{\mu_2}{2\pi} \iint_\Omega (u^4 + 2u^2 v^2 + v^4) Z_F Z_F^* \, dudv$$

where the asterisk $*$ denotes the complex conjugate. Differentiating the above expression with respect to Z_F^* and setting the result to zero, we can deduce the necessary condition for a minimum of the cost function (5.11) as follows:

$$(u^2 Z_F + ju P + v^2 Z_F + jv Q) + \lambda (u^4 Z_F + ju^3 P + v^4 Z_F + jv^3 Q)$$

$$+ \mu_1 (u^2 + v^2) Z_F + \mu_2 (u^4 + 2u^2 v^2 + v^4) Z_F = 0$$

A rearrangement of this equation then yields

$$\left[\lambda (u^4 + v^4) + (1 + \mu_1)(u^2 + v^2) + \mu_2 (u^2 + v^2)^2 \right] Z_F(u,v)$$

$$+ j(u + \lambda u^3) P(u,v) + j(v + \lambda v^3) Q(u,v) = 0$$

Solving the above equation except for $(u,v) \neq (0,0)$, we obtain that

$$Z_F(u,v) = \frac{-j(u + \lambda u^3) P(u,v) - j(v + \lambda v^3) Q(u,v)}{\lambda(u^4 + v^4) + (1 + \mu_1)(u^2 + v^2) + \mu_2 (u^2 + v^2)^2} \qquad (5.15)$$

Therefore, a Fourier transform of the unknown surface height $Z(x,y)$ is expressed as a function of Fourier transforms of given gradients $p(x,y)$ and $q(x,y)$. This Fourier-transform based method can be summarized as follows:

Theorem 5.1 *Given a gradient field $(p(x,y), q(x,y))$; the corresponding surface height function $Z(x,y)$ can be computed by taking the inverse Fourier transform of $Z_F(u,v)$ in (5.15), where $Z_F(u,v)$, $P(u,v)$, and $Q(u,v)$, respectively, are Fourier transforms of $Z(x,y)$, $p(x,y)$, and $q(x,y)$.*

Algorithm 5.1 Fourier-Transform Based Method

---

input gradients $p(x,y), q(x,y)$; parameters λ, μ_1, and μ_2
for $0 \leqslant x,y \leqslant N-1$ **do**
  if $(|p(x,y)| < pq_{max}$ & $|q(x,y)| < pq_{max})$ **then**
    P1(x,y)=p(x,y); P2(x,y)=0;
    Q1(x,y)=q(x,y); Q2(x,y)=0;
  else
    P1(x,y)=0;   P2(x,y)=0;
    Q1(x,y)=0;   Q2(x,y)=0;
  end if
end for
Calculate Fourier transform in place: P1(u,v), P2(u,v);
Calculate Fourier transform in place: Q1(u,v), Q2(u,v);
for $0 \leqslant u,v \leqslant N-1$ **do**
  if $(u \neq 0$ & $v \neq 0)$ **then**
    $\Delta = \lambda\left(u^4 + v^4\right) + (1+\mu_1)\left(u^2+v^2\right) + \mu_2\left(u^2+v^2\right)^2$;
    $H1(u,v) = [(u+\lambda u^3)P2(u,v) + (v+\lambda v^3)Q2(u,v)]/\Delta$;
    $H2(u,v) = [-(u+\lambda u^3)P1(u,v) - (v+\lambda v^3)Q1(u,v)]/\Delta$;
  else
    $H1(0,0) =$ average height; $H2(0,0) = 0$;
  end if
end for
Calculate inverse Fourier transform of H1(u,v) and H2(u,v) in place: H1(x,y), H2(x,y);
for $0 \leqslant x,y \leqslant N-1$ **do**
  $Z(x,y) = H1(x,y)$;
end for

Our algorithm (see Algorithm 5.1) specifies the implementation details for this Fourier-transform based method. The constant pq_{max} eliminates gradient estimates which define angles with the image plane close to $90°$, and a value such as $pq_{max} = 12$ is an option. Real parts are stored in arrays P1, Q1, and H1, and imaginary parts in arrays P2, Q2, and H2. The initialization in line 19 can be by an estimated value for the average height of the visible scene. Parameters λ, μ_1 and μ_2 should be chosen based on experimental evidence for the given scene.

5.5 Wavelet-Transform Based Method

Wavelet theory has proved to be a powerful tool, and has begun to play a serious role in a broad range of applications, including numerical analysis, pattern recognition, signal and image processing. The wavelet transform is a generalization of the Fourier transform. Wavelets have advantages over traditional Fourier methods in analyzing physical situations where the function contains discontinuities and sharp spikes. In order to take the advantages of the wavelet transform, we present a wavelets based method for depth recovery from gradients.

5.5.1 *Daubechies Wavelet Basis*

Let $\phi(x)$ and $\psi(x)$ are the Daubechies *scaling function* and *wavelet*, respectively. They both are implicitly defined by the following, two-scale relation:[22]

$$\phi(x) = \sum_{k \in \mathbb{Z}} a_k \phi(2x - k) \tag{5.16}$$

and the equation

$$\psi(x) = \sum_{k \in \mathbb{Z}} (-1)^k a_{1-k} \phi(2x - k) \tag{5.17}$$

where $\mathbb{Z} = \{\cdots, -1, 0, 1, \cdots\}$, and a_k are called the Daubechies wavelet filter coefficients.

Connection coefficients (see, for example, Beylkin,[23] Mallat[24]) play an important role in representing the relation between the scaling function and differential operators. For $k \in \mathbb{Z}$, the connection coefficients with Mth order *vanishing moments* (that is, $\int_{-\infty}^{+\infty} x^k \phi(x) dx = 0$, for $0 \leqslant k \leqslant M$) are defined by

$$\Gamma_k^0 = \int \phi(x) \phi(x - k) dx \tag{5.18}$$

$$\Gamma_k^1 = \int \phi^{(x)}(x) \phi(x - k) dx \tag{5.19}$$

$$\Gamma_k^2 = \int \phi^{(x)}(x) \phi^{(x)}(x - k) dx \tag{5.20}$$

Then we have the following properties:

(i) $\Gamma_0^1 = 0$,
(ii) for the scaling function $\phi(x)$, which has Mth order vanishing moments, $\Gamma_k^1 = \Gamma_k^2 = 0$, $k \notin [-2M+2, 2M-2]$, and
(iii) $\Gamma_k^0 = \begin{cases} 1, & k = 0, \\ 0, & \text{otherwise.} \end{cases}$

Table 5.1　Connection coefficients for $N = 3$.

Γ_k^1	Γ_k^2
$\Gamma_{-4}^1 = 0.00034246575342$	$\Gamma_{-4}^2 = -0.00535714285714$
$\Gamma_{-3}^1 = 0.01461187214612$	$\Gamma_{-3}^2 = -0.11428571428571$
$\Gamma_{-2}^1 = -0.14520547945206$	$\Gamma_{-2}^2 = 0.87619047619052$
$\Gamma_{-1}^1 = 0.74520547945206$	$\Gamma_{-1}^2 = -3.39047619047638$
$\Gamma_0^1 = 0.0$	$\Gamma_0^2 = 5.26785714285743$
$\Gamma_1^1 = -0.74520547945206$	$\Gamma_1^2 = -3.39047619047638$
$\Gamma_2^1 = 0.14520547945206$	$\Gamma_2^2 = 0.87619047619052$
$\Gamma_3^1 = -0.01461187214612$	$\Gamma_3^2 = -0.11428571428571$
$\Gamma_4^1 = -0.00034246575342$	$\Gamma_4^2 = -0.00535714285714$

The connection coefficients for Daubechies' wavelet with 3rd order vanishing moments are shown in Table 5.1:

5.5.2　*Iteration Formula for Wavelet-Transform Based Method*

In order to discretize the functional (5.5), the tensor product of the third-order Daubechies' scaling functions is used to span the solution space. The surface height is described as a linear combination of a set of scaling basis functions. After discretization, the problem of depth recovery from gradients becomes a discrete minimization problem. To solve the minimization problem, a perturbation method will be used. The surface height is finally decided after finding the weight coefficients.

We assume that the size of the domain of the surface $Z(x,y)$ equals $N \times N$. Suppose further that the surface $Z(x,y)$ is represented by a linear combination of a set of third-order Daubechies' scaling basis functions in the following format:

$$Z(x,y) = \sum_{m=0}^{N-1} \sum_{n=0}^{N-1} z_{m,n} \phi_{m,n}(x,y) \tag{5.21}$$

where $z_{m,n}$ are the weight coefficients, $\phi_{m,n}(x,y)$ are the tensor products of the third-order Daubechies scaling functions, with

$$\phi_{m,n}(x,y) = \phi(x-m)\phi(y-n) \tag{5.22}$$

For the known gradient values $p(x,y)$ and $q(x,y)$, we assume that

$$p(x,y) = \sum_{m=0}^{N-1} \sum_{n=0}^{N-1} p_{m,n} \phi_{m,n}(x,y) \tag{5.23}$$

$$q(x,y) = \sum_{m=0}^{N-1} \sum_{n=0}^{N-1} q_{m,n}\phi_{m,n}(x,y) \qquad (5.24)$$

where the weight coefficients $p_{m,n}$ and $q_{m,n}$ can be determined by

$$p_{m,n} = \iint p(x,y)\phi_{m,n}(x,y)dxdy \qquad (5.25)$$

$$q_{m,n} = \iint q(x,y)\phi_{m,n}(x,y)dxdy \qquad (5.26)$$

Substituting (5.21), (5.23) and (5.24) into (5.5), we have that

$$W = \iint \left[\left(\sum_{m,n=0}^{N-1} z_{m,n}\phi_{m,n}^{(x)}(x,y) - \sum_{m,n=0}^{N-1} p_{m,n}\phi_{m,n}(x,y) \right)^2 \right] dxdy$$

$$+ \iint \left[\left(\sum_{m,n=0}^{N-1} z_{m,n}\phi_{m,n}^{(y)}(x,y) - \sum_{m,n=0}^{N-1} q_{m,n}\phi_{m,n}(x,y) \right)^2 \right] dxdy$$

$$= W_1 + W_2, \qquad (5.27)$$

where

$$\phi_{m,n}^{(x)}(x,y) = \frac{\partial \phi_{m,n}(x,y)}{\partial x}, \phi_{m,n}^{(y)}(x,y) = \frac{\partial \phi_{m,n}(x,y)}{\partial y}$$

In order to derive the iterative scheme for computing the surface height function $Z(x,y)$, let $\Delta z_{i,j}$ represent the updates of $z_{i,j}$ in the iterative equation, and $z'_{i,j}$ be the value after the update. Then we have that

$$z'_{i,j} = z_{i,j} + \Delta z_{i,j} \qquad (5.28)$$

Substituting $z'_{i,j}$ into W_1, W_1 will be changed by ΔW_1, that is,

$$W'_1 = W_1 + \Delta W_1$$

$$= \iint \left[\left(\sum_{m,n=0}^{N-1} z_{m,n}\phi_{m,n}^{(x)}(x,y) - \sum_{m,n=0}^{N-1} p_{m,n}\phi_{m,n}(x,y) \right) + \Delta z_{i,j}\phi_{i,j}^{(x)}(x,y) \right]^2 dxdy$$

$$= W_1 + 2\Delta z_{i,j} \sum_{m,n=0}^{N-1} z_{m,n} \iint \phi_{m,n}^{(x)}(x,y)\phi_{i,j}^{(x)}(x,y)dxdy$$

$$- 2\Delta z_{i,j} \sum_{m,n=0}^{N-1} p_{m,n} \iint \phi_{m,n}(x,y)\phi_{i,j}^{(x)}(x,y)dxdy$$

$$+ \Delta z_{i,j}^2 \iint \phi_{i,j}^{(x)}(x,y)\phi_{i,j}^{(x)}(x,y)dxdy \qquad (5.29)$$

By using the tensor product (5.22) and the definitions of the connection coefficients (5.18), (5.19) and (5.20) yields

$$
\iint \phi_{m,n}^{(x)}(x,y)\phi_{i,j}^{(x)}(x,y)\,dx\,dy
$$

$$
= \iint \phi^x(x-m,y-n)\phi^{(x)}(x-i,y-j)\,dx\,dy
$$

$$
= \iint \phi^x(x-m)\phi(y-n)\phi^{(x)}(x-i)\phi(y-j)\,dx\,dy
$$

$$
= \int \phi^x(x-m)\phi^{(x)}(x-i)\,dx \int \phi(y-n)\phi(y-j)\,dy
$$

$$
= \int \phi^x(x)\phi^{(x)}(x-i+m)\,dx \int \phi(y)\phi(y-j+n)\,dy
$$

$$
= \Gamma_{i-m}^2 \Gamma_{j-n}^0 \tag{5.30}
$$

Using the same way, we obtain that

$$
\iint \phi_{m,n}(x,y)\phi_{i,j}^{(x)}(x,y)\,dx\,dy = \Gamma_{i-m}^1 \Gamma_{j-n}^0 \tag{5.31}
$$

$$
\iint \phi_{i,j}^{(x)}(x,y)\phi_{i,j}^{(x)}(x,y)\,dx\,dy = \Gamma_0^2 \tag{5.32}
$$

Substituting (5.30), (5.31) and (5.32) into (5.29) gives

$$
W_1' = W_1 + 2\Delta z_{i,j} \sum_{m,n=0}^{N-1} z_{m,n}\Gamma_{i-m}^2 \Gamma_{j-n}^0
$$

$$
- 2\Delta z_{i,j} \sum_{m,n=0}^{N-1} p_{m,n}\Gamma_{i-m}^1 \Gamma_{j-n}^0 + \Delta z_{i,j}^2 \Gamma_0^2 \tag{5.33}
$$

Using the same derivation, we have that

$$
W_2' = W_2 + \Delta W_2
$$

$$
= W_2 + 2\Delta z_{i,j} \sum_{m,n=0}^{N-1} z_{m,n}\Gamma_{i-m}^0 \Gamma_{j-n}^2
$$

$$
- 2\Delta z_{i,j} \sum_{m,n=0}^{N-1} q_{m,n}\Gamma_{i-m}^0 \Gamma_{j-n}^1 + \Delta z_{i,j}^2 \Gamma_0^2 \tag{5.34}
$$

Substituting (5.33) and (5.34) into (5.27), it is shown that the energy change is given by

$$\Delta W = \Delta W_1 + \Delta W_2$$

$$= 2\Delta z_{i,j} \sum_{m,n=0}^{N-1} z_{m,n} \left(\Gamma_{i-m}^2 \Gamma_{j-n}^0 + \Gamma_{i-m}^0 \Gamma_{j-n}^2 \right)$$

$$- 2\Delta z_{i,j} \sum_{m,n=0}^{N-1} p_{m,n} \Gamma_{i-m}^1 \Gamma_{j-n}^0 - 2\Delta z_{i,j} \sum_{m,n=0}^{N-1} q_{m,n} \Gamma_{i-m}^0 \Gamma_{j-n}^1 + 2\Delta z_{i,j}^2 \Gamma_0^2$$

In order to make the cost function decrease as fast as possible, ΔW must be maximized. From $\partial \Delta W / \partial \Delta z_{i,j} = 0$, we have that

$$\Delta z_{i,j} = \frac{1}{2\Gamma_0^2} \sum_{k=-2N+2}^{2N-2} \left[(p_{i-k,j} + q_{i,j-k}) \Gamma_k^1 - (z_{i-k,j} + z_{i,j-k}) \Gamma_k^2 \right] \qquad (5.35)$$

Substituting (5.35) into (5.28) leads to the following iterative scheme:

$$z_{i,j}^{t+1} = z_{i,j}^t + \Delta z_{i,j} \qquad (5.36)$$

where t is the iteration index. By taking zero as the initial values, we can iteratively solve the depth recovery from gradients problem using the iterative scheme (5.36).

5.6 Experimental Results

To investigate the performance of the algorithms described in the previous sections, we have done several computer simulations on both synthetic and real images.

5.6.1 *Test on Noiseless Gradients*

The two-scan method was tested on a synthetic vase image, which is generated mathematically by the following explicit surface equation:

$$Z(x,y) = \sqrt{f^2(y) - x^2}$$

where

$$f(y) = 0.15 - 0.1y(6y+1)^2(y-1)^2(3y-2)^2,$$
$$-0.5 \leqslant x \leqslant 0.5, \quad 0.0 \leqslant y \leqslant 1.0$$

The image of this synthetic vase object is shown on the left of Figure 5.1. The 3D plot of the reconstructed surface using the proposed two-scan algorithm is shown in the middle of Figure 5.1. By comparing the 3D plots of the true surface (middle)

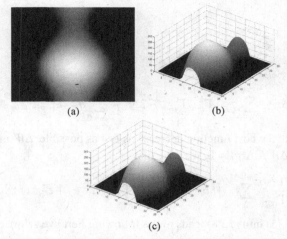

Fig. 5.1   Results of a synthetic vase object. (a) Original image. (b) 3D plot of the vase object. (c) Reconstruction result using the proposed two-scan method.

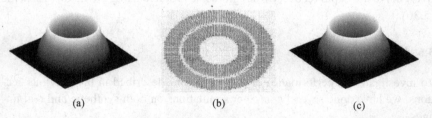

Fig. 5.2   Results for a torus object. (a) Original surface. (b) Gradient vector fields. (c) Reconstructed surface using wavelet-transform based method.

and the reconstructed surface (right), we can see that they look very similar to each other.

The wavelet-transform based method was applied to a torus image. The original torus image is illustrated on the left of Figure 5.2. The gradient fields of the torus surface is shown in the middle of Figure 5.2. The reconstructed surface height from this gradient fields by the proposed wavelet based method is shown on the right of Figure 5.2. It can be seen that the shape of the torus object is correctly reconstructed.

Figure 5.3 shows three captured images of a Beethoven plaster statue, using a static camera but different light sources. The gradients were generated using the albedo-independent photometric stereo method with three light sources (3S

Fig. 5.3   Image triplet of a Beethoven statue.

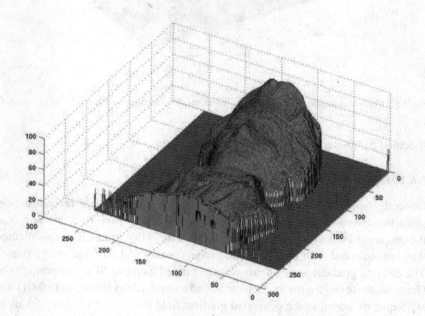

Fig. 5.4   Recovered surface using the Frankot-Chellappa method.

PSM) as specified in.[8] Figure 5.4 illustrates both recovered surfaces. The left-hand surface was calculated using the Frankot-Chellappa algorithm[17] as specified in,[8] and the right-hand surface was calculated using our Fourier-transform based method with $\lambda = 0.5$, $\mu_1 = 0$, and $\mu_2 = 0$. By comparing the reconstructed surfaces shown in Figure 5.5, we see that the Fourier-transform based method (with the

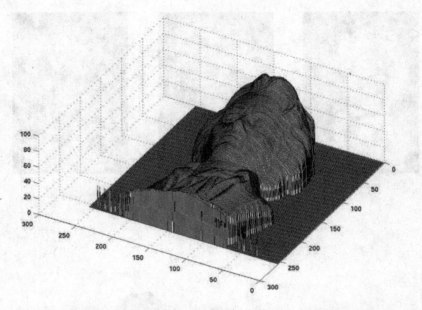

Fig. 5.5   Recovered surface using our Fourier-transform based method, with $\lambda = 0.5$, and $\mu_1 = \mu_2 = 0$.

specified parameters) improves the recovered shape.

5.6.2 *Test on Noisy Gradients*

Generally speaking, local methods may provide an unreliable reconstruction, since the errors can propagate along the scan paths. Therefore, we only test the proposed Fourier-transform based method for noisy gradients. This method was implemented with one synthetic image (peaks) and one real image (vase). The discrete gradient vector fields were generated using an SFS algorithm.[4] The Gaussian noise (with a mean, set to zero, and a standard deviation, set to 0.01) was subsequently added to the generated gradient field in order to test the sensitivity to noise.

Figure 5.6 and Figure 5.7 show the reconstructed surfaces when the parameters are given by some specific values. Figure 5.6 shows the original 3D height plot of a synthetic peaks surface (left), the 3D plot of reconstructed surfaces using the Frankot-Chellappa algorithm (middle), and the 3D plot of the reconstructed surface using our Fourier-transform based method with $\lambda = 0$, $\mu_1 = 0.1$, and $\mu_2 = 1$. By comparing the true heights, we can see that the noise is reduced.

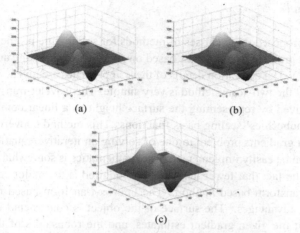

Fig. 5.6   Results of a synthetic image. (a) Original surface. (b) Reconstructed surface using the Frankot-Chellappa algorithm. (c) Reconstructed surface using our Fourier-transform based method with $\lambda = 0, \mu_1 = 0.1$ and $\mu_2 = 1$.

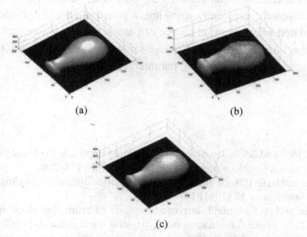

Fig. 5.7   Results of a vase object. (a) Original surface. (b) Reconstructed surface using the Frankot-Chellappa algorithm. (c) Reconstructed surface our Fourier-transform based method with $\lambda = 0, \mu_1 = 0.1, \mu_2 = 10$.

Figure 5.7 shows the 3D plot of the original vase surface (left). The reconstructed surfaces with $\lambda = 0$, $\mu_1 = \mu_2 = 0$ and $\mu_1 = 0.1, \mu_2 = 10$ are shown in the middle Figure 5.7 and on the right of Figure 5.7.

5.7 Conclusion

This chapter proposed three classes of methods for solving the problem of depth recovery from gradient vector fields, based on previous work by the authors (see, for example,[19]). The derivation details of these approaches are given. The derivation process of the two-scan method is very simple. The wavelet-transform based method is derived by representing the surface height as a linear combination of third-order Daubechies' scaling basis functions. This method converts the depth recovery from gradients problem to one of solving an iterative equation. Hence, the method can be easily implemented. The mathematics is somewhat more complicated, but the fact that fewer iterations are required is the major advantage of the wavelet-transform based method. The Fourier-transform based method has some distinct advantages. The surface of the object is constructed in one pass utilizing all of the given gradient estimates, and the robustness of the Fourier-transform based method to noisy gradient fields can be improved by choosing associated weighting parameters. The choice of parameters heavily affects the surface reconstructed from gradients. Therefore, the criterion for the choice of the parameters and the relation between the parameters and noise should be a future topic of research. Generally speaking, a constrained minimization problem can be formulated for an optimal choice of parameters. This is a problem which many researchers have been trying to solve for several years, and so far there is no systematic way derived for choosing parameters.

References

1. B. K. P. Horn and M. J. Brooks, The variational approach to shape from shading, *Computer Vision Graphics Image Processing* **33** (1986) 174–208.
2. K. Ikeuchi and B. K. P.Horn, Numerical shape from shading and occluding boundaries, *Artificial Intelligence* **17** (1981) 141–184.
3. C. -H. Lee and A. Rosenfeld, Improved methods of estimating shape from shading using the light source coordinate system, *Artificial Intelligence* **26** (1985) 125–143.
4. P. L. Worthington and E. R. Hancock, New constraints on data-closeness and needle map consistency for shape-from-shading, *IEEE Trans. Pattern Analysis Machine Intelligence* **21** (1999) 1250–1267.
5. W. A. P. Smith and E. R. Hancock, Recovering facial shape using a statistical model of surface normal direction, *IEEE Trans. Pattern Analysis Machine Intelligence* **28** (2006) 1914–1930.
6. R. J. Woodham, Photometric method for determining surface orientation from multiple images, *Optical Engineering* **19** (1980) 139–144.
7. K. Ikeuchi, Determining surface orientations of specular surfaces by using the pho-

tometric stereo method, *IEEE Trans. Pattern Analysis Machine Intelligence* **3** (1981) 661–669.

8. R. Klette, A. Koschan, and K. Schlüns, *Computer Vision – Räumliche Information aus digitalen Bildern*, Vieweg, Braunschweig, 1996.

9. M. L. Smith, T. Hill, and G. Smith, Surface texture analysis based upon the visually acquired perturbation of surface normals, *Image Vision Computing* **15** (1997) 949–955.

10. M. L. Smith, The analysis of surface texture using photometric stereo acquisition and gradient space domain mapping, *Image Vision Computing* **17** (1999) 1009–1019.

11. J. Fan and L. B. Wolff, Surface curvature and shape reconstruction from unknown multiple illumination and integrability, *Computer Vision Image Understanding* **65** (1997) 347–359.

12. N. E. Coleman, Jr. and R. Jain, Obtaining 3-dimensional shape of textured and specular surfaces using four-source photometry, *Computer Vision Graphics Image Processing* **18** (1982) 439–451.

13. G. Healey and R. Jain, Depth recovery from surface normals, in Proc. *Int. Conf. Pattern Recognition* **2** (1984) 894–896.

14. Z. Wu and L. Li, A line-integration based method for depth recovery from surface normals, *Computer Vision Graphics Image Processing* **43** (1988) 53–66.

15. R. Klette and K. Schlüns, Height data from gradient fields, in Proc. *Machine Vision Applications Architectures Systems Integration*, SPIE 2908 (1996), pp. 204–215.

16. A. Robles-Kelly and E. R. Hancock, A graph-spectral method for surface height recovery, *Pattern Recognition* **38** (2005) 1167–1186.

17. R. T. Frankot and R. Chellappa, A method for enforcing integrability in shape from shading algorithms, *IEEE Trans. Pattern Analysis Machine Intelligence* **10** (1988) 439–451.

18. B. K. P. Horn, Height and gradient from shading, *Int. J. Computer Vision* **5** (1990) 37–75.

19. T. Wei and R. Klette, Depth recovery from noisy gradient vector fields using regularization, in Proc. *Computer Analysis Images Patterns*, LNCS 2756, Springer, Berlin, (2003), pp. 116–123.

20. A. Robles-Kelly and E. R. Hancock, Shape-from-shading using the heat equation, *IEEE Trans. Image Processing* **16** (2007) 7–21.

21. M. Castelán, A. Robles-Kelly and E. R. Hancock, A coupled statistical model for face shape recovery from brightness images, *IEEE Trans. Image Processing* **16** (2007) 1139–1151.

22. I. Daubechies, Orthonormal bases of compactly supported wavelets, *Commun. Pure Appl. Mathematics* **41** (1988) 909–996.

23. G. Beylkin, On the representation of operators in bases of compactly supported wavelets, *SIAM J. Numerical Analysis* **29** (1992) 1716–1740.

24. S. G. Mallat, *A Wavelet Tour of Signal Processing*, Academic Press, 1999.

Chapter 6

Convolutional Compactors for Guaranteed 6-Bit Error Detection

F. Börner*, A. Leininger† and M. Gössel*

*Institute of Computer Science, University of Potsdam,
August-Bebel-Str. 89, D-14482 Potsdam, Germany
fboerner@cs.uni-potsdam.de
mgoessel@cs.uni-potsdam.de

†Infineon Technologies AG
Am Campeon 1–12, D-85579 Neubiberg, Germany
andreas.leininger@infineon.com

In this paper a new convolutional compactor with a single output is introduced. It is characterized by very good error masking properties: All odd errors, all 2-bit, 4-bit and 6-bit errors at the inputs of the compactor are not masked. Additionally, the depth of the compactor, i.e. the necessary number of flip flops, is only $O(\log^2 n)$, which makes the proposed convolutional compactor of special interest when a very large number of chip internal scan chains or test outputs are to be compressed.

Contents

6.1 Introduction

To reduce the huge amount of test and diagnosis data of large circuits under test recently convolutional compactors were investigated.[1,4,7,8] These compactors are called convolutional since their structure is the same as for convolutional encoders in their observer canonical form.[5]

Motivated by low pin count testing, this paper focuses on convolutional compactors with a single output. A convolutional compactor with a single output consists of a set of m (overlapping) XOR-trees and a Shift Register of m Flips-Flops. The outputs of the XOR-trees are connected to the inputs of the Shift Register (Figure 6.1).

In a more abstract notation a convolutional compactor is a finite linear definite automaton.[6] The state of the compactor, i.e. the content of the m flip-flops depends only on the last m inputs of the compactor. Therefore convolutional compactors are very well-adapted to the compaction of data with X-states. In a convolutional compactor a possible X-value is shifted through the flip-flops of the MISRs until it reaches the outputs. No trace is left in the state or the contents of the flip-flops of the convolutional compactor after the X-value reaches the outputs of the MISRs.

This is the main advantage of a convolutional compactor compared to a compactor where the outputs of (non-overlapping) XOR-trees are connected to the inputs of a multi-input linear feedback shift register MILFSR as described by Savir.[9] In the case of a MILFSR with feedback, x-values may corrupt the entire signature of the MILFSR.

For the convolutional compactor with a single output, it was proven that all odd errors and all two-bit errors at the inputs are not masked.[8] It was demonstrated by extensive simulation experiments that a very high percentage of 4-bit errors can also be detected by this compactor. But the detection of 4-bit errors and 6-bit errors cannot be guaranteed.

In this paper it is shown how to systematically configure a convolutional compactor based on generalized construction rules. Approaches for optimization to ease the implementation are only mentioned briefly. In particular, a convolutional compactor with n inputs is introduced, $n \geqslant 6$, which does not mask 2-bit, 4-bit, 6-bit and odd input errors at its output or tolerates a corresponding number of unknown values. Thereby the necessary number of flip-flops or the depth of the convolutional compactor is only $m \in O(\log^2 n)$. For example, for 1000 outputs of the CUT which are to be compacted the sequential depth or the necessary number of flip-flops is only about 100 instead of 1000.

A short introduction to convolutional compactors is given in in section 6.2. Then the basic terminology and the theoretical approach to the theorems used later for constructing the compactors are explained in section 6.2 and section 6.3. Section 6.4 to 6.6 show how the convolutional compactor is constructed based on these theorems. In section 6.7 conclusions are drawn.

6.2 A Convolutional Compactor with a Single Output

6.2.1 *The Observer Canonical Form of a Convolutional Compactor*

Figure 6.1 shows a feedback free convolutional compactor in its observer canonical form.[5] The compactor has n inputs and a single output out. The compactor consists of m (overlapping) parity trees $P_0, \ldots, P_{m-1}$ and an m-stage MISR (Multi-Input Signature Register) of m flip-flops $FF_0, \ldots, FF_{m-1}$.

For time $t = 0, 1, \ldots$ we assume that the input signals are the test responses

$$\vec{a}(t) = \begin{pmatrix} a_0(t) \\ \vdots \\ a_{n-1}(t) \end{pmatrix}$$

of n scan-paths of the circuit under test CUT. The output signals of the parity trees $P_0, P_1, \ldots, P_{m-1}$ are denoted by

$$\vec{b}(t) = \begin{pmatrix} b_0(t) \\ \vdots \\ b_{m-1}(t) \end{pmatrix}.$$

These $b_i(t)$ are the input signals of the MISR and the output signal of the MISR, denoted by $c(t)$, is the output of the compactor.

Formally the parity trees P_i are described by a binary $(m \times n)$ parity–matrix $\mathbf{W} = (w_{i,j})$ with $w_{i,j} \in \{0, 1\}$ for all $0 \leqslant i \leqslant m-1, 0 \leqslant j \leqslant n-1$. The jth input of the compactor is connected to a corresponding input of the ith parity tree P_i if and only if $w_{i,j} = 1$. In Figure 6.1 the parity trees are described by horizontal lines. If $w_{i,j} = 1$, i.e. if the jth input is XORed to the ith parity tree, then the crosspoint of the jth input line and the ith parity tree is marked with an $\oplus$.

This is the observer canonical form of a feedback free convolutional compactor with a single output. For a given number n of inputs, different such compactors differ only by their $(m \times n)$-parity-matrices $\mathbf{W}$.

Thereby m determines the sequential depth or the number of flip-flops of the MISR.

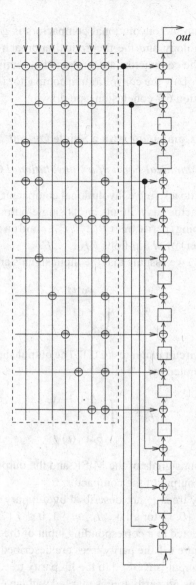

Fig. 6.1   A Convolutional Compactor.

6.2.2   *The Input-Output-Behavior of a Convolutional Compactor*

We want to design parity–matrices **W** with special error detecting properties. Therefore we first study the mathematical dependencies between inputs and output of the compactor. The signals $a_j(t)$, $b_i(t)$, $c(t)$ and the entries $w_{i,j}$ of the parity matrix $\mathbf{W}$ are elements of the two-element field $F_2 = \{0,1\}$ with addition $\oplus$ modulo 2 and the usual multiplication.

For the n-dimensional input vectors $\vec{a}(t)$ at time t and the corresponding m-dimensional output vectors $\vec{b}(t)$ of the parity trees we have $\vec{b} = \mathbf{W}\vec{a}$. If we define the matrix $\mathbf{A}$

$$\mathbf{A} = (\vec{a}(0), \vec{a}(1), \vec{a}(2), \ldots, \vec{a}(t), \ldots)$$

with the input vectors $\vec{a}(t)$ of the compactor as columns and the matrix $\mathbf{B}$

$$\mathbf{B} = (\vec{b}(0), \vec{b}(1), \vec{b}(2), \ldots, \vec{b}(t), \ldots)$$

with the vectors $\vec{b}(t)$ as columns, then we have

$$\mathbf{B} = \mathbf{W} \cdot \mathbf{A}. \tag{1}$$

We assume that the initial state of the MISR is $\underline{0}$, i.e. the initial content of all flip-flops is 0. Then the components of the m-dimensional output vectors $\vec{b}(t)$ of the m parity trees are compacted by the MISR according to the equation

$$c(t) = b_0(t) \oplus b_1(t-1) \oplus \ldots \oplus b_{m-1}(t-m+1), \tag{2}$$

where we put $b_i(s) = 0$ for $s < 0$.

Feedback free convolutional compactors are special cases of finite linear definite automata ([63]) which have a finite impulse response. Therefore we can describe their behavior with the help of the so called D-transforms of the input and output sequences. The D-transform of an infinite sequence

$$(x(t))_{t \geqslant 0} = (x(0), x(1), \ldots, x(t), \ldots)$$

is the formal power series

$$T_D[x(t)] := T_D[(x(t)_{t \geqslant 0}] = x(0) \oplus x(1)D \oplus \ldots \oplus x(t)D^t \oplus \ldots,$$

i.e. an element of $F_2[[D]]$. (D is sometimes called the *delay operator*.) We say that the *monomial* D^k *belongs to* $T_D[x(t)]$ if $x(k) = 1$.

If the sequence $(x(t))$ is finite, then we consider the D-transform $T_D[x(t)]$ as a formal polynomial in D, i.e. as element of $F_2[D]$. In the sequel, we will not distinguish sharply between a sequence $(x(t))$ and its D-transform $T_D[x(t)]$.

We denote the D-transform of the sequence $(a_j(t))_{t \geqslant 0}$ (the jth row of the input matrix $\mathbf{A}$) by

$$g_j^{\mathbf{A}}(D) := T_D[a_j(t)],$$

and the D-transform of the output sequence $(c^{\mathbf{A}}(t))_{t \geqslant 0}$ for the input matrix $\mathbf{A}$ is denoted by

$$h^{\mathbf{A}}(D) := T_D[c^{\mathbf{A}}(t)].$$

Moreover, to the jth column $\vec{w}_{*,j}$ of the parity matrix $\mathbf{W}$ ($0 \leqslant j \leqslant n-1$) we assign the formal D-polynomial

$$f_j(D) := w_{0,j} \oplus f_{1,j}D \oplus \ldots \oplus w_{m-1,j}D^{m-1}.$$

Then equations (1) and (2) lead to

$$h^{\mathbf{A}}(D) = f_0(D)g_0^{\mathbf{A}}(D) \oplus f_1(D)g_1^{\mathbf{A}}(D) \oplus \ldots \oplus f_{n-1}(D)g_{n-1}^{\mathbf{A}}(D) \qquad (3)$$

$$= (f_0(D), \ldots, f_{n-1}(D)) \begin{pmatrix} g_(^{\mathbf{A}}D) \\ \vdots \\ g_{n-1}^{\mathbf{A}}(D) \end{pmatrix}. \qquad (4)$$

Consequently, the D-transform of the output sequence $(c^{\mathbf{A}}(t))$ is the convolution of the n-tuple $(f_0(D), \ldots, f_{n-1}(D))$ with the D-transforms $g_j^{\mathbf{A}}(D)$ of the rows of the input matris $\mathbf{A}$. Equation 3 completely describes the input–output behaviour of our compactor.

6.2.3 *Linear Superposition and Errors*

In this subsection it is investigated how errors in the input sequence influence the output sequence of the convolutional compactor. Since the convolutional compactor is a linear automaton (with initial state $\underline{0}$), the deviation of the actual output sequence from the expected correct output sequence is determined by the deviation of the sequence of the actual input vectors from the correct sequence of input vectors only and not by the concrete values of the inputs.

We consider the case when the (correct) input matrix $\mathbf{A}$ is changed into an erroneous input matrix $\mathbf{A} \oplus \mathbf{E}$ by an *input error matrix* $\mathbf{E} = (\vec{e}(0), \vec{e}(1), \ldots, \vec{e}(t), \ldots)$ of the same size as $\mathbf{A}$. So

$$\vec{e}(t) = \begin{pmatrix} e_0(t) \\ \vdots \\ e_{n-1}(t) \end{pmatrix}$$

are column-vectors of dimension n.

We denote the D-transform of the jth row of $\mathbf{E}$ with

$$g^{\mathbf{E}}(D) = e_j(0) \oplus e_j(1)D \oplus \ldots \oplus e_j(t)D^t \oplus \ldots.$$

Then the D transform of the jth row of $\mathbf{A} \oplus \mathbf{E}$ is $g^{\mathbf{A}}(D) \oplus g^{\mathbf{E}}(D)$. Let $h^{\mathbf{A} \oplus \mathbf{E}}(D)$ denote the D-transform of the output sequence for the input $\mathbf{A} \oplus \mathbf{E}$. Then equation (3) yields

$$
\begin{aligned}
h^{\mathbf{A} \oplus \mathbf{E}}(D) &= f_0(D)(g_0^{\mathbf{A}}(D) \oplus g_0^{\mathbf{E}}(D)) \oplus f_1(D)(g_1^{\mathbf{A}}(D) \oplus g_1^{\mathbf{E}}(D)) \oplus \ldots \\
&\quad \ldots \oplus f_{n-1}(D)(g_{n-1}^{\mathbf{A}}(D) \oplus g_{n-1}^{\mathbf{E}}(D)) \\
&= h^{\mathbf{A}} \oplus f_0(D)g_0^{\mathbf{E}}(D) \oplus f_1(D)g_1^{\mathbf{E}}(D) \oplus \ldots \oplus f_{n-1}(D)g_{n-1}^{\mathbf{E}}(D)
\end{aligned}
$$

According to equation (3), the convolution

$$
h^{\mathbf{E}}(D) = f_0(D)g_0^{\mathbf{E}}(D) \oplus f_1(D)g_1^{\mathbf{E}}(D) \oplus \ldots \oplus f_{n-1}(D)g_{n-1}^{\mathbf{E}}(D)
$$

is the D-transform of the output sequence $(c^{\mathbf{E}}(t))$ for the input $\mathbf{E}$. So we have

$$
h^{\mathbf{A} \oplus \mathbf{E}}(D) = h^{\mathbf{A}}(D) \oplus h^{\mathbf{E}}(D), \tag{5}
$$

and the deviation of the (incorrect) output sequence $h^{\mathbf{A} \oplus \mathbf{E}}(D)$ from the correct sequence $h^{\mathbf{A}}(D)$ is the sequence $h^{\mathbf{E}}(D)$, and this sequence is independent from the original input matrix A. This *principle of superposition* (equation (5)), which is a consequence of the linearity of the automata, reduces the error discussion to the consideration of the error matrix $\mathbf{E}$ (represented by the D-transforms $g_j^{\mathbf{E}}(D)$) and the output $h^{\mathbf{E}}(D)$.

Errors of the input vectors of the convolutional compactor are described by their corresponding binary error matrix $\mathbf{E}$. In this paper, by an *error* we mean an error matrix $\mathbf{E} \neq \mathbf{0}$, i.e. $g_j^{\mathbf{E}}(D) \neq 0$ for at least one $j \in \{0, \ldots, n\}$. An error $\mathbf{E}$ is *masked* if $h^{\mathbf{E}}(D) = 0$, and it is *detected* if

$$
h^{\mathbf{E}}(D) = f_0(D)g_0^{\mathbf{E}}(D) \oplus f_1(D)g_1^{\mathbf{E}}(D) \oplus \ldots \oplus f_{n-1}(D)g_{n-1}^{\mathbf{E}}(D) \neq 0. \tag{6}
$$

Our aim is the construction of parity matrices $\mathbf{W}$ *(represented by the polynomials* $f_j(D)$*) such that as many errors as possible can be detected by the inequality (6).*

In the sequel, we will suppress the superscript $\mathbf{E}$ and write $h(D)$ for $h^{\mathbf{E}}(D)$ and $g_j(D)$ instead of $g^{\mathbf{E}}(D)$.

6.3 Classification of Errors

In order to detect errors, we first want to introduce a certain classification of possible errors. For simplicity, we will assume that $\mathbf{E}$ is of finite size, i.e. $0 \leqslant t \leqslant s$ for some (possibly large) number s. In this case the $g_j(D)$ are polynomials of degree $< s$.

Definition 6.1 A *k-bit error* $(k \geqslant 1)$ is a matrix $\mathbf{E}$ where exactly k entries $e_j(t)$ are equal to 1. The columns of $\mathbf{E}$ with at least one value 1 are called the *error columns* of $\mathbf{E}$. An *one column error* is a matrix $\mathbf{E}$ with exactly one error column.

More generally we define: Let $k_1, k_2, \ldots, k_l \geqslant 1$. Then a $[k_1, k_2, \ldots, k_l]$-*error* corresponds to an error matrix $\mathbf{E}$ with exactly l error columns, where the rth error column $(1 \leqslant r \leqslant l)$ contains exactly k_r values 1. Between error columns there may be an arbitrary number of zero-columns.

Therefore a $[k_1, k_2, \ldots, k_l]$-error corresponds to an error matrix $\mathbf{E}$ with altogether $k_1 + k_2 + \ldots + k_l$ entries 1. The one-column errors with k errors are the $[k]$-errors. In practice, a one column error means that all errors occur at the same time t at the inputs of the compactor.

6.4 Construction of Parity Matrices for Error Detection

The construction of parity matrices is based on some theorems concerning the detection of special types of errors, the construction of parity matrices for specific errors and a recursive determination of larger parity matrices from smaller ones. In this paper, the proofs of these theorems are usually shortened or omitted. For the detailed proofs we refer to Börner.[2]

6.4.1 *First Observations*

We start with a general negative result.

Theorem 6.1 *Let $n \geqslant 2$. Then for each choice of $\mathbf{W}$ (or of $f_0(D), \ldots, f_{n-1}(D)$) there exist error matrices $\mathbf{E} \neq \mathbf{0}$ that are masked.*

Proof If $f_j(D) = 0$ for some j then all errors with $g_j(D) \neq 0$, $g_i(D) = 0$ for $i \neq j$ are masked. If $f_j(D) \neq 0$ for all j then choose $g_0(D) = f_1(D)$, $g_1(D) = f_0(D)$ and $g_j(D) = 0$ for $j \geqslant 2$. $\qquad\square$

Therefore we cannot detect all possible errors with our convolutional compactor. But the next result is positive.

Theorem 6.2 *If every column of the parity matrix $\mathbf{W}$ has an odd number of ones, then all error matrices $\mathbf{E}$ with an odd number of errors are detected, i.e. all k-bit errors with odd k are detected.*

Proof We call a D-polynomial *odd* if the number of the occuring monomials D^k is odd, otherwise we call it *even*. By our assumption, all polynomials $f_j(D)$ are odd. If $g_j(D)$ is odd, then $f_j(D)g_j(D)$ is also odd, and if $g_j(D)$ is even, then $f_j(D)g_j(D)$ is also even. If the number of 1s in $\mathbf{E}$ is odd, then an odd number of the $g_j(D)$ must be odd. Consequently, $h(D) = f_0(D)g_0(D) \oplus \ldots \oplus f_{n-1}(D)g_{n-1}(D)$ must be odd and therefore $h(D) \neq 0$. $\qquad\square$

If not every column of a given parity matrix **W** has an odd number of ones, then it is easy to insert an additional row such that the number of ones in every column of the modified parity matrix is odd. We call such an additional row an *add-odd row*. If we add an add-odd row to **W**, then a single flip-flop has to be added to the convolutional compactor.

In particular, with the matrix $\mathbf{W} = (1, 1, \ldots, 1)$ we can already detect every k-bit error for odd k. (This corresponds to the simple parity observation of the input signals at time t.) More general, a row $(w_{i,0}, w_{i,1}, \ldots, w_{i,n-1})$ of **W** with $w_{i,j} = 1$ for all j is called a *complete row*. The existence of such complete rows in **W** has consequences for the error detection.

Theorem 6.3 *If the first row of the parity matrix **W** is a complete row, then all $[k_1, k_2, \ldots, k_l]$-errors with odd k_1 are detected.*

*If the last row of the parity matrix **W** is a complete row, then all $[k_1, k_2, \ldots, k_l]$-errors with odd k_l are detected.*

Proof We only prove the first statement. W.l.o.g. we assume that the first error column in **E** is the first column in **E**. Then exactly k_1 of the polynomials $g_j(D)$ have a monomial 1. By our assumption, all $f_j(D)$ have a monomial 1. Therefore there occur exactly k_1 monomials 1 in $h(D) = f_0(D)g_0(D) \oplus \ldots \oplus f_{n-1}(D)g_{n-1}(D)$, all other monomials have degree $\geqslant 1$. But as k_1 is odd, an odd number of 1s adds to 1, so $h(D) \neq 0$. $\qquad\square$

If both, the first and the last row of the parity matrix **W**, are complete rows, then the overall parity must be implemented only one time and has to be connected with the first and the last flip-flop of the MISR.

By Definition 6.1, a $[k_1, \ldots, k_l]$-error **E** remains such an error if we permute the rows of **E**. Consequently:

Theorem 6.4 *Let $k_1, \ldots, k_l \geqslant 1$ be fixed natural numbers. If all $[k_1, k_2, \ldots, k_l]$-errors can be detected by use of the parity matrix **W**, then these errors are also detected by a parity matrix **W**′, if **W**′ is derived from **W** by a permutation of its columns.*

6.4.2 *Detection of Even One–Column Errors*

Because of Theorem 6.2 we have no problems with the detection of k-bit errors for odd k. So now we concentrate on k-bit errors for even k. First of all we have to consider the even one–column errors. We will design special parity matrices only for the detection of such special errors. Later these matrices will appear as parts of matrices **W** with more general error detection capabilities.

Definition 6.2 Let $k \geqslant 1$. A parity matrix $\mathbf{W}$ is an $[\leqslant 2k]$-*even-error-detection matrix*, (in short, a $[\leqslant 2k]$-*eed-matrix*), if $\mathbf{W}$ detects all one-column $[2l]$ errors for all l with $1 \leqslant l \leqslant k$. (Sometimes we abbreviate this by writing $\mathbf{W}$ with index $[\leqslant 2k]$ as $\mathbf{W}_{[\leqslant 2k]}$.)

The smallest value m for which a $[\leqslant 2k]$-eed-matrix of size $(m \times n)$ exists, is denoted by $\mathrm{minsize}_n(k)$. (If n is fixed, we also denote this value by $m_{[\leqslant 2k]}$.)

Consequently a $[\leqslant 8]$-eed matrix can detect all $[2]$-, $[4]$-, $[6]$- and $[8]$-one-column errors.

Since we are interested in a convolutional compactor with a small number of flip-flops we try to find such $[\leqslant 2k]$-eed matrices of size $(m \times n)$ where m is near to $\mathrm{minsize}_n(k)$. If $\mathbf{E}$ describes an one-column error, then we can w.l.o.g. assume that the first column of $\mathbf{E}$ is the error column. Then for all $j \in \{0, \ldots, n-1\}$, the $g_j(D)$ are either $g_j(D) = 0$ or $g_j(D) = 1$. Therefore $h(D) = \sum\{f_j(d) \mid g_j(D) = 1\}$. Consequently, $\mathbf{W}$ is $[\leqslant 2k]$-eed if $\sum_{j \in J} f_j(D) \neq 0$ for all $J \subseteq \{0, \ldots, n-1\}$ with $|J| = 2l$ for some l with $1 \leqslant l \leqslant k$. This leads to the following easy but useful criterion for such matrices.

Theorem 6.5 *An $(m \times n)$-matrix $\mathbf{W}$ is an $[\leqslant 2k]$-eed matrix if and only if for all index sets $I, J \subseteq \{0, \ldots, n-1\}$ with $|I| = |J| = k$ and $I \neq J$ holds*

$$\sum_{i \in I} f_i(D) \neq \sum_{j \in J} f_j(D). \tag{7}$$

For the least possible m holds

$$\mathrm{minsize}_n(k) \geqslant \left\lceil \log \binom{n}{k} \right\rceil. \tag{8}$$

Proof The first part of the Theorem follows from the preceding remarks, for details we refer to Börner.[2] The lower bound for $\mathrm{minsize}_n(k)$ follows from the fact that all sums $\sum_{j \in J} f_j(D)$ for $|J| = k$ have to be different. There are $\binom{n}{k}$ possible choices for J, so the number 2^m of possible column vectors of dimension m must be $\geqslant \binom{n}{k}$. $\qquad \square$

On the other hand, it is obvious that

$$\mathrm{minsize}_n(k) \leqslant n-1, \tag{9}$$

since the XOR-sums of arbitrary $2l$ pairwise different columns of the $((n-1) \times n)$-matrix

$$\begin{bmatrix} 0 & 1 & & & \\ 0 & & 1 & & \\ 0 & & & 1 & \\ & & \cdots & & \\ 0 & & & & 1 \end{bmatrix} \tag{10}$$

are never equal to zero.

In general every **H**-matrix of a $2k$ error detecting code can be used as a $\mathbf{W}_{[\leqslant 2k]}$ matrix. But since odd errors are not necessarily to be detected, different matrices are also possible and a $\mathbf{W}_{[\leqslant 2k]}$ matrix can even contain a single 0-column.

If and only if all columns of a parity matrix are different, the parity matrix is a $\mathbf{W}_{[\leqslant 2]}$ matrix which detects all even [2]-errors. Such a matrix can be constructed from the columns $i_{(2)}$, $0 \leqslant i \leqslant n$ in binary representation (i.e. by a H-matrix of a Hamming code with an additional 0-column). This parity matrix has $\text{minsize}_n(1) = \lceil \log n \rceil$ columns.

If we add to this matrix an add-odd row to make the number of ones in every column odd then this parity matrix will detect all [2]-errors and all odd errors within the error matrix **E**. An example of such a parity matrix for $n = 4$ is the following (3×4)-matrix:

$$\begin{bmatrix} 0 & 1 & 0 & 1 \\ 0 & 0 & 1 & 1 \\ 1 & 0 & 0 & 1 \end{bmatrix} \tag{11}$$

A consequence of 6.5 is the following Theorem.

Theorem 6.6 *Let* **W** *be a* $[\leqslant 2k]$*-eed matrix. If the parity-matrix* $\mathbf{W}'$ *is obtained from* **W** *by the following operations,*

- *insertion of additional rows,*
- *interchanging of rows,*
- *adding a row to another row,*
- *adding a complete row* $(1, 1, \ldots, 1)$ *to an arbitrary row,*
- *permutation of columns,*

then $\mathbf{W}'$ *is also a* $[\leqslant 2k]$*-eed matrix.*

Proof   All these operations do not influence the criterion in Theorem 6.5.   □

If a $[\leqslant 2k]$-eed matrix **W** is of minimal size, i.e. $m = \text{minsize}_n(k)$, then **W** cannot contain a row with only zero entries. Consequently, using the operations of Theorem 6.6, we can apply a Gaussian elimination algorithm to $\mathbf{W}$ to obtain a matrix $\mathbf{W}'$ in a special "diagonal form", without losing the $[\leqslant 2k]$-eed property. If we note that even the addition of a complete row is allowed, then we obtain the following Theorem.

Theorem 6.7 *If $m = \text{minsize}_n(k)$, then there exists a $[\leqslant 2k]$-eed matrix of size $(m \times n)$ of the following form:*

$$
\begin{bmatrix}
0 & 1 & 0 & 0 & \dots & 0 & \\
0 & 0 & 1 & 0 & \dots & 0 & \\
0 & 0 & 0 & 1 & \dots & 0 & \text{submatrix } R \\
\vdots & & & & \vdots & & \\
0 & 0 & 0 & 0 & \dots & 1 &
\end{bmatrix}
\tag{12}
$$

Here every column of the $(m \times (n - m - 1))$ (and possibly empty) submatrix R must have at least $2k$ ones.

6.4.3 *Optimal $(m_{[\leqslant 2k]} \times n)$-Matrices for Small n*

Using the normal form of Theorem 6.7, we can determine optimal $[\leqslant 2k]$-eed matrices for small values of n.

We start with $k = 2$. (The case $k = 1$ was already solved in subsection 6.4.2.) Because of $2k \leqslant n$ the smallest value for n is $n = 4$. In this case (9) and (8) yield $3 = \lceil \log \binom{4}{2} \rceil \leqslant \text{minsize}_4(2) \leqslant 3$, therefore $\text{minsize}_4(2) = 3$. According to (12), a possible $[\leqslant 4]$-eed-matrix is:

$$
\begin{bmatrix}
0 & 1 & 0 & 0 \\
0 & 0 & 1 & 0 \\
0 & 0 & 0 & 1
\end{bmatrix}
\tag{13}
$$

Similarly, we obtain $\text{minsize}_5(2) = \text{minsize}_6(2) = 4$.

The corresponding $[\leqslant 4]$-eed matrices of the form (12) are:

$$
\begin{bmatrix}
0 & 1 & 0 & 0 & 0 \\
0 & 0 & 1 & 0 & 0 \\
0 & 0 & 0 & 1 & 0 \\
0 & 0 & 0 & 0 & 1
\end{bmatrix}
\quad and \quad
\begin{bmatrix}
0 & 1 & 0 & 0 & 0 & 1 \\
0 & 0 & 1 & 0 & 0 & 1 \\
0 & 0 & 0 & 1 & 0 & 1 \\
0 & 0 & 0 & 0 & 1 & 1
\end{bmatrix}
$$

It is easy to verify that these matrices are in fact $[\leqslant 4]$-eed matrices. In these three cases we have $\text{minsize}_n(k) = \lceil \log \binom{n}{k} \rceil$. But this lower bound is not always sharp. For $n = 8$ we have $\lceil \log \binom{8}{2} \rceil = 5$, but there is no $[\leqslant 4]$-eed matrix of the form (12) with 8 columns and 5 rows. In this case we obtain $\text{minsize}_8(2) = 6$.

In this manner we can proceed for $k = 3$. In particular we obtain $\text{minsize}_6(3) = 5$, $\text{minsize}_7(3) = 6$. The corresponding $[\leqslant 6]$-eed-matrices for these values are of the form (10), i.e. the submatrix R in (12) is empty. For $n = 8$ we obtain $\text{minsize}_8(3) = 6$, and it is easy to verify that the matrix

$$\begin{bmatrix} 0 & 1 & 0 & 0 & 0 & 0 & 0 & 1 \\ 0 & 0 & 1 & 0 & 0 & 0 & 0 & 1 \\ 0 & 0 & 0 & 1 & 0 & 0 & 0 & 1 \\ 0 & 0 & 0 & 0 & 1 & 0 & 0 & 1 \\ 0 & 0 & 0 & 0 & 0 & 1 & 0 & 1 \\ 0 & 0 & 0 & 0 & 0 & 0 & 1 & 1 \end{bmatrix} \tag{14}$$

is a $[\leqslant 6]$-eed matrix.

6.4.4 *Recursive Determination of Larger $[\leqslant 2k]$-eed Matrices*

The next Theorem shows, how a $2k$-eed matrix with $2n$ columns can be constructed from matrices with n columns.

Theorem 6.8 *Let $W_{[\leqslant 2k]}$ be a $[\leqslant 2k]$-eed-matrix of size $(m_0 \times n)$, and let U be a $[\leqslant 2\lfloor k/2 \rfloor]$-eed-matrix of size $(m_1 \times n)$ with an additional add-odd row. Then the matrix*

$$V = \begin{bmatrix} \begin{array}{c|c} W_{[\leqslant 2k]} & W_{[\leqslant 2k]} \\ \hline \textit{(zero-matrix)} & U \end{array} \end{bmatrix}$$

is a $[\leqslant 2k]$-eed matrix of size $((m_0 + m_1) \times 2n)$.

Proof We have to verify the validity of the criterion in Theorem 6.5. For details we refer to Börner.[2] □

Let $k = 2$. Then we can start with a small matrix $W_{[\leqslant 4]}$, e.g. with the (3×4) matrix of (13). As matrix U we choose the $[\leqslant 2]$-eed matrix with an add-odd row from (11). Then we obtain the following (6×8)-matrix:

$$\begin{bmatrix} 0 & 1 & 0 & 0 & 0 & 1 & 0 & 0 \\ 0 & 0 & 1 & 0 & 0 & 0 & 1 & 0 \\ 0 & 0 & 0 & 1 & 0 & 0 & 0 & 1 \\ 0 & 0 & 0 & 0 & 0 & 1 & 0 & 1 \\ 0 & 0 & 0 & 0 & 0 & 0 & 1 & 1 \\ 0 & 0 & 0 & 0 & 1 & 0 & 0 & 1 \end{bmatrix}$$

This matrix is again a $[\leqslant 4]$-eed matrix, i.e. the sum of two or four different columns is never a zero-column.

We can iterate this construction. Each step doubles the number n of columns to $2n$, but the number of rows grows only by $1 + \lceil \log n \rceil$. In this way we obtain matrices where the number of columns is a power of 2. If the desired number of columns is not a power of 2, then we simply can omit some columns, without weakening the property to be a $[\leqslant 4]$-eed matrix. (This follows once more from 6.5.) A careful count of the numbers of rows in the new matrices leads to the bound

$$\text{minsize}_n(2) \leqslant \frac{(\lfloor \log n \rfloor)^2 + \lfloor \log n \rfloor}{2} + 1 + r(n) \tag{15}$$

Here we abbreviate $r(n) = \lceil \log(n - 2^{\lfloor \log n \rfloor}) \rceil$ if $n \neq 2^{\lfloor \log n \rfloor}$, and $r(n) = -1$ otherwise.

It is a bit surprising, that we obtain the same bound also for $k = 3$. In this case we start e.g. with the (6×8) matrix $W_{[\leqslant 6]}$ of (14). Because of $\lfloor 2/2 \rfloor = \lfloor 3/2 \rfloor = 1$ we can use the same matrices U in the construction of Theorem 6.8 as in the case $k = 2$. Therefore we have for $n \geqslant 6$:

$$\text{minsize}_n(3) \leqslant \frac{(\lfloor \log n \rfloor)^2 + \lfloor \log n \rfloor}{2} + 1 + r(n) \tag{16}$$

6.5 Detection of 2- and 4-Bit Errors

Now we use our results for the detection of several classes of errors. We start with the following parity matrix.

$$\mathbf{W} = \begin{array}{|ccccccc|l} 1\,1\,1\,1\,1\,1\,1 & \text{complete row} \\ \hline 0\,1\,0\,1\,0\,1\,0 & \\ 0\,0\,1\,1\,0\,0\,1 & [\leqslant 2]\text{-eed-matrix} \\ 0\,0\,0\,0\,1\,1\,1 & (\lceil \log n \rceil \text{ rows}) \\ \hline 0\,1\,1\,0\,1\,0\,0 & \text{add-odd row} \end{array} \tag{17}$$

(In our example we have $n = 7$ inputs, but the following holds for all values $n \geqslant 2$.) The add-odd row assures that each column in $\mathbf{W}$ has an odd number of ones. Therefore by Theorem 6.2, all odd errors will be detected with this parity matrix. A 2-bit error can occur in the form of a $[1, 1]$ error or as a $[2]$-error. A $[1, 1]$-error will be detected by the complete row in $\mathbf{W}$ by Theorem 6.3. A $[2]$ error will be detected because the $\lceil \log n \rceil$ rows in the middle of $\mathbf{W}$ form a $[\leqslant 2]$-eed matrix. So we have:

Theorem 6.9 *If the matrix* **W** *of (17) is used as the parity matrix in the convolutional compactor with n inputs of Fig. 6.1, then all odd errors and all 2-bit errors are detected. The compactor needs* $\lceil \log n \rceil + 2$ *flip-flops.*

But there are undetectable 4-bit errors for the simple matrix of (17). Therefore we now turn to a design of a parity matrix that can detect all odd errors and all 2- and 4-bit errors.

We need an additional notation: Let **V** be a matrix. Then the *mirrored matrix* **V*** is the matrix with the same rows as in **V**, but in reverse order.

Theorem 6.10 *Consider a matrix* **W** *with the structure as shown in Fig. 6.2.*

1 1 1 1 1 1 1	complete row of ones
0 1 0 1 0 1 0	
1 0 0 1 1 0 0	$[\leqslant 2]$-eed matrix **V**, ($\lceil \log n \rceil$ rows)
1 1 0 0 0 0 1	
1 0 0 0 1 1 1	add–odd row
0 1 0 0 0 0 1	
0 0 1 0 0 0 1	$[\leqslant 4]$-eed matrix,
0 0 0 1 0 0 1	($\leqslant \dfrac{(\lfloor \log n \rfloor)^2 + \lfloor \log n \rfloor}{2} + 1 + \lceil \log(n - 2^{\lfloor \log n \rfloor}) \rceil$ rows)
0 0 0 0 1 0 1	
0 0 0 0 0 1 1	
1 1 0 0 0 0 1	
1 0 0 1 1 0 0	mirrored matrix **V***, ($\lceil \log n \rceil$ rows)
0 1 0 1 0 1 0	
1 1 1 1 1 1 1	complete row of ones

Fig. 6.2 Parity Matrix for 4-Bit Error Detection.

(The example has only n = 7 inputs, but the structure is valid for all values $n \geqslant 4$.*)*

If the matrix **W** *of Fig. 6.2 is used as the parity matrix of the convolutional compactor with n inputs in Fig. 6.1, then the compactor detects all odd errors and all 2-bit and 4-bit errors. This compactor needs*

$$\frac{(\lfloor \log n \rfloor)^2 + \lfloor \log n \rfloor}{2} + 2\lceil \log n \rceil + 4 + r(n)$$

flip-flops. (The number of different parity trees is lower, because some rows occur twice in **W**, *and the corresponding trees must be realized only once.)*

Proof The add-odd row assures that each column in **W** has an odd number of ones. Consequently, by Theorem 6.2 all odd errors will be detected with this parity

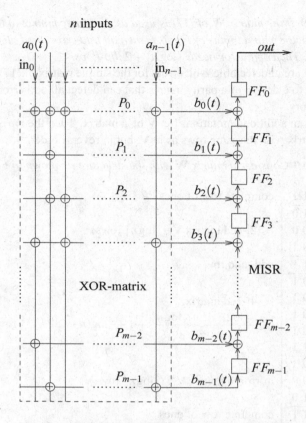

Fig. 6.3   A Convolutional Compactor for 4-Bit Error Detection.

matrix. The first and last rows of this matrix are complete rows of ones. Therefore, by Theorem 6.3 all $[k_1, \ldots, k_l]$-errors with odd k_1 or odd k_l are detected. Every $[2]$-error and every $[4]$-error will be detected by the $[\leqslant 4]$-eed matrix in the middle of $\mathbf{W}$.

This shows already that all 2-bit errors and all 4-bit errors will be detected, with the possible exception of the $[2,2]$-errors. The detectability of these $[2,2]$-errors is a consequence of the symmetric arrangement of the $[\leqslant 2]$-eed matrices $\mathbf{V}$ and $\mathbf{V}^\star$, we refer to Börner.[2]   □

We want to mention, that our parity matrix detects a lot more errors. E.g. all $[2,4]$-errors and all $[4,2]$-errors are also detectable. Moreover, there are many possible modifications of our structure. Figure 6.3 shows a convolutional compactor for 7 inputs, based on the parity matrix of Fig. 6.2. If the parity matrix contains

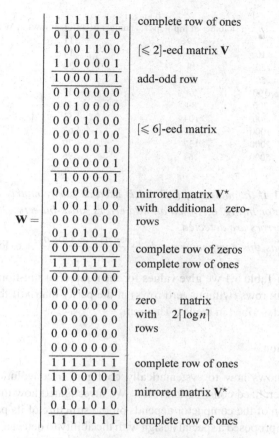

$$\mathbf{W} = \begin{array}{|l|l|}
\hline
1\ 1\ 1\ 1\ 1\ 1\ 1 & \text{complete row of ones} \\
\hline
0\ 1\ 0\ 1\ 0\ 1\ 0 & \\
1\ 0\ 0\ 1\ 1\ 0\ 0 & [\leqslant 2]\text{-eed matrix } \mathbf{V} \\
1\ 1\ 0\ 0\ 0\ 0\ 1 & \\
\hline
1\ 0\ 0\ 0\ 1\ 1\ 1 & \text{add-odd row} \\
\hline
0\ 1\ 0\ 0\ 0\ 0\ 0 & \\
0\ 0\ 1\ 0\ 0\ 0\ 0 & \\
0\ 0\ 0\ 1\ 0\ 0\ 0 & [\leqslant 6]\text{-eed matrix} \\
0\ 0\ 0\ 0\ 1\ 0\ 0 & \\
0\ 0\ 0\ 0\ 0\ 1\ 0 & \\
0\ 0\ 0\ 0\ 0\ 0\ 1 & \\
\hline
1\ 1\ 0\ 0\ 0\ 0\ 1 & \\
0\ 0\ 0\ 0\ 0\ 0\ 0 & \text{mirrored matrix } \mathbf{V}^\star \\
1\ 0\ 0\ 1\ 1\ 0\ 0 & \text{with additional zero-} \\
0\ 0\ 0\ 0\ 0\ 0\ 0 & \text{rows} \\
0\ 1\ 0\ 1\ 0\ 1\ 0 & \\
\hline
0\ 0\ 0\ 0\ 0\ 0\ 0 & \text{complete row of zeros} \\
1\ 1\ 1\ 1\ 1\ 1\ 1 & \text{complete row of ones} \\
\hline
0\ 0\ 0\ 0\ 0\ 0\ 0 & \\
0\ 0\ 0\ 0\ 0\ 0\ 0 & \\
0\ 0\ 0\ 0\ 0\ 0\ 0 & \text{zero matrix} \\
0\ 0\ 0\ 0\ 0\ 0\ 0 & \text{with } 2\lceil \log n \rceil \\
0\ 0\ 0\ 0\ 0\ 0\ 0 & \text{rows} \\
0\ 0\ 0\ 0\ 0\ 0\ 0 & \\
\hline
1\ 1\ 1\ 1\ 1\ 1\ 1 & \text{complete row of ones} \\
\hline
1\ 1\ 0\ 0\ 0\ 0\ 1 & \\
1\ 0\ 0\ 1\ 1\ 0\ 0 & \text{mirrored matrix } \mathbf{V}^\star \\
0\ 1\ 0\ 1\ 0\ 1\ 0 & \\
\hline
1\ 1\ 1\ 1\ 1\ 1\ 1 & \text{complete row of ones} \\
\hline
\end{array}$$

Fig. 6.4   Parity Matrix for 6 Bit Error Detection.

identical rows, then the parity tree according to such a row must be realized only once.

6.6 Detection of 6-Bit Errors

It is possible to extend our approach also to the detection of all 6-bit errors. Fig. 6.4 shows the structure of a corresponding parity matrix. Similar arguments as in the last section show that this matrix detects all odd errors, all 2-bit errors, all 4-bit errors, all $[2,4]$-errors and all $[4,2]$-bit errors. Every $[6]$-error is recognized with the help of the $[\leqslant 6]$-eed matrix in the middle of $\mathbf{W}$. For the remaining $[2,2,2]$- and $[2,1,1,2]$-errors we refer to Börner.[2]

Table 6.1

n	number of flip-flops	number of different rows in $\mathbf{W}$
7	29	11
10	37	14
20	48	20
50	62	29
100	75	37
200	89	46
500	104	56
1000	120	67
2000	137	79
5000	164	96

Theorem 6.11 *If the matrix of Fig. 6.4 is used as parity matrix of the convolutional compactor in Fig. 6.1, then all odd errors, all 2-bit errors, all 4-bit errors and all 6-bit errors are detected.*

For n inputs this compactor needs $\dfrac{(\lfloor \log n \rfloor)^2 + \lfloor \log n \rfloor}{2} + 6\lceil \log n \rceil + 6 + r(n)$ *flipflops.*

Finally, in Table 6.1 we give values for the number of flip-flops and the number of different rows (without zero rows) in the parity matrix if the compactor is structured as described in the last Theorem.

6.7 Conclusions

This paper shows how to systematically configure a convolutional compactor based on generalized construction rules. We investigated how the capability for error-detection of the compactor depends on the structure of its parity matrix. In particular, we proposed a special design with the ability to detect all odd errors and all 2-bit, 4-bit and 6-bit errors at the inputs. The number of flipflops in the shift register of the compactor is in $O(\log^2 n)$, Therefore this compactor is of interest for the test of circuits with a large number of scan-pathes.

References

1. M. Arai, S. Fukumoto, K. Iwasaki, Analysis of error masking and X-masking probabilities for convolutional compactors, in Proc. *IEEE International Test Conference*, 2005.
2. F. Börner, Kompaktierung von Daten aus Scanpfaden mit einem Faltungskompaktor, *Report 003/2005*, Institut für Informatik, Universität Potsdam, 2005.
3. M. Gössel, *Lineare Automaten und Schieberegister.* Akademie-Verlag, Berlin, 1971.
4. Y. Han, Y. Hu, H. Li, X. Li and A. Chandra, Response compaction for test time and test pins reduction based on advanced convolutional codes, in Proc. *IEEE Int. Symp. on Defect and Fault Tolerance in VLSI Systems* (DFT 2004), 298–305.

5. R. Johannesson, K. Sh. Zigangirov, *Fundamentals of Convolutional Coding*, IEEE Press, 1999.
6. Z. Kohavi, *Switching and Finite Automata Theory*, McGraw-Hill, 1979.
7. G. Mrugalski, A.Pogiel, J. Rajski, C. Wang, Fault diagnosis in designs with convolutional compactors, in Proc. *IEEE International Test Conference* (ITC 2004), 498–506.
8. J. Rajski, J. Tyszer, C. Wang, S.M. Reddy, Convolutional compaction of test responses, in Proc. *IEEE International Test Conference* (ITC 2003), 745–754.
9. J. Savir, On shrinking wide compressors, in Proc. *13th VLSI Test Symposium* (VTS 1995), 108–117.

Chapter 7

Low-Energy Pattern Generator for Random Testing

Bhargab B. Bhattacharya*, Sharad C. Seth, and Sheng Zhang

Department of Computer Science and Engineering
University of Nebraska-Lincoln Lincoln
NE 68588-0115, USA

A new built-in self-test (BIST) scheme for scan-based circuits is proposed for reducing energy consumption during testing. In a random testing environment, a significant amount of energy is wasted in the LFSR and the circuit-under-test (CUT) by useless patterns that do not contribute to fault dropping. Another major source of energy drainage is the loss due to random switching activity in the CUT and in the scan path between applications of two successive vectors. In this work, a pseudorandom test sequence is generated by a linear feedback shift register (LFSR), and useless patterns are identified by fault simulation. Switching activity of the remaining (useful) patterns in the CUT including the scan path is estimated, and their optimal reordering that minimizes switching activity, is determined. Two components of switching activity, namely intrinsic (independent of test ordering) and variable (dependent on test ordering) are identified. An efficient technique of computation of switching activity occurring in the scan path is reported. Next, we design a mapping logic, which modifies the state transitions of the LFSR such that only the useful vectors are generated by the LFSR according to the desired sequence. Earlier techniques of energy reduction were based on blocking of useless vectors at the inputs to the CUT after being produced by the LFSR. Similarly, test reordering was used for low-energy BIST design, but only in the context of deterministic testing. The novelty of this approach lies in the fact that it not only inhibits the LFSR from generating the useless patterns, but also enforces the patterns to appear in a desired sequence. Further, it reduces test application time without affecting fault coverage. Experimental results on ISCAS-89 benchmark circuits reveal a significant amount of energy savings in the LFSR during testing. The design is simple and is applicable to a general scan-based test scheme like the STUMPS architecture.

---

*While on leave from Indian Statistical Institute, Kolkata, India

.

Contents

7.1 Introduction

With the emergence of mobile computing and communication devices, such as laptops, cell-phones, portable multimedia and video equipments, design of low-energy VLSI systems has become a major concern in circuit synthesis.[1,2] A significant component of the energy consumed in CMOS circuits is caused by the total amount of *switching activity* (*SA*) at various circuit nodes during operation. The energy dissipated at a circuit node is proportional to the total number of $0 \rightarrow 1$ and $1 \rightarrow 0$ transitions the logic signals undergo at that node multiplied by its capacitance (which depends on its fanout, and its transistor implementation). The allowable energy is limited by the battery and the thermal capacity of the package. The higher device densities and clock rates of the deep submicron technology further exacerbate the thermal dissipation problem.

It has been observed that energy consumption in an IC may be significantly higher during testing due to increased switching activity than that needed during normal (system) mode, which can cause excessive heating and degrade circuit reliability.[3] The average-power optimization has to do with extending the battery life in mobile applications. The maximum-power optimization, on the other hand, can be in terms of sustained or instantaneous power. Maximum sustained power, over a specified limit, can cause excessive heating of the device and even burnout. Maximum instantaneous power, on the other hand, may cause excessive (inductive) voltage drop in the power and ground lines because of current swing. Thus, the logic states at circuit nodes may erroneously change.[4]

Conventional built-in self-test (BIST) schemes with random patterns may need an excessive amount of energy because of the length of the test set and randomness of the consecutive vectors applied to the CUT. Further, in a scan-based system, a significant amount of energy may be wasted during just the scan operations. Although extensive research has been done for designing low-power circuits while operating in the system mode,[1,2] many issues of reducing power/energy consumption are still open while the circuit is in the test mode. In this paper, we focus on the factors that influence the energy consumption during testing and propose a new scheme for scan-BIST architecture that minimizes it during *pseudorandom testing*. The test application time is also significantly reduced and hence, we may lessen the chances of burnout due to sustained maximum power. Designing a pseudorandom test pattern generator (TPG) that produces vectors in a (desired) *deterministic* sequence, in order to minimize switching activity, is an open problem.

7.2 Related Works

BIST techniques are widely used in IC testing for their cost-effectiveness, convenience and ease of implementation.[5,6] They provide on-chip test pattern generators (TPG) for applying a pseudorandom test sequence to the CUT, and multiple input signature analyzers (MISR) to capture the test responses. The TPG is usually implemented by a linear feedback shift register (LFSR). They are widely used for testing scan-based systems employing, for example, the STUMPS architecture.[6] To improve fault coverage in a BIST scheme, several approaches are known - use of weighted random patterns,[7] reseeding or redefining transition functions of the LFSR,[8,9] synthesis for testability,[10] arithmetic BIST[11,12] and embedding of deterministic test cubes for hard-to-detect faults generated by ATPG tools in the pseudorandom sequence using deterministic BIST,[13] bit flipping,[14] or bit fixing.[15,16]

With the emergence of deep submicron technology, minimizing instantaneous peak power, sustained peak power, or total energy consumption during testing (both for manufacturing and routine field tests) has become an important goal of BIST design. The existing techniques include distributed BIST and test scheduling,[3,17] toggle suppression and blocking useless patterns,[18] designing low-power TPG,[19,20] and low-transition random pattern generator,[21] optimal vector selection,[22] designing new scan path architecture,[23] use of Golomb coding for low-power scan testing with compressed test data,[24] and BIST for data paths.[25] For deterministic testing, energy reduction can be achieved by reordering scan chains and test vectors to minimize switching activity.[26,27] Compaction of test vectors for low power in scan-based systems is available in the literature.[28] Several commercial BIST design tools are also available.[29] Other low power design schemes for scan-based BIST[30] and for system-on-chip[31] have been reported recently.

7.3 Proposed Scheme

We assume a *test-per-scan* BIST scheme as in STUMPS architecture (Figure 7.1) and further assume that a netlist description of the scan-based CUT is known. A *modulo-m* bit-counter keeps track of the number of scan shifts, where m is the length of the longest scan path. Test energy reduction by toggle suppression and pattern blocking approaches were considered earlier.[18] In the latter, random patterns generated by the LFSR that do not contribute to fault dropping (useless patterns) are suppressed at the input to the CUT, using a pattern counter and blocking logic. Since the number of useful patterns is known to be a very small fraction of all generated patterns, a significant amount of energy is still wasted in the LFSR while cycling through these useless patterns. Further, test-vector reordering can be used to save energy in a deterministic testing environment, but is challenging to implement in a pseudorandom setting. In this paper, we propose a new technique of BIST design that prevents the LFSR to cycle through the states generating useless patterns, as well as reorders the useful patterns in a desired sequence for application to the CUT.

To estimate the consumed energy, we compute the total switching activity as the number of $0 \rightarrow 1$ and $1 \rightarrow 0$ transitions in all the circuit nodes including the LFSR, CUT, and the scan path over a complete test session. The various steps of the proposed method are now summarized below:

(i) A pseudorandom test sequence is generated by an LFSR, and its fault coverage in the CUT is determined through forward and reverse fault simulation; let S denote the test sequence up to the last useful vector (beyond which fault coverage does not improve significantly); we assume single stuck-at fault model.

(ii) Identify the set U of useless patterns in S that do not contribute to fault dropping.

(iii) For all ordered pairs of test vectors in the reduced set $S_r = (S \setminus U)$, determine the switching activity (SA) in the scan path and the CUT.

(iv) Reorder the vectors in S_r to determine an optimal order S' to minimize energy.

(v) Modify the state table of the LFSR such that it generates the new sequence S'.

(vi) Synthesize a mapping logic (ML) with minimum cost, to augment the LFSR; the state transitions of the LFSR are modified under *certain conditions* to serve two purposes: (a) to prevent it from cycling through the states generating useless patterns, and (b) to reorder S_r to S'; for all other conditions, the LFSR runs in accordance to its original state transition function. Figure 7.2 shows a simple MUX-based design that can be used for this purpose (explanation of the scheme will be provided shortly).

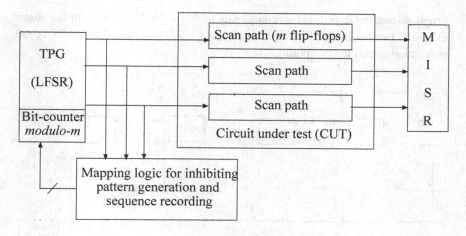

Fig. 7.1 Proposed low-energy BIST design.

It may be noted that determining an optimal order was implemented earlier only for a set of pre-computed *deterministic test vectors*,[26,27] and hence these approaches are unsuitable for random testing. The proposed work aims at *changing the natural order of the test vectors of a *pseudorandom sequence* generated by an LFSR to a *desired order*, and *skipping* the useless patterns by adding a minimum amount of additional mapping logic. Since the useless test patterns are not generated by the LFSR, no blocking logic is needed at the inputs to the CUT.[18] Thus, power is saved both in the LFSR and the CUT. A similar idea of skipping certain LFSR states is used earlier in order to embed a set of deterministic tests,[32] whereas, our objective here is to compact random tests from a low-power perspective. Further, we identify two components of switching activity in the scan-path and the CUT: the *intrinsic* part (the portion that does not depend on vector ordering, but depends only on the test set), and the *variable* part (the portion that depends on vector ordering). Based on this, we propose a very efficient technique for computing the *SA* in the scan path (as listed in Step (iii) earlier) for all ordered vector pairs. This is especially important for a circuit with a large scan path, where simulation of *scan-shift switching activity* (*SSSA*) is extremely time consuming. *SSSA* occurs in the scan flip-flops during shifting in a test vector while shifting out the CUT response of the preceding test vector. For example, for a scan chain with 1000 FF's, and with 500 test vectors, in order to determine *SSSA* for all ordered pairs of vectors, simulation for $(1000*500*500) = 25,000,00,00$ cycles (approx.) is needed. In general, for T tests, and a linear scan chain of length f, the complexity of naive simulation is $O(f*T^2)$, which is unacceptably high. The proposed method improves this complexity to $O(f * T + T^2)$. Thus, for the above example, our method will need around $(1000 * 500) + (500 * 500) = 75,00,00$ simulation cycles for computing *SSSA*.

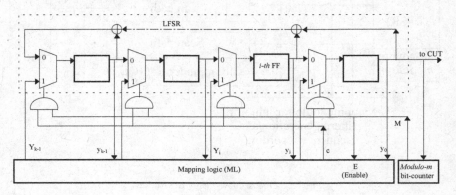

Fig. 7.2   Modification to the LFSR.

The following simple example of a TPG (Figure 7.3) illustrates the underlying idea of state-skipping technique.

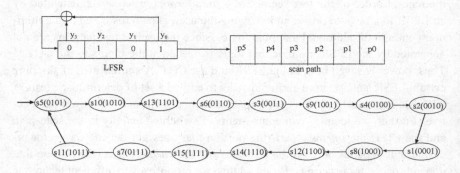

Fig. 7.3   An example LFSR and its state diagram.

The least significant bit of the LFSR is shifted serially into the scan path generating the following test sequence:

Let us assume that the above test sequence S is applied to the CUT. The total energy consumed in the LFSR, the CUT, and the scan-chain is determined by the switching activity summed over all successive test applications. The number of transitions between applications of two consecutive vectors depends on a complex distance function determined by the CUT, and the test and response func-

Table 7.1   Test vectors and their start- and end-states.

Test sequence S	$p0$	$p1$	$p2$	$p3$	$p4$	$p5$	start-state of LFSR	end-state of LFSR
$t1$	1	0	1	0	1	1	$s5$	$s9$
$t2$	0	0	1	0	0	0	$s4$	$s14$
$t3$	1	1	1	1	0	1	$s15$	$s13$
$t4$	0	1	1	0	0	1	$s6$	$s1$
$t5$	0	0	0	1	1	1	$s8$	$s11$

tion. We will demonstrate shortly that some component of SA is *intrinsic*, and the rest is *variable*. Hence, in a test session, switching activity can be represented as a directed complete graph called activity graph (see Figure 7.4b, where a bi-directional edge means two directed edges in opposite directions). Each node in the graph represents a test vector, and the directed edge (e_{ij}) represents application of test vector t_j following the test vector t_i. The weight $w(e_{ij})$ on the edge e_{ij} denotes the variable component of switching activity corresponding to the or-dered pair of tests (t_i, t_j). The intrinsic component may be represented as a node weight, and being invariant over a test session, may be ignored as far as determi-nation of optimal ordering is concerned. As an example, let us consider an activity graph as in Figure 7.4b. The edge weights are represented as a cost matrix in Fig-ure 7.4a, which in general is asymmetric, because the variable component of SA strongly depends on ordering of test pairs. Thus, for the above test sequence S ($t1 \rightarrow t2 \rightarrow t3 \rightarrow t4 \rightarrow t5$), the variable component of switching activity is $(8 + 10 + 10 + 9) = 37$. Let us now assume that $t3$ is found to be a useless test pattern by fault simulation. Thus, if $t3$ is blocked at the input to the CUT, it saves some energy. Further, if $t3$ is not generated by the LFSR, a significant amount of energy is saved, because the LFSR has to cycle through six states in order to generate a test vector. Thus, the node $t3$ along with all incident edges on it can be deleted. An optimal ordering of test vectors that minimizes the energy consumption can now be found by determining a min-cost Hamiltonian path, which in this example is S' ($t1 \rightarrow t2 \rightarrow t5 \rightarrow t4$), the path cost being equal to 23 ($= 8 + 7 + 8$), as shown in Figure 7.4c.

The activity in the LFSR, not shown here, will also reduce significantly, if it does not generate the useless patterns. To achieve these two goals, we design our mapping logic as follows.

In the new sequence S', for the ordered pair ($t1 \rightarrow t2$), no action is required, as $t2$ is generated by the LFSR as a natural successor vector of $t1$. So, for s9 (end-state of $t1$), we set the Y-outputs of the mapping logic (see Figure 7.2) to don't cares (d), and the control line C to 0 (see Table 7.2). However, we need an addi-

	t1	t2	t3	t4	t5
t1	0	8	9	11	9
t2	11	0	10	12	7
t3	8	6	0	10	6
t4	10	9	6	0	9
t5	11	8	7	8	0

Fig. 7.4   (a) Cost matrix, (b) Activity graph, and (c) Optimal H-path in the reduced graph.

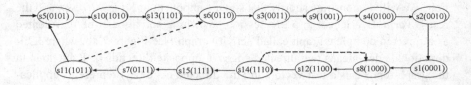

Fig. 7.5   State skipping transitions for suppressing useless patterns and reordering useful patterns.

Table 7.2   Truth table of mapping logic.

Present State of the LFSR (inputs to mapping logic)					Outputs of the mapping logic					
	y_3	y_2	y_1	y_0		Y_3	Y_2	Y_1	Y_0	C
s9:	1	0	0	1		d	d	d	d	0
s14:	1	1	1	0	s8:	1	0	0	0	1
s11:	1	0	1	1	s6:	0	1	1	0	1
all other states:						d	d	d	d	d

tional transition from $s14$ (end-state of $t2$) to $s8$ (start-state of $t5$), and similarly from $s11$ (end-state of $t5$) to $s6$ (start-state of $t4$). For these combinations, the Y-outputs are determined by the corresponding start-states, and C is set to 1. For all other remaining combinations, all outputs are don't cares. These transitions will not only generate the useful test patterns in a desired sequence, but also prevent the LFSR to cycle through the states that generate useless patterns (in this example, test $t3$). Further, the output M of the *modulo-m* bit counter assumes 1 only when scan path (whose length is m) is filled, i.e., at the end-states of the test vectors. Thus, in order to generate the altered sequence S', we need to skip the natural next state of the LFSR and jump to the start state of the desired next test pattern. These are called state skipping transitions and are shown with dotted lines in Figure 7.5. The truth-table description of the mapping logic (ML) for this example is shown in Table 7.2, from which the required functions can be synthesized.

Table 7.3 General description of the mapping logic.

	Present State of the LFSR (inputs to mapping logic)		Outputs of the mapping logic				
	y_{k-1}	$\cdots$	y_0	Y_{k-1}	$\cdots$ $\cdots$	Y_0	C
Case (i)*	End state of t_i'			d	d d	d	0
Case (ii)	End state of t_i'			start-state(t_{i+1}')			1
	all other states			d	d d	d	d

* Case(i) is applicable if the consecutive test pair (t_i', t_{i+1}') of S' appears in consecutive order in the original test sequence S as well; otherwise, case (ii) is applicable.

In general, the mapping logic can be described as follows:

Given a seed, let S denote the original test sequence generated by the LFSR, and $S' = \{t_1', t_2', \ldots, t_i', t_{i+1}', \ldots\}$ denote the optimally ordered reduced test sequence consisting of useful vectors only. Let y_i denote the output of the i-th flip-flop of the LFSR, and Y_i denote the output of the mapping logic feeding the i-th flip-flop through a MUX (see Figure 7.2). The mapping logic is a combinational circuit with k inputs $\{y_0, y_1, \ldots y_{k-1}\}$, and $k+1$ outputs $\{Y_0, Y_1, \ldots Y_{k-1}, C\}$, where k is the length of the LFSR, and C is a control output. For every test t_i' in S', there is a corresponding row in the truth table given in Table 7.3

Thus, the next-state of the LFSR follows the transition diagram of the original LFSR when either $C = 0$, or $M = 0$, and is determined by the outputs of the mapping logic if and only if $C * M = 1$. Since these additional transitions emanate only from the end-states of test patterns, their occurrences can be signaled by the output M of the bit-counter, and also when $C = 1$. In order to prevent the switching activity from occurring in ML for every scan shift cycle, the inputs to ML can be made transparent to it by using an enable signal E controlled by M. Thus, the y-inputs become visible to ML if and only if $M = 1$. Further, all the patterns generated by the LFSR are distinct and hence each state can have at most one additional outgoing and one incoming transition. This mapping logic enforces the LFSR to generate only the useful vectors in a desired sequence. Most of the entries in the specification of mapping logic will be don't cares, as shown in Table 7.3. Termination of the test session can be signaled when the end-state of the last useful pattern in S' is reached. Determination of optimal reordering of test patterns is equivalent to solving a traveling salesman problem (TSP) in a directed graph, which being an NP-hard problem, needs heuristic techniques for quick solution.[33] However, finding a suitable test order which minimizes the overall switching activity and which can be synthesized in a cost-effective way, is a major problem.

Upadhyaya and Chen[32] used the skipping of LFSR states in a scenario, where the order of test application was immaterial. In our case, some component of power dissipation depends strongly on the order. The synthesis of mapping logic takes care of don't care states to minimize cost.

7.4 Modeling and Computation of Switching Activity

We assume fully isolated scan-path architecture (see Figure 9.21, p. 368 of reference[35]), in which the scan register (SCAN) is completely separated from the CUT by a buffer register (BUFFER), and SCAN has read/write access to the CUT only through BUFFER. This eliminates the switching activity rippling through the CUT while shifting in a test pattern. When all the bits of a test pattern are shifted in, the content of SCAN is copied to BUFFER, and then applied to the CUT. During the system mode, the CUT outputs are captured in BUFFER, and then copied to SCAN. In the next shift cycle, the response vector is shifted out while shifting in the next test pattern to SCAN.

For application of tests, primary inputs (PI) are scanned in, along with the current state. Similarly, primary outputs (PO) are scanned out, along with the next state. This requires extending the length of the SCAN register by $\max\{\#PI, \#PO\}$, where $\#PI$ and $\#PO$ indicate the number of primary inputs and the number of primary outputs respectively.

A *test cycle* corresponds to applying a particular test vector to the CUT and capturing its response. In the context of fully isolated scan, the test cycle can be broken down as follows. Assume that the test vector t_j is applied following the vector t_i, and let r_i denote the response vector when t_i is applied to the CUT.

Step	Action	Number of cycles	SA occurs in
1	Scan-in t_j in SCAN, and scan out last response (r_i)	m	LFSR, SCAN, MAP
2	Copy SCAN to BUFFER, and evaluate the response (r_j) of the CUT	1	BUFFER, CUT
3	Capture the response (r_j) in BUFFER	1	BUFFER
4	Copy BUFFER to SCAN	1	SCAN

Notice that unlike the full serial integrated scan designs, the isolated scan adds to more clock cycles (in Steps 2 and 4) per test cycle. Now we analyze the dependence of these steps on the test vectors. At the beginning of Step 1, the states of SCAN and BUFFER are determined by the response (r_i) of the preceding test vector (t_i). Therefore the SA in Step 1 depends on the ordered pair (t_i, t_j). The same is true for Step 2 because the BUFFER holds the response vector of the previously applied test vector. However, in Step 3 when the CUT response is captured in BUFFER, the state we overwrite is determined completely by the response of current test vector. Similarly, in Step 4, the scan-register state being overwritten depends only on the current test vector (t_j). In other words, the switching activity in Steps 1 and 2 depends on vector ordering (variable component), whereas, that in Steps 3 and 4 depends on the current test vector, and invariant on vector ordering (intrinsic). We will show in the next subsection, that a major component of SA occurring during scan shift (Step 1) although being dependent on consecutive vector pairs, remains invariant over a complete test session. Hence, this portion may also be treated as intrinsic as far as TSP optimization is concerned. Thus, the variable part of SA for a vector pair (t_i, t_j) contributed by Steps 1 and 2 will be treated as a weight on the corresponding directed edge e_{ij} in the activity graph, and the SA in Steps 3 and 4, as well as the intrinsic part of Step 1, will be associated as a weight to the node t_j of the activity graph.

Efficient computation of scan-shift switching activity ($SSSA$)

The computation of $SSSA$ is most complex, as for all pairs of useful vectors, the test patterns are to be shifted in the scan chain while shifting out the response of the previous pattern. This is particularly true when the length of the scan path is large. An example was given in Section 7.3.

Assume the length of the scan chain $= \max\{(\#PI + \#FF), (\#PO + \#FF)\}$. Let t_i and t_j denote two consecutive test vectors (t_j follows t_i), and let r_i denote the response vector of t_i. Since the test vectors (response vectors) are applied (captured) via the scan chain, we consider for the sake of computation of $SSSA$, that both a test and its response vector have the same number of bits ($= f$, the length of the scan chain); if $\#PI \neq \#PO$, then the content of the scan chain can be determined accordingly from the knowledge of test and response vectors, and the difference of $\#PI$ and $\#PO$. Let us now assume that the test vectors t_j is applied after t_i, i.e., in the scan chain, the contents of r_i will be replaced by t_j as shown in the Figure 7.6.

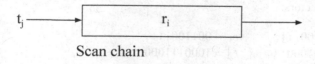

Scan chain

Fig. 7.6   Scan chain.

Thus, $SA(t_i, t_j) = \#$ transitions for shifting r_i out while shifting t_j in. In this section, we show that although $SSSA$ depends on vector ordering, a major component of it is intrinsic in nature, which remains invariant over a test session, and is independent of test vector ordering.

Method

For each test t_i and its response vector r_i compute the following:

Let v denote a bit vector; scan v from left-to-right and observe the maximal run weights.

Example: let v = 0011001110
Run weight W: 2 2 2 3 1

This means that the string v starts with a 0-run of length 2, followed by a 1-run of length 2, and so on. Let $x(v)$ denote the number of transitions in the string v. Thus, $x(v) = 4$. We also index the run weight W with the increasing value of x by looking into them from the left side. Thus, for the above example, we say, $w(0) = 2$, $w(1) = 2$, $w(2) = 2$, $w(3) = 3$, $w(4) = 1$.

For a bit vector v, we define a parameter $IT(v)$, namely *intrinsic transition*, as follows:

$$IT(v) = 1.w(1) + 2.w(2) + 3.w(3) + 4.w(4) + . + x.w(x)$$

(It may be noted that $w(0)$ is not needed in computation.)

For the above example, $IT(v) = (1*2) + (2*2) + (3*3) + (4*1) = 19$.

The following theorem can now be easily proved.

Theorem 7.1 *The number of transitions for shifting r_i out while shifting t_j in is*

$$SA(t_i, t_j) = IT(t_j) + IT(r_i), \text{ if } LSB(t_j) = MSB(r_i);$$
$$= IT(t_j) + IT(r_i) + f, \text{ if } LSB(t_j) \neq MSB(r_i),$$

where, f is the length of the scan chain, and LSB (resp. MSB) denotes the least (resp. most) significant bit.

Example: Let $f = 10$; consider the following two test and corresponding response vectors:

$t_1 = 0111000111$ $r_1 = 0011001110$
$t_2 = 0111010001$ $r_2 = 1000111000$

$IT(t_1) = (1*3) + (2*3) + (3*3) = 18$; $IT(r_1) = (1*2) + (2*2) + (3*3) + (4*1) = 19$;
$IT(t_2) = (1*3) + (2*1) + (3*1) + (4*3) + (5*1) = 25$; $IT(r_2) = (1*3) + (2*3) + (3*3) = 18$.

Computation of $SA(t_1, t_2)$

$$
\begin{aligned}
SA(t_1, t_2) &= \text{\# transitions for shifting } r_1 \text{ out while shifting } t_2 \text{ in} \\
&= IT(t_2) + IT(r_1) + f, \text{ since } LSB(t_2) \neq MSB(r_1); \\
&= 25 + 19 + 10 = 54.
\end{aligned}
$$

Computation of $SA(t_2, t_1)$

$$
\begin{aligned}
SA(t_2, t_1) &= \text{\# transitions for shifting } r_2 \text{ out while shifting } t_1 \text{ in} \\
&= IT(t_1) + IT(r_2), \text{ since } LSB(t_1) = MSB(r_2); \\
&= 18 + 18 = 36.
\end{aligned}
$$

Complexity: For each test vector t_i and its response vector r_i, we calculate $IT(t_i)$ and $IT(r_i)$. This needs $O(f * T)$ time, where T is the total number of useful tests. Then, for each ordered pair (t_i, t_j), $SA(t_i, t_j)$ can be computed in constant time. Since, we have $T(T-1)$ ordered pairs, the total complexity of computing SA for all ordered pairs is $O(f * T + T^2)$.

It may be noted in this context that $SA(t_1, t_2)$ apparently depends strongly on the vector pair (t_1, t_2), but it is not true. A major component of $SSSA$ contributed by the term $\{IT(t_2) + IT(r_1)\}$ is indeed intrinsic in nature, because the contribution due to $IT(t_2)$ or $IT(r_1)$ will appear elsewhere even if t_2 is not immediately preceded by t_1 in the test sequence. Therefore, the total contributions to SA due to $\{IT(t_2) + IT(r_1)\}$ remains invariant over a complete test session (considering a Hamiltonian cycle in the TSP; a little variation might occur because a test sequence is a Hamiltonian path in the activity graph). The variable component of $SSSA$ depends only on the match/mismatch of MSB(previous response vector) and the LSB(current test vector), the computation of which is straightforward. Hence, for every pair of useful vectors, this can be performed quite easily.

7.5 Experimental Results

Experiments were carried out on several ISCAS-89 scan benchmark circuits, and results are reported in Tables 7.4 through 7.7. A 25-bit LFSR is used to generate 20,000 pseudorandom test vectors, and the useful vectors (i.e., those contributing to fault dropping, under the stuck-at fault model) are identified by running the HOPE fault simulator.[39] The reduced sequence S_r thus consists of only the useful vectors. Table 7.4 shows for each circuit, the numbers of FF's, PI's, and PO's, the number of useful patterns obtained after forward and reverse fault simulation, the last useful pattern position in the random sequence, and fault coverage. Since the modified LFSR generates only the useful patterns, a significant amount of test application time is saved as shown in the last column of Table 7.4. Next, for

Table 7.4   Useful patterns and savings in test application time.

Circuit	Number of FF's	Number of PI's	Number of PO's	Number of useful patterns	Last useful pattern position	Fault coverage	Savings in test application time (%)
s27.scan	3	4	1	7	23	100.000	69.56
s208.scan	8	11	2	31	2992	100.000	98.96
s208.1.scan	8	10	1	38	9612	100.000	99.60
s298.scan	14	3	6	38	313	100.000	87.86
s344.scan	15	9	11	26	182	100.000	85.71
s349.scan	15	9	11	26	182	99.429	85.71
s382.scan	21	3	6	33	394	100.000	91.62
s386.scan	6	7	7	74	1756	100.000	95.79
s400.scan	21	3	6	34	394	98.585	91.37
s420.scan	16	19	2	47	16244	93.953	99.71
s420.1.scan	16	18	1	59	15552	89.011	99.62
s444.scan	21	3	6	38	273	97.046	86.08
s510.scan	6	19	7	63	1479	100.000	95.74
s526.scan	21	3	6	72	10864	99.820	99.34
s526n.scan	21	3	6	72	10864	100.000	99.34
s641.scan	19	35	24	53	5691	98.056	99.07
s713.scan	19	35	23	53	5691	91.910	99.07
s820.scan	5	18	19	121	19856	99.529	99.39
s832.scan	5	18	19	119	19856	97.931	99.40
s838.scan	32	35	2	62	9755	85.764	99.36
s838.1.scan	32	34	1	57	15893	64.662	99.64
s953.scan	29	16	23	103	18306	99.907	99.44
s1196.scan	18	14	14	141	18473	97.907	99.24
s1238.scan	18	14	14	147	18473	92.989	99.20
s1423.scan	74	17	5	75	19449	98.878	99.61
s1488.scan	6	8	19	150	5089	100.000	97.05
s1494.scan	6	8	19	148	5089	99.203	97.09
s5378.scan	179	35	49	267	19950	98.682	98.66
s9234.1.scan	211	36	39	290	19567	85.347	98.52
s13207.1.scan	638	62	152	443	19863	95.751	97.77
s15850.1.scan	534	77	150	307	19984	91.616	98.46
s35932.scan	1728	35	320	71	197	89.809	63.96
s38417.scan	1636	28	106	620	19879	94.403	96.88
s38584.1.scan	1426	38	304	719	19885	95.160	96.38

each pair of useful test vectors, switching activity in the CUT and the scan path is computed. We assume a single linear scan chain. As mentioned in Section 7.4, a major component of SA that occurs in the BUFFER and the scan path, is intrinsic in nature, i.e., it remains invariant over a complete test session and is independent of vector ordering. The intrinsic components of SA due to scan shift and capturing the responses in the BUFFER are shown in Table 7.6. To determine an optimum reordering of useful test vectors, we therefore, consider only the variable component of SA occurring in the CUT and the scan path for every ordered pair of useful

Table 7.5   Energy savings in the LFSR.

Circuit original LFSR	*SA* in the modified LFSR	*SA* in the mapping logic	Cost of mapping (# literals)	*SA* in the modified logic (ML)	Total *SA* in savings (%) LFSR (3+5)	Energy $(2-6)/2$
s27.scan	2316	1037	61	185	1222	47.24
s208.scan	784901	7763	342	1937	9700	98.76
s208.1.scan	1338742	9469	421	2565	12034	99.1
s298.scan	80006	9952	447	2728	12680	84.15
s344.scan	44787	9310	298	1700	11010	75.42
s349.scan	44787	9369	296	1459	10828	75.82
s382.scan	172935	13281	435	2627	15908	90.8
s386.scan	485417	14720	885	6939	21659	95.54
s400.scan	172935	13254	438	2713	15967	90.77
s420.scan	7954725	21767	529	3724	25491	99.68
s420.1.scan	7467193	24633	619	4563	29196	99.61
s444.scan	103058	12649	422	2640	15289	85.16
s510.scan	311036	22271	754	5939	28210	90.93
s526.scan	3468273	25972	836	6804	32776	99.05
s526n.scan	3468273	25827	825	6590	32417	99.07
s641.scan	3818782	38766	608	4470	43236	98.87
s713.scan	3818782	38786	607	4421	43207	98.87
s820.scan	4769896	38213	1422	14141	52354	98.95
s832.scan	4769896	38178	1413	14456	52634	98.95
s838.scan	15114313	56628	794	5951	62579	99.59
s838.1.scan	13984169	46522	639	4818	51340	99.63
s953.scan	10852686	52480	1100	9426	61906	99.43
s1196.scan	7824730	67033	1791	21846	88879	98.86
s1238.scan	7824730	68701	1791	21427	90128	98.85
s1423.scan	18304013	77295	779	5718	83013	99.55
s1488.scan	1832949	29089	1593	17395	46484	97.46
s1494.scan	1832949	28850	1557	17215	46065	97.49
s5378.scan	47336298	733053	3161	48422	781475	98.35
s9234.1.scan	61658727	896855	3292	52314	949169	98.46
s13207.1.scan	173384645	3746163	4975	94014	3840177	97.79
s15850.1.scan	149195549	2533658	3799	65599	2599257	98.26
s35932.scan	6235825	1594065	841	6523	1600588	74.33
s38417.scan	41575505	12348460	7301	152168	12500628	69.93
s38584.1.scan	360243110	12719030	8580	187308	12906338	96.42

vectors. Determination of such activity in the scan path is simple as shown in Section 7.4, and that in the CUT can be computed by true-value simulation in $O(T^2)$ iterations, where T is the number of useful vectors. We then construct a complete directed graph G (activity graph) where each node represents a useful test pattern; a directed edge with a weight d_{ij} between from node t_i to node t_j is drawn if the variable component of switching activity occurring in the CUT and the scan path is d_{ij} when the test pattern t_j is applied following a test t_i. A minimum-cost Hamiltonian path in G (or equivalently solving a TSP), corresponds to a reordered

Table 7.6 Intrinsic (order-independent component) switching activity due to useful patterns.

Circuit	Scan path	Capture	Total intrinsic SA
s27.scan	145	93	238
s208.scan	3599	998	4597
s208.1.scan	5402	1795	7197
s298.scan	6979	2533	9512
s344.scan	8800	2327	11127
s349.scan	8720	2524	11244
s382.scan	12178	2779	14957
s386.scan	3927	3977	7904
s400.scan	12337	3038	15375
s420.scan	19091	2673	21764
s420.1.scan	29173	4967	34140
s444.scan	12839	3445	16284
s510.scan	17559	4511	22070
s526.scan	24584	7315	31899
s526n.scan	25156	7605	32761
s641.scan	73604	10768	84372
s713.scan	70670	10640	81310
s820.scan	24823	13424	38247
s832.scan	23725	12171	35896
s838.scan	101053	6394	107447
s838.1.scan	109426	9705	119131
s953.scan	110958	10489	121447
s1196.scan	64924	29069	93993
s1238.scan	66057	22573	88630
s1423.scan	246292	18631	264923
s1488.scan	30821	30096	60917
s1494.scan	27610	25163	52773
s5378.scan	5683786	299625	5983411
s9234.1.scan	8884279	631840	9516119
s13207.1.scan	132363299	1440921	133804220
s15850.1.scan	75741475	1362104	77103579
s35932.scan	140635082	569135	141204217
s38417.scan	870769045	5770204	876539249
s38584.1.scan	1021004396	5403604	1026408000

test sequence S' with minimum overall switching activity in the scan path and the CUT. We run a very efficient heuristic TSP solver[35] to find a nearly-optimal ordering. Based on the ordering, the mapping logic (ML) for the LFSR is synthesized using ESPRESSO[36] and SIS.[37] In Table 7.5, we report switching activity (SA) in the original LFSR in column 2. Reordering of test patterns and modification to the LFSR, yield the new values of SA in the LFSR (column 3), and in the mapping logic (column 4). The last column in Table 7.5 indicates a significant amount of total energy savings in the LFSR. The overhead of mapping logic in terms of literals as computed by SIS is also shown in the table. It is apparent that the logic cost is determined by the size of the LFSR, the number of useful patterns, and how many of them are out-of-consecutive-order relative to the original test sequence. A general objective would be to determine a test sequence that reduces switching activity, and at the same time, requires fewer changes in the natural sequence generated by the original LFSR. For small-sized circuits, the relative overhead is high compared to the cost of the CUT, as we have used a 25-bit LFSR. However, for large circuits, e.g., s35932, overhead of mapping logic is low.

Table 7.7 depicts the reduction of order-dependent component of switching activity by optimal ordering of useful patterns. In columns 2 and 3, we report the SA in the scan path and the CUT if the useful test vectors are applied in the natural order as generated by the LFSR. Our simulation considers only application of useful vectors to the CUT via scan path, not the useless vectors, which are already skipped by the modified LFSR. The optimal activity after reordering of test vectors as determined by the TSP solver is shown in columns 5, 6 and 7. The corresponding energy savings are shown in columns 8, 9 and 10. The CPU time on SUN Ultra 10 (233 MHz) for solving the TSP is reported in the last column.

Table 7.5 shows that most of the energy consumed during testing is wasted in the LFSR while cycling through states generating useless patterns, and hence blocking them at the inputs to the CUT is not enough. The proposed method inhibits the LFSR from producing those useless patterns and thus a significant amount of energy is saved during pseudorandom pattern generation even if the additional energy consumed in the mapping logic is accounted for. Table 7.6 depicts a pessimistic observation that a major component of SA in the scan path is intrinsic in nature, which cannot be reduced by test vector reordering. Table 7.7 shows the energy savings of the variable component achieved by changing the order.

7.6 Prediction of Energy Savings in the LFSR and the CUT

Comparison of the last column of Table 7.4 to that of Table 7.5 reveals that the energy savings in the LFSR is roughly proportional to the fraction of useless patterns. On the other hand, statistical distribution of switching activity in a CUT may be used to estimate a priori the potential energy savings obtainable by test vector reordering. For each pair of useful test patterns we compute by simulation,

Table 7.7   Reduction of order-dependent component of switching activity by optimal ordering of useful patterns.

Circuit	SA in the natural LFSR order			SA in the optimal TSP order			Energy savings (%)			CPU time (sec.)
	Scan path	CUT	Total	Scan path	CUT	Total	Scan path	CUT	Total	
s27.scan	49	67	116	14	36	50	71.4%	46.3%	56.9%	58
s208.scan	958	958	1916	114	759	873	88.1%	20.8%	54.4%	90
s208.1.scan	378	1856	2234	162	1177	1339	57.1%	36.6%	40.1%	99
s298.scan	340	2317	2657	360	1393	1753	−5.9%	39.9%	34.0%	104
s344.scan	468	2166	2634	468	1580	2048	0.0%	27.1%	22.2%	86
s349.scan	156	2289	2445	156	1467	1623	0.0%	35.9%	33.6%	84
s382.scan	567	2687	3254	567	1978	2545	0.0%	26.4%	21.8%	107
s386.scan	325	3920	4245	312	2687	2999	4.0%	31.5%	29.4%	195
s400.scan	621	2747	3368	564	1973	2537	9.2%	28.2%	24.7%	107
s420.scan	875	2688	3563	350	2167	2517	60.0%	19.4%	29.4%	125
s420.1.scan	952	4638	5590	646	3403	4049	32.1%	26.6%	27.6%	146
s444.scan	486	3077	3563	459	2271	2730	5.6%	26.2%	23.4%	104
s510.scan	625	4312	4937	500	3066	3566	20.0%	28.9%	27.8%	169
s526.scan	999	6370	7369	1026	4811	5837	−2.7%	24.5%	20.8%	190
s526n.scan	999	6520	7519	999	5098	6097	0.0%	21.8%	18.9%	189
s641.scan	1296	8965	10261	216	6890	7106	83.3%	23.1%	30.7%	138
s713.scan	1296	8762	10058	108	6760	6868	91.7%	22.8%	31.7%	133
s820.scan	648	12798	13446	672	10737	11409	−3.7%	16.1%	15.1%	351
s832.scan	600	11543	12143	624	7672	8296	−4.0%	33.5%	31.7%	351
s838.scan	1943	6206	8149	1072	5338	6410	44.8%	14.0%	21.3%	173
s838.1.scan	1320	9170	10490	924	7803	8727	30.0%	14.9%	16.8%	152
s953.scan	2704	10198	12902	2652	8162	10814	1.9%	20.0%	16.2%	255
s1196.scan	2144	24582	26726	1536	18307	19843	28.4%	25.5%	25.8%	456
s1238.scan	2304	22684	24988	1344	17631	18975	41.7%	22.3%	24.1%	481
s1423.scan	2821	19689	22510	546	16739	17285	80.6%	15.0%	23.2%	169
s1488.scan	75	28602	28677	75	13713	13788	0.0%	52.1%	51.9%	422
s1494.scan	425	26889	27314	425	15745	16170	0.0%	41.4%	40.8%	417
s5378.scan	30388	298565	328953	4280	276382	280662	85.9%	7.4%	14.7%	1283
s9234.1.scan	38750	646794	685544	39000	559996	598996	−0.6%	13.4%	12.6%	1418
s13207.1.scan	167480	1435071	1602551	167480	1346817	1514297	0.0%	6.1%	5.5%	2985
s15850.1.scan	110124	1338785	1448909	110124	1208780	1318904	0.0%	9.7%	9.0%	1912
s35932.scan	67584	582776	650360	67584	503881	571465	0.0%	13.5%	12.1%	181
s38417.scan	498212	5745334	6243546	496470	5521566	6018036	0.3%	3.9%	3.6%	5611
s38584.1.scan	598580	5389292	5987872	600310	5169447	5769757	−0.3%	4.1%	3.6%	7803

its switching activity (distance) in the CUT when these two patterns are applied consecutively. A plot of normalized distance vs. frequency of vectors is then drawn to estimate potential energy savings. For example, the plots for circuits *s832* and *s9234.1* are shown in Figure 7.7. Reordering of test vectors for *s832* yields 33.5% energy savings compared to 13.4% in *s9234.1* (see Table 7.7) as the variance of the former is larger than that of the second.

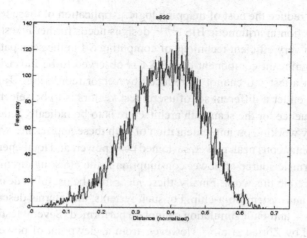

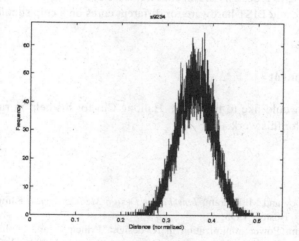

Fig. 7.7   Distribution of switching activity for *s832* and *s9234*.

7.7 Conclusion and Future Problems

A new BIST design is proposed for saving energy both in the LFSR and CUT in a random testing environment. To improve fault coverage further, traditional techniques of embedding deterministic test vectors for hard-to-detect faults,[14–16] can be adopted in the proposed scheme. An open problem in this area is to find a maximal set of consecutive useful vectors in a pseudorandom sequence by changing the order of fault simulation. Preservation of their ordering in the final sequence will in turn, reduce the cost of mapping logic. Application of these techniques to energy reduction in arithmetic BIST[11,12] designs needs further investigation. We also report a very efficient technique of computing *SA* in the scan path for vector pairs. A significant component of the *SA* is observed to be intrinsic in nature, which given a test set, cannot be reduced by vector reordering. To reduce this component, either a different set of useful test vectors is to be selected from the random sequence, or the scan path architecture is to be radically redesigned. We are currently working on implementation of both these approaches with a goal to reduce instantaneous peak power, sustained peak power, and total energy demand. Another intrinsic source of power consumption is the clocking circuitry which is not considered in this work. Finally, these ideas may be useful in designing BIST for IP cores and system-on-a-chip. For such systems, the internal descriptions may not be known, and fault simulation cannot be performed. Several testing schemes are reported by Zorian et al.[38] However, from a viewpoint of power/energy reduction, new BIST designs are needed for core-based systems. Reusability of mapping logic and BIST hardware for different cores on a chip should preferably be ensured.

Acknowledgment

The authors would like to thank Dr. Hailong Cui for his help in running some experiments for this work.

References

1. J. M. Rabey and M. Pedram, *Low Power Design Methodologies*, Kluwer Academic Publishers, Boston, 1995.
2. M. Pedram, Power minimization in IC design: Principles and applications, *ACM Transactions on Design Automation of Electronic Systems*, **1** (1996) 3–56.
3. Y. Zorian, A distributed BIST control scheme for complex VLSI devices, in Proc. *VLSI Test Symposium* (VTS–1993), 4–9.
4. S. Wang and S.K. Gupta, ATPG for heat dissipation minimization during test application, *IEEE Transaction on Computers*, **47** (1998) 256–262.
5. H. -J. Wunderlich, BIST for Systems-on-a-Chip, *Integration, the VLSI Journal* **26** (1998) 55–78.
6. P. Bardell, W. H. McAnney, and J. Savir, *Built-in Test for VLSI: Pseudorandom Techniques*, John Wiley and Sons, NY, 1987.
7. H. -J. Wunderlich, Multiple distributions for biased random test patterns, *IEEE Transactions on Computer Aided Design of Integrated Circuits* **9** (1990) 584–593.
8. S. Hellerbrand, S. Tarnik, J. Rajski, and B. Courtois, Generation of vector patterns

through reseeding of multiple-polynomial linear feedback shift registers, *IEEE Transaction on Computers* **44** (1995) 223–233.

9. S. Devadas and K. Keutzer, An algorithmic approach to optimizing fault coverage for BIST logic synthesis, in Proc. *International Test Conference* (ITC–1998), 164–173.

10. M. Chatterjee, D. K. Pradhan, and W. Kunz, LOT: Logic Optimization with Testability. New transformations for logic synthesis, *IEEE Transactions on Computer Aided Design of Integrated Circuits* **17** (1998) 386–399.

11. J. Rajski and J. Tyser, *Arithmetic Built-in Self-Test for Embedded Systems*, Prentice Hall PTR, NJ, 1998.

12. R. Dorsch and H. -J. Wunderlich, Accumulator based deterministic BIST, in Proc. *International Test Conference* (ITC–1998), 412–421.

13. G. Kiefer and H. -J. Wunderlich, Deterministic BIST with multiple scan chains, *Journal of Electronic Testing: Theory and Applications* **14** (1999) 85–93.

14. H. -J. Wunderlich and G. Keifer, Bit-flipping BIST, in Proc. *IEEE/ACM international conference on Computer-aided design*, (ICCAD–1996), 337–343.

15. N. A. Touba and E. J. McCluskey, Altering a pseudo-random bit sequence for scan-based BIST, in Proc. *IEEE International Test Conference*, (ITC–1996), 167–175.

16. N. A. Touba and E. J. McCluskey, Bit-fixing in pseudorandom sequences for scan BIST, *IEEE Transactions on Computer Aided Design of Integrated Circuits* **20** (2001) 545–555.

17. R. M. Chou, K. K. Saluja, and V. D. Agrawal, Scheduling tests for VLSI systems under power constraints, *IEEE Transactions on VLSI Systems* **5** (1997) 175–185.

18. S. Gerstendoerfer and H. -J. Wunderlich, Minimized power consumption for scan-based BIST, *Journal of Electronic Testing: Theory and Applications* **16** (2000) 203–212.

19. P. Girad, L. Guiller, C. Landrault, S. Pravossoudovitch, and H. -J. Wunderlich, A modified clock scheme for a low power BIST test pattern generator, *Manuscript*.

20. S. Wang and S.K. Gupta, DS-LFSR: A new BIST TPG for low heat dissipation, in Proc. *IEEE International Test Conference* (ITC–1997), 848–857.

21. S. Wang and S.K. Gupta, LT-RTPG: A new test-per-scan BIST TPG for low heat dissipation, in Proc. *IEEE International Test Conference*, (ITC–1999), 85–94.

22. F. Corno, M. Rebaudengo, M. Sonza Reorda, and M. Violante, Optimal vector selection for low power BIST, in Proc. *IEEE Int. Symp. on Defect and Fault-Tolerance in VLSI Systems* (1999), 219–226.

23. N. Nicolici, B. M. Al-Hashimi, Multiple scan chains for power minimization during test application in sequential circuits, *IEEE Transaction on Computers* **51** (2002) 721–734.

24. A. Chandra and K. Chakrabarty, Low-power scan testing and test data compression for system-on-a-chip, *Transactions on Computer Aided Design of Integrated Circuits* **21** (2002) 597–604.

25. D. Gizopoulos, N. Kranitis, A. Paschalis, M. Psarakis, and Y. Zorian, Low power/energy BIST scheme for datapaths, in Proc. *IEEE VLSI Test Symposium* (VTS–2000), 23–28.

26. S. Chakravarty and V. P. Dabholkar, Two techniques for minimizing power dissipation in scan circuits during test application, in Proc. *Asian Test Symposium* (ATS–1994), 324-329.

27. V. P. Dabholkar, S. Chakravarty, I. Pomeranz, and S. M. Reddy, Techniques for minimizing power dissipation in scan and combinational circuits during test application, *Transactions on Computer Aided Design of Integrated Circuits* **17** (1998) 1325–1333.

28. R. Sankaralingam, R. Rao Oruganti, and N. A. Touba, Static compaction techniques to control scan vector power dissipation, in Proc. *IEEE VLSI Test Symposium* (VTS–2000), 35–40.

29. http://www.mentor.com/dft/logic-bist.html

30. M. Shah, Efficient scan-based BIST scheme for low-power testing of VLSI chips, in Proc. *Int. Symp. on Low-Power Electronics and Design*, pp. 376–381, 2006.

31. Z. Ling and K. Jishun, A new BIST solution for system-on-chip, in Proc. *11th. Pacific Rim Int. Symp. on Dependable Computing*, 2005.

32. S. J. Upadhyaya and L-C. Chen, On-chip test generation for combinational circuits by LFSR modification, in Proc. *International Conference on Computer-Aided Design* (ICCAD–1993), 84–87.

33. D. S. Johnson, et al., Experimental analysis of heuristics for the ATSP, in *The Traveling Salesman Problem and its Variations* (Eds. Gutin and Punnen), Kluwer Academic Publisher, 2002.

34. M. Abramovici, M. A. Breuer, and A. D. Friedman, Digital Systems Testing and Testable Design, IEEE Press, 1999.

35. http://lancet.mit.edu/galib-2.4/

36. R. K. Brayton, G. D. Hatchel, C. T. McMullen, and A. L. Sangiovanni-Vincentelli, *Logic Minimization Algorithms for VLSI Synthesis*, Kluwer Academic Publishers, Boston, 1984.

37. E. M. Sentovich, K. J. Singh, L. Lavagno, C. Moon, A. Saldanha, H. Savoj, P. R. Stephan, and A. L. Sangiovanni-Vincentelli, SIS: A system for sequential circuit synthesis, *Tech. Rep. UCB/ERL M92/41*, Electronic Research Lab., 1992.

38. Y. Zorian, E. J. Marinissen, and S. Dey, Testing embedded-core based system chips, in Proc. *IEEE International Test Conference* (ITC–1998), 130-143; (also in *Computer* (June 1999), 52-60).

39. H. K. Lee and D. S. Ha, HOPE: An efficient parallel fault simulator for synchronous sequential circuits, *Transactions on Computer Aided Design of Integrated Circuits* **15** (1996) 1048–1058.

40. G. Hetherington and T. Fryars, N. Tamarapalli, M. Kassab, A. Hassan, and J. Rajski, Logic BIST for large industrial designs: Real issues and case studies, in Proc. *International Test Conference* (ITC–1999), 358–367.

Chapter 8

New Methodologies for Congestion Estimation and Reduction

Taraneh Taghavi and Majid Sarrafzadeh

Computer Science Department
University of California at Los Angeles
Los Angeles, CA 90095, USA

In this paper we summarize different methodologies which have been proposed for congestion estimation and reduction in floorplanning and placement steps for large-scale circuits. Moreover, we formulate the problem of congestion and wirelength minimization in a given initial floorplan to alleviate congestion in the early design cycle. A number of flow-based approaches, to concurrently move congested regions to non-congested regions, have been proposed. These techniques are highly inaccurate due to interference between the block movements. Furthermore, large changes in the floorplan will not preserve region and wirelength constraints. To solve this problem, we propose a novel approach by using a minimum perturbation zero slack assignment for distributing the congestion while preserving the wirelength quality of the design. The proposed technique employs an effective congestion estimator and uses a variation of the ZSA for congestion distribution. Our experimental results show that by applying our method we can reduce congestion by 40% while having less than 11% increase in the wirelength on average.

Contents

8.1 Introduction

Minimizing the total routed wirelength is one of the fundamental goals in VLSI placement stage.[1] Half-perimeter wirelength has emerged as the most typical objective in placement because it adequately models the routed wirelength, especially for two-terminal and three-terminal nets. In general it is believed that there is a positive correlation between half-perimeter wirelength and routed wirelength. Many successful placement tools are based on half-perimeter wirelength minimization.[2,3]

So far, most of the congestion estimation methods perform post-placement congestion estimation. However in the presence of IP blocks, alleviating congestion after placement may result in abrupt increase in wirelength; therefore, congestion needs to be estimated early enough to guide placement to avoid generating highly-congested and hence un-routable designs.

Due to the increasing size of the circuits, many industrial integrated circuits have very high utilization factor. They have a huge number of logic blocks inside the chip area and so congestion is very likely to happen in many areas of the chip after placement. If one area of the chip is very congested, it is very likely that most of the wires may need to have a detour around the congested area although no IP blocks are present in that area. Hence, wirelength estimation may need to know the congestion map of the circuit, and consider calculating the detour for the congested areas. As a result, congestion and wirelength estimation are dependent and should be performed concurrently, early in design flow.

In this paper, our main contribution is to study different methodologies for congestion estimation and reduction. Furthermore, we propose a novel congestion reduction method early in design cycle to improve the quality of floorplanning and placement.

The remainder of this paper is organized as follows. Sections 8.2 and 8.3 present an overview of the previous research on congestion estimation and reduction. In Section 8.4, we discuss our new methodology for early congestion estimation and reduction. Section 8.5 represents the experimental results. Concluding remarks are described in Section 8.6.

8.2 Congestion Estimation

Congestion can be modeled as the summation of linear[4] or quadratic[5] function of difference between routing demand and routing resource. Existing congestion reduction techniques include incorporating congestion into cost function of simulated annealing,[5] combining a regional router into placement tool,[6] and performing a cost placement processing step.[7] While congestion reduction at late or post placement stage is empirically effective, congestion estimate achieved in early placement stage would be equally valuable.

8.2.1 *Overview of Published Work in Congestion Estimation*

As VLSI system complexity continues to increase, physical design is getting more and more difficult.[8] Traditional placement tools focus on minimizing total wirelength to obtain better routability and smaller layout area.[9–11] Despite the pervasive use of half perimeter wirelength objective, there is a mismatch between wirelength and congestion objectives in placement;[12] a placement with less total wirelength does not necessarily mean a better layout after routing. Congestion, an important objective indicating routability, has not drawn enough research attention in placement related studies. Dealing with congestion is widely addressed in routing algorithms. However, in most cases, a portion of routing violation cannot be removed given fixed cell locations. It is of value to consider routability in placement stage where the effort on congestion reduction would be more effective.[13]

In order to increase routability of a circuit, a routability model should be considered and the congestion should be estimated based on this model. Routability is usually modeled on a global routing grid on the whole die area. In,[14] the authors categorize routability modeling into two major categories: topology-free, where no explicit routing is done, and topology-based, where routing trees are explicitly constructed on some routing grid.

8.2.1.1 *Topology-Free Modeling*

Several methods have been proposed for topology-free modeling of routing congestion. In RISA modeling proposed by Cheng,[5] the routing supply for each bin in the routing grid structure is modeled according to how the existing wiring of power or clock nets, regular cells, and macros (macros are referred to as mega cells) are placed, and the routing demand of a net is modeled by its weighted bounding box length.[5]

The routing resource demand model proposed in this paper is based on net bounding box. The wire crossing at a specific cut line through a bounding box depends on not only pin count of the net but also the location of the cut line. Optimal Steiner tree is not used for estimating routing demand because routers might not route nets in similar patterns and it is more expensive to calculate than bounding box. The probability of having a wire at location (x,y) within a net bounding box can be approximated by adding up and normalize K optimal Steiner trees of K sets of randomly located M pins.

While the reduction on the congestion clearly highlights the advantage of the model, the proposed approach discards the extensive research work on wirelength minimization, and it significantly degrades the placement speed.

Methods which use bounding box model for routing estimation constrain the router to use box boundaries for routing estimation.[15] Obviously, these methods are used for simplicity and the results in[16] show that these models fail to match with real routers' behavior. Using probabilistic distribution for routing estimation can also fail to match with real routers' behavior because it does not model routers' preference for via minimization.[17,18] Lou et al. propose a net-based stochastic uniform routing distribution model to compute the expected track usage and developed an algorithm for congestion estimation.[19] In,[17] the authors modified Lou's algorithm with calibration data and used it for wirelength estimation.

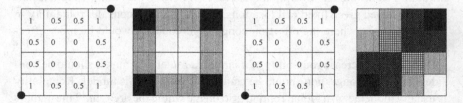

Fig. 8.1   Routing estimation using BBOX router (a), its congestion map (b), Lou's model (c), its congestion map (d).[15]

In,[15] a complete theoretical analysis of different routing congestion estimation methods have been proposed. There, has been stated that these algorithms differ mainly in their methods of using routing resources and routing multi-terminal net. Let the terminals of a 2-terminal net be located on the corner of a grid. Suppose we use the bounding box router and Lou's model as routing models. Furthermore, assume that for each bin, routing congestion is calculated as the summation of routing usage over its horizontal and vertical edges. Figure 8.1 shows the congestion distribution for bounding box (a) and Lou's models (c). The resulted congestion maps are shown in Figure 8.1.b and Figure 8.1.d, respectively. A congestion map

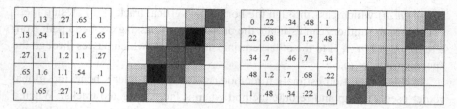

0	.13	.27	.65	1
.13	.54	1.1	1.6	.65
.27	1.1	1.2	1.1	.27
.65	1.6	1.1	.54	.1
0	.65	.27	.1	0

0	.22	.34	.48	1
.22	.68	.7	1.2	.48
.34	.7	.46	.7	.34
.48	1.2	.7	.68	.22
1	.48	.34	.22	0

Fig. 8.2   Routing estimation using Lou's model (a), its congestion map (b), model in[17] (c), its congestion map (d).[15]

visually plots the congestion in the design by assigning different colors to different congestion levels (A darker color means higher congestion level). As shown in this figure, these two models behave differently for congestion prediction.

In order to compare Lou's model with the model in,[17] a more complex example must be considered. Suppose two 2-terminal nets, as shown in Figure 8.2, are routed using these two models. In this figure, dark circles represent the terminals of one net and the shaded circles show terminals of the other net. The congestion distributions using Lou's model and the model in[17] are shown in Figure 8.2.a. and Figure 8.2.c. and their congestion maps are shown in Figure 8.2.b. and Figure 8.2.d, respectively. Because the Lou's behavior for minimization, the congestion value of each central bin in this model is more than the equivalent bin in the other model.

The estimation of both peak congestion and congestion distribution has been done at early top-down placement stages.[20] Specifically, they estimate the maximum congestion prior to placement stage. Also they give a congestion distribution picture of the chip layout at coarse levels of hierarchical placement flow. Both estimates are made based on Rent's rule, a well known stochastic model for real circuits.

In order to analyze peak congestion over all the bin boundaries of the layout, they assume that the circuit is an ideal circuit which strictly obeys Rent's rule. This ideal circuit is placed using a hierarchical placement flow which is based on recursively bi-partitioning. On each hierarchical level of the top-down placement, each sub-circuit is quadric-sectioned into four smaller sub-circuits. A quadric-section step consists of a vertical bi-partitioning followed by a horizontal one.[8]

In another paper,[15] the authors presented a stochastic closed loop congestion estimation algorithm for a placed netlist based on router's behavior. Furthermore, an efficient congestion reduction technique is proposed which is based on contour plotting.

They have calculated the equations for number of v-bend paths from bin (i, j)

to bin (n,m) which pass through the horizontal/vertical track (x,y). Using these equations, the authors can estimate the probability of passing a v-bend path from a terminal at bin (i,j) to the other terminal at bin (n,m) which passes through the specific track (x,y).

In this congestion estimation method, considering router's behavior, global router uses a rip-up and reroute method when it face congested tracks. When a track with a routing demand less than T_a (congestion avoidance threshold) is the parts of a layer which cannot be used for routing are called routing blockages. In order to present an accurate congestion prediction model, routing blockages must also be considered during prediction. If a bin is blocked, its routing demand is distributed among its neighboring bins based on the model.[19]

8.2.1.2 *Topology-Based Modeling*

Topology-based methods usually are more accurate than topology-free methods since the routing topologies they use have a fairly strong correlation with the topologies a global router generates. These methods, for each net, a Steiner tree topology is generated on the given routing grid.[14] Despite their accuracy, topology-based methods are of higher complexity and therefore, most of the research on this topic focuses on efficiency of topology generation.

A multi-partitioning technique using pre-determined Steiner trees for estimating wiring demand was introduced by Mayrhofer and Lauther.[22] In this method, the restriction on the number of partitions confines the performance of the approach.

In another paper,[23] the authors presented two algorithms with logarithmic complexity for this problem. The first algorithm is a congestion-driven two-bend algorithm for two-pin nets (*LZ*-router). The second algorithm they proposed can support incremental updates for building a rectilinear Steiner arborescence tree (*A*-tree) for a multi-pin net (*IncA*-tree).

In their first algorithm, the *LZ*-router incorporates using a complex data structure for finding better routes through applying a binary search on the possible routes for a two-pin net. They have shown that, if the bin structure is $g_x \times g_y$, the complexity of *LZ*-router to route two-pin nets with coordinates of (i,j) and $(i+x,j+y)$ is $O(\log(|x|+|y|)\log(g_x+g_y))$.

In their second algorithm, if the grid structure consists of $(2^m+1) \times (2^m+1)$ grids, the circuit is hierarchically quadric-sectioned until each partition contains only one single unit. A sub-tree in constructed over each partition which connects all the pins inside that partition. The lower left corner of the partition is the root for this sub-tree. As it is shown in Figure 8.3, by recursively quadric-

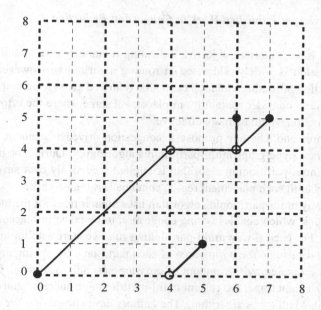

Fig. 8.3　An illustration of IncA-tree algorithm.[14]

sectioning, an *A*-tree can be built up such that each pin at location (x, y) is connected to the origin at location $(0, 0)$ with $\max(\log x, \log y)$ edges. They have shown that for an *n*-pin net with bounding box length L, the complexity of updating a non-root pin move is $O(L \log L)$ times the complexity of *LZ*-router which ends up with $O((\log L)^2 \log(g_x + g_y))$. For moving the root the complexity is $O(n(\log L)^2 \log(g_x + g_y))$. Due to the logarithmic nature of the above algorithm's complexity, the runtime overhead of this congestion cost updating grows slowly. Since *IncA*-tree may generate routes with longer wire comparing to *A*-tree, it may overestimate the congestion. So, it is very appropriate to use to guide the placement tool rather than final measurement of the placement congestion.[14]

8.3　Congestion Reduction

While the wirelength minimization is an important goal of placement algorithms, congestion alleviation is needed during the placement to avoid ending up with un-routable circuits.

8.3.1 *Overview of Published Work in Congestion Reduction*

Congestion is an important objective indicating routability of a circuit. Dealing with congestion is widely addressed in routing algorithms. However, in most cases, a portion of routing violation cannot be removed given fixed cell locations. It is of value to consider routability in placement stage where the effort on congestion reduction would be more effective.[13]

Mayrhofer and Lauther propose a congestion-driven placement technique based on a new hypergraph multi-partitioning algorithm.[22] During the placement, the L-way multi-partitioning algorithm is applied recursively over several levels of hierarchy until each placement region contains less than L cells.

The L-way multi-partitioning algorithm uses Steiner trees for the modeling of net topologies, which enables wiring congestion to be taken into account during placement. For each L-way multi-partitioning run, we start with an initial partitioning and determine the wiring area of each partition. The wiring area of each partition is estimated by the amount of crossing nets and local nets. In the iterative improvement stage, the current multi-partitioning is incrementally modified by Fuduccia-Mattheyses algorithm. The authors have shown that the time complexity of one pass of the multi-partitioning heuristic is $O(L^4 p)$, where p is the total number of pins of a circuit. Thus the time complexity of the algorithm is linear in the total number of pins.

Parakh et al. integrate congestion estimation into the quadratic placement framework to alleviate congestion while minimizing wirelength.[6] In Gordian, placement generated by minimizing the wirelength metric is hierarchically partitioned into regions, and placed with new center of gravity constraints. Before each successive placement, internal route estimation and region-based global route are performed on each region to estimate supply-demand ratios. These ratios are used to influence the quadratic placer into growing (or shrinking) regions.

The computation of the growth matrix is based on the congestion analysis of the current regions. For each region we examine both routing supply and routing demand. Routing supply is computed by the region size and the horizontal/vertical metal pitch. Routing demand of a region consists of two parts: internal routing demand and external routing demand. The former is estimated by a line-probe cost evaluation method. The latter is obtained using a region router which uses an A^* algorithm to route the global nets on the region-based graph. The difference between the routing demand and the routing supply of a region determines the corresponding value in the growth matrix.

Wang et al.[24] combine congestion cost into the objective function during placement optimization. Experiments show that the traditional cost function, wirelength, works the best in this case. A two step approach is then proposed to produce a congestion minimized placement. The first step is a traditional wirelength minimization stage which can also reduce the congestion globally. After that, a post processing stage is used to reduce local congested spots. This two-stage minimization flow is found to be much more effective than minimizing congestion in one step or to simultaneously minimize wirelength and congestion.

A similar cell inflation technique is used in partitioning based[25] and quadratic placement approaches.[26] Change et al.[27] use global routing guided information for congestion control. This approach is generally computationally expensive. Incremental A-tree routing and efficient data structure help to reduce the running time cost. Zhong and Dutt[28] propose a systematic approach to tackle multiple constraints in placement, with application on congestion control. The essential idea is to allow violations on the congestion constraints at the intermediate placement stages.

White space allocation is another way to alleviate congestion in placement orthogonal to above congestion management techniques. The authors proposed a white space allocation approach that dynamically assigns white space according to the congestion distribution of the placement.[29]

8.4 Congestion Reduction in Floorplanning

In this section, we propose a new congestion-driven floorplanning method which uses IP block movement to alleviate congestion. Our floorplanning framework is based on minimum cut block placement with irregular bin grid. We iteratively apply our floorplanning framework until we satisfy the density constraints. At each of the iterations, we modify the resultant floorplan to remove congestion. We impose a bin grid on our circuit and fix the IP blocks which are bigger than one bin or located on the borders of the bins. We show that flow network algorithms greedily handle this problem by moving the overflow of the congested regions to other non-congested regions.[30] But, they ignore the effect of conflict of the expanded congested regions and may result in producing new congested areas. Thus, flow algorithms can just be applied to remove the congestion for non-interfering congested regions. To globally solve this problem, we propose a novel approach by applying a modified version of zero slack assignment formulation, namely minimum perturbation zero slack assignment. Zero slack assignment algorithm has been used in the area of timing budget management previously. We show how to modify this algorithm to use it for congestion minimization purposes. We construct a graph over the congested regions and form a maximum independent set problem to find the maximum non-interfering congested regions. To remove the

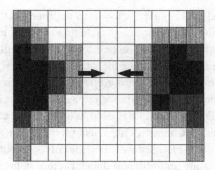

Fig. 8.4 A window showing part of congestion map.

congestion on these regions, we use a non-interfering minimum cost flow algorithm. Furthermore, we apply a minimum perturbation zero slack assignment for congestion minimization on the set of interfering congested regions. We apply a minimum perturbation zero-slack assignment formulation on this set of congested regions. Moreover, to pick the best IP blocks to move, we formulate our problem as a 0-1 knapsack problem which can be solved by an effective heuristic.

8.4.1 *Motivating Examples*

In order to remove the congestion, we need a global view of the congestion map of a design. Let us consider Figure 8.4. In this figure a small window of the congestion map of a design is shown. Here we can see two congested regions on the right and left sides of the window. In the first glance it seems that to remove the congestion, two congested regions should be expanded toward the middle part of the window which is not congested. Now let us consider the whole congestion map of the same design in Figure 8.5. As we see in this figure, the above statement is not true anymore, since the two sides of those congested regions have plenty of white space and they should be expanded in the opposite directions.

Flow algorithms try to distribute congestion by moving the cells out of congested areas. The main problem with this category of algorithms is that they greedily try to expand congested regions but ignore the newly generated congested regions due to the interfering of the expanded regions. Figure 8.6 depicts this problem. As it can be seen in this figure, there are two congested regions which make a new crossing congested region when get expanded. To fix this problem in flow algorithm, new congestion estimation is needed after each round of augmenting flow inside the flow algorithm which is very inefficient. Moreover, flow algorithms produce large changes in floorplan which cannot deteriorate the quality

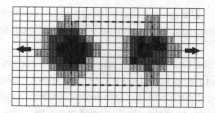

Fig. 8.5   The whole congestion map of a design; the rectangle with dashed border is the same window in Figure 10.4.

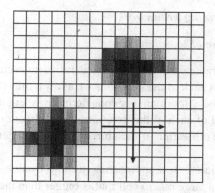

Fig. 8.6   The interfering flow happens in the crossing of two arrows.

of wirelength optimization.

These facts imply that flow algorithms are unable to handle congestion minimization for interfering regions and furthermore cannot preserve wirelength quality of the design. This fact motivates us to propose new approaches to address the problem of congestion reduction for the whole design in floorplanning step.

8.4.1.1 *Previous Work on Floorplanning Congestion Reduction*

Recently, some congestion driven methods have been used in floorplanning stage. The authors proposed an empirical model for estimating wire congestion and embed it into their objective function to reduce wire congestion.[31] The authors used a two-stage simulated annealing for congestion reduction.[32] In another research, the authors presented a probabilistic analysis congestion model with considering irregular bin grid.[33] In another paper the authors used Lagrangian relaxation to handle soft modules.[34] Their method has been corrected and improved.[35] The authors use a nonlinear programming method for modular shaping to minimize the wire congestion.[36] The authors have incorporated congestion and timing constraints into the optimization cost function of a hierarchical min-cut floorplanning framework.[37]

In the current research, we have proposed new congestion reduction techniques in floorplanning step based on moving IP block instead of standard cells. We have used a min-cut partitioning framework for our floorplanning and applied a novel approach based on zero slack assignment to reduce congestion on the over-utilized bins in the bin grid.

8.4.2 *Our Floorplanning Framework*

8.4.2.1 *Min-Cut Based Floorplanning*

Our floorplanning framework is based on iterative min-cut based hierarchical floorplanning.[38,39] At each of the iterations we apply a min-cut based partitioning approach to find the best floorplanning considering wirelength minimization as an objective. At the end of each of the iterations we apply our congestion reduction method to reduce the utilization of each bin in the bin grid generated after min-cut partitioning.

Min-cut based hierarchical approaches run into trouble in mixed-size placement, when there is a large macro cell that is bigger than the bin size at a certain hierarchical level. The size of both sub-bins has to be equal to have a regular bin structure. However, the macro is too large to fit into any of the sub-bins, even though the actual area of the macro is equal to the half of the bin are that is being cut. Since each cell has to be assigned to only one bin, we have to put the macro either to the left or to the right sub-bin. In order to deal with macro cells, we give up the regularity of bin structure, meaning that bins can have different sizes. We use hMetis[40] as the partitioner. If there is more than one macro in the bin being partitioned, we pre-assign macros so that they can fit in the sub-bins they will belong to, and perform partitioning for the rest of standard cells. If a macro can fit in neither of the sub-bins, it is pre-assigned to a sub-bin that minimizes the violation. When a bin contains only one cell/macro, the bin is no longer partitioned but still can move around during simulated annealing to minimize wirelength.

After each bipartition, bin based simulated annealing takes place to find a good location for each partition to be placed in, minimizing the total wirelength. Because the bin structure is irregular due to unbalance partitioning, we have to take care of different bin sizes during simulated annealing. There are three types of moves in bin-based simulated annealing: horizontal switch, vertical switch, and diagonal switch. These moves switch two adjacent bins. If the bin structure is regular, we can freely choose any type of move. However, we now have constraints for these moves: Diagonal switches are allowed only when the two bins have the same size (both width and height); vertical switches are allowed only when the widths of both bins are the same, and horizontal switches occur only when the heights are the same. The moves that do not satisfy these constraints are automatically rejected. When we accept a move, the size and the position of bins have to be updated accordingly to keep the bin structure correct. By restricted bin-based simulated annealing, macro cells can still move around to find a better location to improve the quality. Also, the solution space is limited because of the constraints, resulting in speed-up.

8.4.3 *Problem Definition*

The main goal is to minimize the utilization of the congested areas after each round of floorplanning without deteriorating the quality of the wirelength of the floorplanned solution. After each step of min-cut floorplanning, we fix the IP blocks which are bigger than one bin or located on the borders of two adjacent bins in the bin grid. To alleviate the congestion we move the IP blocks instead of standard cells since we get more congestion reduction by just moving fewer number of IP blocks rather than huge number of standard cells.

A good metric for estimating the congestion of a design is to measure its target utilization as was used in ISPD 2006 placement design contest. The target utilization (or density) can be defined as a constraint for the placement or floorplanning tool. The target utilization should be set higher than the design utilization which is a characteristic of the designed circuit. To compute the utilization of each bin, a bin grid is imposed over the whole circuit. The bin overflow is defined as

$$BOF = \sum MovableAreaBin - BinFreeSpace \times TargetDensity \qquad (1)$$

and total overflow is defined as . The white space is defined as

$$WS = BinFreeSpace \times TargetDensity - \sum MovableAreaBin \qquad (2)$$

for each bin.

Congestion Reduction Floorplanning Problem (CRF): Given a bin grid with some bins with overflow and some bins with white space such that $\sum WS \geqslant \sum BOF$ over all bins, we want to assign the excessive overflow to the bins with white space with the possible minimum perturbation.

To solve the problem of **CRF**, we detect the congested regions by plotting a contour around each of them. To form the non-interfering congested regions, we construct a graph with each vertex corresponding to a congested region. Each edge in this graph is connecting two congested regions if their contours overlap with each other. We find the non-interfering congested regions by a maximum independent set algorithm. We construct a graph for all the bins belonging to the regions of this set and apply flow algorithm to remove the congestion. For the rest of the congested regions (the interfering ones), we apply a modified ZSA algorithm.

We remove congested areas by distributing IP blocks such that the total overflow which is defined in (1) becomes almost zero. To actually find the best set of IP blocks to move we formulate a 0-1 knapsack problem and apply an effective heuristic used to solve 0-1 knapsack problem.

8.4.4 *Non-Interfering Congested Regions Detection*

We try to find the disjoint congestion regions using a maximum independent set algorithm. Each congested region consists of several over-utilized bins. To detect the congested areas, we apply the contour plotting method.[15] A contour plot is a set of level curves of different heights for a function of two variables. A level curve of height h for a function $f(x,y)$ is the set of all points (x,y) such that $f(x,y) = h$. There are several methods for contour plotting which are all discussed in.[41]

After the congested regions are detected by contour plotting, we construct a graph which each vertex corresponds to a congested region and there is an edge between two vertices whenever their corresponding regions overlap with each other. Applying maximum independent set problem will give us the maximum number of vertices which their corresponding congested regions do not overlap each other. It has been shown that a good heuristic for maximum independent set problem is the greedy approach. More precisely, Turan[42] shows that every graph with n vertices and an average degree of $\square$ contains an independent set of size at least $\frac{n}{\delta+1}$. An elegant proof of Turan's theorem, due to Erdos,[43] is easily converted into the following algorithm for finding an independent set S of size times the maximum independent set in at most $O(m)$ steps where m is the number of edges. The details of this algorithm are shown in Algorithm 1.

After finding the maximum independent set S, we find all the bins corresponding to the cells inside S. The problem of congestion reduction for non-interfering congested regions can be viewed as a flow network problem. As we have shown in section 8.4.1, the flow algorithms have several problems with interfering congested regions due to their greedy nature. Moreover, if the flow transportation takes place in the whole chip area, it may lead into large changes in the floorplan which has a large potential to ruin the wirelength quality. Thus, these algorithms can just minimize congestion on non-interfering congested areas. We form a graph for flow network over all the bins belonging to the non-interfering congested regions of set S. We apply the minimum cost flow algorithm on this graph. It should be noticed that to avoid conflicting other congested regions, the overflow of a bin we limit the movement of the overflow of a bin to the bins in its neighborhood.

Algorithm 1: Maximum Independent Set heuristic

Input: Graph $G(V, E)$
Output: Set $S \subseteq V$ = maximum independent set
Set $S = 0$
repeat
    choose the vertex v with smallest degree d
    Add v to S
    Delete v and all its neighbors (and incident edges)
until G is empty.

For the bins in the interfering congested regions, we use a modified version of Zero Slack Assignment (ZSA) algorithm namely minimum perturbation zero slack assignment.

8.4.5 *Minimum Perturbation Zero Slack Assignment for Interfering Congested Regions*

The concept of slack assignment has been completely researched in the area of timing budget management to increase circuit performance. Timing budget management intuitively translates to relaxing the timing constraints for as many components as possible without violating the system timing constraints. In the timing budget management area, the notion of the slack of a module refers to the upper bound in the delay increase of that module without violating the timing constraints of the whole circuit. In our congestion reduction problem, slack of a bin is referred to the upper bound of utilization increase in each bin without violating total target density of the circuit. Obviously, slack of each bin is less than its current white space.

There have been several papers on reasonably formulating this problem using different slack assignment methods as in.[44-46] In this section we first briefly review the well-known Zero Slack assignment method, and then conform it to our own congestion reduction problem.

Given a graph, ZSA starts with nodes of minimum positive slack and performs slack assignment such that their slacks become zero.[46] More specifically, at each iteration, ZSA identifies a path on which all nodes have minimums lack s_{min}, then

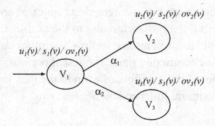

Fig. 8.7   Zero Slack Assignment; a(v), r(v), and s(v) are arrival time, required time and slack, respectively.[12]

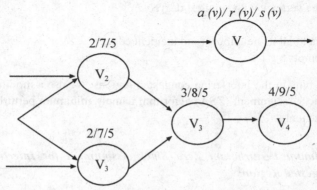

Fig. 8.8   Constructed graph for minimum perturbation slack assignment; u(v), s(v), and ov(v) are utilized space, total space and overflow (slack), respectively.

assigns each node an additional budget $\frac{s_{min}}{N_{min}}$ where N_{min} is the number of nodes on the path. In Figure 8.7, for example, path $\{v_1, v_3, v_4\}$ is first identified, and each node on the path is assigned an additional delay 5/3. Slacks of all nodes in the figure except v_2 become zero, while slack of v_2 is updated as 5/3. After assigning additional delay of 5/3 again to v_2, the algorithm terminates with effective slack of 20/3 (note that maximum effective slack of Figure 8.7 is 10).

To solve our problem with zero slack assignment we construct a directed acyclic graph (DAG) over the bins that are not included in the maximum independent set we formed in the previous section. Each node of the directed acyclic graph G_1 corresponds to one of those bins. There is a directed edge from a bin with extra white space to all the bins with overflow. An example of such constructed graph is shown in Figure 8.8.

Since we do not want to deteriorate the wirelength quality after congestion reduction, we need to incorporate perturbation minimization parameter into the modeling of our problem by slack assignment algorithm. For this purpose, each edge needs to have a weight corresponding to the Manhattan distance of its two endpoints. In the traditional zero slack assignment algorithm, the objective is to maximize the weighted sum of the slacks. To adjust our formulation with the traditional slack assignment algorithm, we define the weight of each edge $w(u, v)$ proportional to $\frac{1}{distance(u,v)}$. This implies that the further two nodes (bins) are, the less the slack (white space) is traversing between them and so the less the IP blocks are moved between those nodes (bins). We normalize edge weights such that the summation of edge weights for all outgoing edges of a vertex should be equal to 1. This ensures that the portion of the white space of a bin which is going to be assigned to several other bins is distributed among them proportional to their Manhattan distance from that specific bin. So, our problem formulation can be represented as:

$$\text{Maximize } \sum_{v \in Adj(u)} slack(u,v) \times \alpha(u,v) \quad \forall u$$
$$\text{Subject to } \sum_{v \in Adj(u)} \alpha(u,v) = 1$$

After constructing the graph G_1, we apply the zero slack assignment algorithm which tries to assign the white space of the underutilized bins to the over utilized bins considering minimum perturbation constraint. After this assignment is done, some of the IP blocks inside each bin should move in the reverse direction from the over utilized bins to the underutilized bins. The total area of the IP blocks which are moving from one bin to another bin equals to the white space (slack) allocated from the latter bin to the former bin. The problem of which IP blocks to choose is discussed in the following section.

8.4.6 *Knapsack Problem Formulation*

After we determine the amount of overflow to move out of each bin, we need to pick the best IP blocks to move out. It matters which IP blocks we are choosing, since it has an effect on the wirelength. Intuitively, if we pick the biggest IP blocks, we minimize perturbation and according the wirelength increase due to the movement of those IP blocks .We use knapsack problem formulation to solve this problem. Knapsack problem is a combinatorial optimization problem which by given a set of items, each with a cost and a value, determines the number of each item to include in a collection so that the total cost is less than some given cost C and the total value is as large as possible. The 0-1 knapsack problem restricts the number of each kind of item to zero or one. Mathematically the 0-1-knapsack problem can be formulated as:

$$\text{Maximize } \sum_{j=1}^{n} P_j x_j$$
$$\text{Subject to } \sum_{j=1}^{n} w_j x_j \leqslant C, \quad x_j = 0, 1 \; \forall j$$

Martello and Toth[47] proposed a greedy approximation algorithm to solve the knapsack problem. Their version sorts the essentials in decreasing order and then proceeds to insert them into the knapsack, starting from the first element (the greatest) until there is no longer space in the sack for more. Their heuristic gives a $\frac{1}{2}$-approximation algorithm for knapsack problem.

We use the same heuristic for our own purpose. Here the cost constraint C for each bin equals to the area corresponding to the overflow of that bin. We define the value of each IP block equal to its size. The bigger the size is the more valuable an IP block is to move out. This value function definition leads to more preservation of wirelength since we move smaller number of IP blocks which intuitively implies we have smaller perturbation. It is obvious that we can just move the IP blocks whose areas are less than the cost constraint C.

8.5 Experimental Results

We have implemented the technique proposed in Section 8.4 in C language. In order to verify our theoretical results on the real-world circuits, we picked most of our benchmarks from the ISPD 2002 placement benchmark suite[48] and ISPD 2005 placement design contest benchmark suite.[49] The characteristics of these benchmarks are shown in Table 8.1. The utilization is the percentage of area of standard cells over the non-blocked area of chip. We have applied our floorplanning tool in macro-cell placement mode on all of the benchmarks to place all the macro-cells. Then we did a complete placement and routing. For placement and routing our benchmarks, we used the Magma BlastFusion which is a commercial CAD tool. We have reported our final congestion and wirelength with and without applying our method. For measuring the congestion and wirelength we used Magma BlastFusion tool to report congestion and wirelength.

Table 8.2 illustrates the congestion reduction by applying our floorplanning algorithm. As it can be seen in this table, we can obtain 40% decrease in total overflow with less than 11% increase in total wirelength.

8.6 Conclusion and Future Work

In this paper we studied the previous research conducted on congestion estimation and reduction. Moreover, we proposed a new iterative approach for floorplanning with the concurrent wirelength and congestion minimization objective. Simula-

Table 8.1   Specification of the Benchmarks.

Test Circuit	# Cells	# Net	Utilization(%)
Test3	12,997	13,865	61.62
IBM01	12,282	11,507	79.99
IBM02	19,321	18,429	27.15
IBM07	45,135	44,394	80.00
IBM08	50,997	47,944	79.99
IBM09	51,746	50,393	80.00
Adaptec1	211,447	211,447	57.32
Adaptec2	255,023	266,009	55.70
Adaptec3	451,650	466,758	33.64
Adaptec4	496,045	515,951	27.22
Bigblue1	278,164	284,479	44.67
Bigblue2	557,866	577,235	37.84

Table 8.2   Congestion Reduction after our algorithm.

Test Circuit	Overflow (%)			Wirelength		
	Before	After	Dec	Before	After	Inc (%)
IBM01	43.56	3.94	39.62	2.95	3.21	8.81
IBM02	38.77	1.81	36.96	5.60	6.12	9.29
IBM07	22.45	2.48	19.97	12.98	14.56	12.17
IBM08	56.78	3.42	53.36	14.72	15.89	7.95
IBM09	37.96	3.25	34.71	16.16	18.4	13.86
Adaptec1	54.68	6.81	47.87	89.37	96.91	8.44
Adaptec2	45.88	3.67	42.21	112.12	124.5	11.04
Adaptec3	39.06	5.33	33.73	246.35	273.45	11.00
Adaptec4	67.30	9.65	57.65	212.22	240.66	13.40
Bigblue1	24.78	2.89	21.89	114.66	129.06	12.56
Bigblue2	55.33	6.55	48.78	196.95	219.87	11.64
Average	44.23	4.53	39.70	93.10	103.88	10.92

tion results show that our approach can reduce congestion on over-utilized bins by 40% while the increase in wirelength is less than 11%. This work can be used in floorplanning step to feed the placemen step with high quality floorplanned designs and prevent it from ending up with un-routable designs because of over-utilized regions.

References

1. M. Sarrafzadeh, M. Wang, and X. Yang, Modern Placement Techniques: Kluwer Academic Publishers, 2003.
2. C. Sechen and A. Sangiovanni-Vinecentelli, The Timberwolf3.2: A new standard cell

placement and global routing package, in Proc. *Design Automation Conference* (DAC 1986), 432–439.

3. G. Sigl, K. Doll, and F. M. Johannes, Analytical placement: a linear or a quadratic objective function, in Proc. *Design Automation Conference* (DAC 1991), 427–432.

4. M. Wang and M. Sarrafzadeh, On the behavior of congestion minimization during placement, in Proc. *Int. Symp. on Physical Design* (ISPD 2000), 145–150.

5. C. E. Cheng, RISA: Accurate and efficient placement routability modeling, in Proc. *Int. Conf. on Computer-Aided Design* (ICCAD 1994), 690–695.

6. P. N. Parakh, R. B. Brown, and K. A. Sakallah, Congestion driven quadratic placement, in Proc. *Design Automation Conference* (DAC 1998), 275–278.

7. M. Wang, X. Yang, K. Eguro, and M. Sarrafzadeh, Multicenter congestion estimation and minimization during placement, in Proc. *Int. Symp. on Physical Design* (ISPD 2000), 147–152.

8. X. Yang, R. Kastner, and M. Sarrafzadeh, Congestion reduction during placement with provably good approximation bound, *ACM Transactions on Design Automation of Electronic Systems* (TODAES) **8** (2003), 316–333.

9. A. E. Dunlop and B. W. Kernighan, A procedure for placement of standard cell VLSI circuits, *IEEE Trans. on Computer Aided Design* **4** (1985) 92–98.

10. J. M. Kleinhans, G. Sigl, F. M. Johannes, and K. J. Antreich, GORDIAN: VLSI placement by quadratic programming and slicing optimization, *IEEE Transactions on Computer Aided Design* **10** (1991) 356–365.

11. W. J. Sun and C. Sechen, Efficient and effective placement for very large circuits, *IEEE Trans. on Computer Aided Design* **14** (1995) 349–359.

12. A. E. Caldwell, A. B. Kahng, and I. L. Markov, Can recursive bisection alone produce routable placements?, in Proc. *Design Automation Conference* (DAC 2000), 477-482.

13. A. B. Kahng, S. Mantik, and D. Stroobandt, Requirements for models of achievable routing, in Proc. *Int. Symp. on Physical Design* (ISPD 2000), 4–11.

14. J. Cong, J. R. Shinnerl, and M. Xie, Large-scale circuit placement, *ACM Trans. on Design Automation of Electronic Systems* **10** (2005) 1–42.

15. M. Saeedi, M. S. Zamani, and A. Jahanian, Prediction and reduction of routing congestion, in Proc. *Int. Symp. on Physical Design* (ISPD 2006), 72–77.

16. M. Wang, X. Yang, and M. Sarrafzadeh, Congestion minimization during placement, *IEEE Trans. of Computer-Aided Design* **19** (2000) 1140–1148.

17. A. B. Kahng and X. Xu, Accurate pseudo-constructive wirelength and congestion estimation, in Proc. *Int. Workshop on System Level Interconnect Prediction*, 2003, 61–68.

18. J. Westra, C. Bartel, and P. Groeneveld, Probabilistic congestion prediction, in Proc. *Int. Symp. on Physical Design* (ISPD 2004), 204–209.

19. J. Lou, S. Thakur, S. Krishnamoorthy, and H. S. Sheng, Estimating routing congestion using probabilistic analysis, *IEEE Transactions on Computer-Aided Design of Integrated Circuits and Systems* **21** (2002) 32–41.

20. X. Yang, R. Kastner, and M. Sarrafzadeh, Congestion estimation during top-down placement, *IEEE Trans. on Computer-Aided Design of Integrated Circuits and Systems* **21** (2002) 72–80.

21. R. S. Tsay, S. C. Chang, and J. Thorvaldson, Early wireability checking and 2-D congestion-driven circuit placement, in Proc. *IEEE Int. ASIC Conf. and Exhibit* (1992), 50–53.

22. S. Mayrhofer and U. Lauther, Congestion-driven placement using a new multi-partioning heuristic, in Proc. *Int. Conf. on Computer Aided Design* (ICCAD 1990), 332–335.

23. C. -C. Chang, J. Cong, Z. D. Pan, and X. Yuan, Multilevel global placement with congestion control, *IEEE Trans. Computer-Aided Design of Integrated Circuits and Systems* **22** (2003) 395–409.

24. M. Wang, X. Yang, and M. Sarrafzadeh, Dragon2000: Fast standard-cell placement for large circuits, in Proc. *Int. Conf. on Computer-Aided Design* (ICCAD 2000), 260–263.

25. W. Hou, H. Yu, X. Hong, Y. Gai, W. Wu, J. Gu, and W. H. Kao, A new congestion-driven placement algorithm based on cell inflation, in Proc. *Asia and South Pacific Design Automation Conference*, (2001), 605–608.

26. U. Brenner and A. Rohe, An effective congestion driven placement framework, in Proc. *Int. Symp. on Physical Design* (ISPD 2002), 6–11.

27. C.-C. Chang, J. Cong, Z. D. Pan, and X. Yuan, Physical hierarchy generation with routing congestion control, in Proc. *Int. Symp. on Physical Design* (ISPD 2002), 36–41.

28. K. Zhong and S. Dutt, Algorithms for simultaneous satisfaction of multiple constraints and objective optimization in a placement flow with application to congestion control, in Proc. *Design Automation Conference* (DAC 2002), 854–859.

29. X. Yang, B. K. Choi, and M.Sarrafzadeh, Routability-driven white space allocation for fixed-die standard-cell placement, *IEEE Trans. Computer-Aided Design of Integrated Circuits and Systems* **22** (2003) 410–419.

30. M. Wang and M. Sarrafzadeh, Modeling and minimization of routing congestion, in Proc. *Asia and South Pacific Design Automation Conference* (ASP-DAC 2000), 185–190.

31. M. Kang and W. W. M. Dai, General floorplanning with L-shaped, T-shaped and soft blocks based on bounded slicing grid structure, in Proc. *Asia and South Pacific Design Automation Conference* (ASP-DAC 1997), 265–270.

32. C. -W. Sham and E. F. Y. Young, Routability driven floorplanner with buffer block planning, in Proc. *IEEE Trans. Computer-Aided Design of Integrated Circuits and Systems* **22** (2003) 470–480.

33. Y. L. Hsieh and T. M. Hsieh, A new effective congestion model in floorplan design, in Proc. *Design, Automation and Test in Europe Conference and Exhibition* (DATE 2004), volume **2**, 1204–1209.

34. F. Y. Young, C. C. N. Chu, W. S. Luk, and Y. C. Wong, Handling soft modules in general non-slicing floorplan using lagrangian relaxation, *IEEE Transactions on Computer Aided Design* **20** (2001) 687–692.

35. C. Lin, H. Zhou, and C. Chu, A revisit to floorplan optimization by lagrangian relaxation, in Proc. *Int. Conf. on Computer Aided Design* (ICCAD 2006), 164–171.

36. H.-H. Huang, C.-C. Chang, C.-Y. Lin, T.-M. Hsieh, and C.-H. Lee, Congestion-driven floorplanning by adaptive modular shaping, in Proc. *Midwest Symposium on Circuits and Systems* (2005), 1067–1070.

37. A. Ranjan, K. Bazargan, and M. Sarrafzadeh, Fast hierarchical floorplanning with congestion and timing control, in Proc. *IEEE Int. Conf. on Computer Design* (ICCD 2000), 357–362.

38. T. Taghavi, X. Yang, B.-K. Choi, M. Wang, and M. Sarrafzadeh, Dragon2005: Large scale mixed-sized placement tool, in Proc. *Int. Symp. on Physical Design* (ISPD 2005), 245–247.
39. T. Taghavi, X. Yang, B.-K. Choi, M. Wang, and M. Sarrafzadeh, Dragon2006: Blockage-aware congestion-controlling mixed-sized placer, in Proc. *Int. Symp. on Physical Design* (ISPD 2006), 15–24.
40. G. Karypis, R. Aggrawal, V. Kumar, and S. Shekhar, Multilevel hypergraph partitioning: application in VLSI domain, in Proc. *Design Automation Conference* (DAC 1997), 526–529.
41. M. J. Aramini, Implementation of an improved contour plotting algorithm, vol. Masters: UIUC, 1981.
42. P. Turan, An external problem in graph theory, *Mat. Fiz. Lapok* **48** (1941) 436–452.
43. P. Erdos, On the graph theorem of Turan, *Math. Lapok* **21** (1970) 249–251.
44. C. Chen, X. Yang, and M. Sarrafzadeh, Potential slack: an effective metric of combinational circuit performance, in Proc. *Int. Conf. on Computer-Aided Design* (ICCAD 2000), 198–201.
45. S. Ghiasi, S. Choudhuri, E. Bozorgzadeh, and M. Sarrafzadeh, A unified theory for timing budget management, in Proc. *Int. Conf. on Computer-Aided Design* (ICCAD 2004), 653–659.
46. R. Nair, C. L. Berman, P. S. Hauge, and E. J. Yoffa, Generation of performance constraints for layout, *IEEE Trans. on Computer Aided Design* **8** (1989) 860–874.
47. S. Martello and P. Toth, Knapsack problems: algorithms and computer implementations, *Wiley-Interscience Series In Discrete Mathematics And Optimization*, (1990), page 296.
48. http://vlsicad.eecs.umich.edu/BK/ISPD02bench.
49. http://www.sigda.org/ispd2005/contest.htm.

Chapter 9

Multimedia Channel Assignment in Cellular Networks

Bhabani P. Sinha* and Goutam K. Audhya[†]

*ACM Unit, Indian Statistical Institute, Kolkata - 700 108, India
[†]BSNL, Calcutta - 700 001, India

This paper deals with the channel assignment problem in a hexagonal cellular network with two-band buffering, supporting multimedia services. We consider the simplest case of a multimedia cellular network dealing with only two different types of multimedia signals, where each cell has a single demand for each type. We first derive the lower bounds on the minimum bandwidth requirement for assigning multimedia channels to a seven-node subgraph of the hexagonal cellular network. We next estimate the lower bounds on the minimum bandwidth requirement for assigning channels in some real-life situations where the relative values of frequency separation constraints are somewhat restricted. Next, we present an algorithm for solving the multimedia channel assignment problem in its most general form, using Genetic Algorithm (GA). We then propose a technique for a clever re-use of the channels, by exploiting the hexagonal symmetry of the cellular network and using only eighteen distinct frequency bands on a nine-node subgraph of the network. With this concept of re-using the channels, we first find the required frequency separation constraints among the channels to be assigned to the different nodes of the network, and then use our proposed GA-based algorithm for assigning the multimedia channels for the complete network. Experiments with different values of the frequency separation constraints show that the proposed assignment algorithm converges very rapidly and generates near-optimal results with a bandwidth pretty close to our derived lower bound.

Contents

9.1 Introduction

Mobile services constitute the fastest growing area of telecommunication that allows access to information super-highway using wireless network environment. In a wireless environment, the radio frequency spectrum is a scarce resource which must be utilized with the objective of increasing network capacity and minimizing interference. The geographical area under the service domain of a mobile cellular network is divided into a number of cells, typically hexagonal in shape. Whenever a mobile cellular network is established, each cell is assigned a set of frequency channels to provide services to the individual calls of that cell. The Channel Assignment Problem (CAP) for such a network is the task of assigning frequency channels to the calls satisfying some frequency separation constraints with a view to avoiding channel interference and using as small bandwidth as possible.

A lot of research work has already been done on the optimal assignment of channels.[1–9,11–14] The available radio frequency spectrum is divided into non-overlapping frequency bands termed as channels. The frequency bands are assumed to be of equal length and are numbered as $0, 1, 2, 3, \ldots$, from the lower end. The highest numbered channel required in an assignment problem is termed as the required bandwidth. The same frequency channel may be assigned to different cells (reused) if they are at a sufficient distance, without causing any interference. For avoiding interference, the assignment of channels should satisfy certain constraints: i) co-channel constraint, due to which the same channel is not allowed to be assigned to certain pair of calls simultaneously, ii) adjacent channel constraint, for which adjacent channels are not allowed to be assigned to certain pair of calls simultaneously, and iii) co-site constraint, which implies that any pair of channels assigned to calls in the same cell must be separated by a certain number.[13]

The cellular network is often modeled as a graph and the problem of channel assignment is mapped to the problem of graph coloring. In its most general form, the channel assignment problem (CAP) is NP-complete.[15] As a result, researchers have attempted to develop approximation algorithms, or heuristic approaches using genetic algorithms,[6,11,13,14,16] neural networks[17] or simulated annealing to solve the problem. While using genetic algorithms, it is quite possible that an optimal assignment has been achieved, but is not recognized, if we do not have a prior knowledge about the optimal bandwidth requirement. On the other hand, a prior idea about lower bounds on bandwidth will help in proceeding towards the ultimate goal and will also provide us an idea about the performance of the algorithm used. In case of neural network and simulated annealing approaches, the techniques start from known lower bounds and improve the result in each iteration. Therefore, in all these approaches, it is extremely necessary to have an idea of the lower bound on the bandwidth needed for the given channel assignment problem.

In,[18] Gamst presented some lower bound on the bandwidth for channel assignment problems in general for only one type of signal. Tcha, Chung and Choi presented some new results on the lower bound on bandwidth in,[19] improving the results by Gamst. Authors in[1] proposed some new lower bounds on channel bandwidth, taking the regular geometry of the cellular network [8-11] into account. They considered hexagonal cellular networks where every cell has a demand of only one channel with 2-band buffering restriction, i.e., the channel interference does not extend beyond two cells. All these results pertain to only one type of signal communication throughout the network. However, the next generation (4G) wireless network aims at supporting multimedia to meet the demands for variety of services, e.g., voice, video and data at any time at any place. Multimedia clients require larger bandwidth to meet the QoS guarantee for video applications, whereas services like voice or e-mail requires smaller bandwidth. Thus a mobile cellular network, supporting multimedia services, must assign frequency channels of different bandwidths to different types of service calls in a particular cell.

We first develop here a general model for channel assignment in a cellular network for multimedia signal communication. We then derive lower bounds on the required bandwidth for assigning channels for a restricted environment where only two types of multimedia signals will be used. A preliminary version of this approach appeared in.[20] In general, as mentioned before, different types of multimedia signals may require different bandwidths for communication over the network. However, we still assume that the total frequency band is divided into a number of smallest size channels numbered as $0, 1, 2, \ldots$. Hence, a signal of a specific type may need to be assigned one or more such adjacent channels to maintain the QoS. For example, a call request for voice communication may be assigned only one channel, while a call request for video communication may possibly need a number of adjacent channels to be assigned for maintaining the required quality of service. The difference between the center frequencies of the bands (a set of adjacent channels) assigned to different calls will, of course, be appropriately chosen to avoid channel interference. We also assume here that each cell has only a single demand for each type of signal, with 2-band buffering restriction, i.e., channel interference does not extend beyond two cells.

We next estimate the lower bounds on the required bandwidth for assigning channels in some real-life situations where the relative values of the frequency separation constraints are somewhat restricted. We next present an algorithm for solving the multimedia channel assignment problem, in its most general form, using genetic algorithm (GA), under the condition of 2-band buffering and with only two types of multimedia signals where each cell has a single demand for each type of signal. We then show that this general approach can conveniently be applied to our network model, by exploiting the symmetric nature of the hexagonal cellular structure, to assign the frequency channels with a very small execution time. To achieve this, we select a subset of only nine nodes of the network and propose a clever technique of re-using the frequency channels so that by repeatedly using only eighteen bands (two bands for each node for assigning both types of multimedia signals), the required assignment for the whole network can be completed. For this purpose, we first find the required frequency separation constraints among the channels to be assigned to the different nodes of the network, and then use our proposed GA-based algorithm for optimally assigning the multimedia channels for the complete network. Experiments with different values of the frequency separation constraints show that the proposed assignment algorithm converges very rapidly and generates near-optimal results, with a resulting bandwidth pretty close to our derived lower bounds.

The paper is organized as follows. In Section 2, we present the system model for formulating the multimedia channel assignment problem. Notations and terminologies used have been discussed in Section 3. The lower bounds on the required bandwidth under various conditions have been derived and estimated in some real-life situations in Section 4. The GA-based algorithm for solving the multimedia channel assignment problem, in general, has then been described in Section 5. Finally, the proposed channel assignment technique for the hexagonal cellular network with re-use of channels is presented in Section 6, followed by conclusion in Section 7.

9.2 System Model

We first note that when we assign varying number of adjacent channels to different types of multimedia signals to maintain the required quality of service, the frequency gaps between two adjacent call assignments will also be different depending on the pairs of signal types associated with these two calls. For example, let us assume that a voice signal needs only one channel, while a video signal needs a band containing 6 adjacent channels. Suppose we assign channel 1 to call 1 for a voice signal. Now, if the adjacent channel 2 is assigned to call 2 for another voice signal, then the frequency separation between these two calls (in terms of channel numbers) will be 1. Instead, if call 2 was for a video signal and the adjacent channels 2, 3, 4, 5, 6 and 7 are all assigned to call 2, then the center frequency for call 2 will be taken as channel number 4.5 (central point of channels 2 to 7), and hence, the frequency separation between calls 1 and 2 will be the gap between the center frequencies of the bands allocated to the two calls which will be computed as 4.5 - 1 = 3.5. Similarly, if adjacent channels are assigned to two video calls, their separation will be 6.

Let us define a function *midpoint* on a set of consecutive positive integers (including zero) which delivers a value equal to the average of these integer values. Thus, *midpoint* $\{2, 3, 4, 5, 6, 7\} = 4.5$

With all these ideas, we now describe the general model for representing the multimedia channel assignment problem in a cellular mobile network by the following components:

1) A set X of n distinct cells, with labels $0, 1, 2, , n - 1$.
2) A set of distinct channels numbered as $0, 1, 2, \ldots$.
3) t different types of multimedia signals denoted by $T_1, T_2, \ldots, T_t$, where a signal of type T_j requires a bandwidth of BW_{T_j}. That is, BW_{T_j} number of adjacent channels need to be assigned to a signal of type T_j.
4) A demand vector $W = (w_{i1}, w_{i2}, \ldots, w_{it})$ for cell i, where w_{ik} represents the channel demand of cell i for the multimedia signal of type T_k.
5) A channel assignment matrix $\Phi = (\phi_{ijk})$, where ϕ_{ijk} represents the set of channels assigned to call j of type k in cell i ($0 \leqslant i \leqslant n - 1, 1 \leqslant j \leqslant w_{ik}, 1 \leqslant k \leqslant t$) with the required bandwidth, i.e., $|\phi_{ijk}| = BW_{T_k}$.
6) A frequency separation matrix $C = (c_{ij,kl})$ where $c_{ij,kl}$ represents the minimum frequency separation requirement between the center frequencies assigned to a call of type T_j in cell i, and a call of type T_l in cell k, $0 \leqslant i, k \leqslant n - 1$ and $1 \leqslant j, l \leqslant t$.
7) A set of frequency separation constraints specified by the frequency separation matrix as follows:
   $|midpoint\{\phi_{i_1 j_1 k_1}\} - midpoint\{\phi_{i_2 j_2 k_2}\}| \geqslant c_{i_1 k_1, i_2 k_2}, \forall i_1, i_2, j_1, j_2, k_1, k_2$ (except when both $i_1 = i_2$ and $j_1 = j_2$).

Example 1: Suppose there are two calls, one from cell 1 and the other from cell 2 of the cellular network. Let the call from cell 1 is of type T_1 and needs only one channel while that from cell 2 is of type T_2 and needs 6 adjacent channels. If channel 1 is assigned to the call from cell 1, while a band consisting of six channels e.g., 3, 4, 5, 6, 7 and 8 are assigned to the call from cell 2, then $midpoint\{3,4,5,6,7,8\} = 5.5$, and hence, the frequency separation (in terms of channel numbers) between these two calls will be $5.5 - 1 = 4.5$. If the minimum separation $c_{11,22}$ between these two calls is specified as 5 to avoid interference, then either of these two calls should be shifted by assigning different channel numbers, so as to satisfy this minimum separation of 5 channels.

9.3 Notations and Terminologies

We consider here the simplest case of a multimedia cellular network, where there are only two types of multimedia signals and each cell has a unit demand for each of these two types. Let us denote these two types of multimedia signals as T_1 and T_2, where the type T_1 signal will need more bandwidth than type T_2 signal. We also consider a 2-band buffering system in which there is no interference between calls in two cells which are more than distance two apart.

For simplicity of notations in our following discussions, let the two types of multimedia signals be denoted as $T_1 = A$ and $T_2 = B$, respectively. The bandwidth BW_A needed for a type A call is assumed to be larger than the bandwidth BW_B needed for a type B call. Also, the required frequency separations for avoiding interference in a 2-band buffering system will be denoted as follows :

- Let s_0, s_1 and s_2 be the required frequency separations between two type A calls in the same cell, in two cells at distance 1 apart and two cells at distance 2 apart, respectively.
- Let s_0', s_1' and s_2' be the required frequency separations between two calls, one of type A and the other of type B, arising in the same cell, in two cells at distance 1 apart and two cells at distance 2 apart, respectively.
- Let s_0'', s_1'' and s_2'' be the required frequency separations between two calls, both of type B, arising in the same cell, in two cells at distance 1 apart and two cells at distance 2 apart, respectively.

Because $BW_A \geqslant BW_B$, we assume that $s_0 \geqslant s_0' \geqslant s_0'', s_1 \geqslant s_1' \geqslant s_1''$ and $s_2 \geqslant s_2' \geqslant s_2''$. We further assume that $s_0 \geqslant s_1 \geqslant s_2, s_0' \geqslant s_1' \geqslant s_2'$ and $s_0'' \geqslant s_1'' \geqslant s_2''$.

We represent the hexagonal cellular network by a cellular graph where each cell of the cellular network is represented by a node and two nodes have an edge between them if the corresponding cells are adjacent to each other. We consider a seven-node subgraph of the cellular graph as shown in Fig. 9.1, with d as the central node from which every other node of the subgraph is at distance one.

Every node in this subgraph is within distance two from each other, and hence, no frequency reuse is possible within this subgraph. Thus, the bandwidth required for assigning channels in this subgraph will give a lower bound on the bandwidth requirement for the whole cellular network.

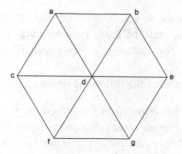

Fig. 9.1   Seven-node subgraph of a hexagonal cellular network.

We denote the channel number $(n_1 s_0 + n_2 s_1 + n_3 s_2 + n_4 s_0' + n_5 s_1' + n_6 s_2' + n_7 s_0'' + n_8 s_1'' + n_9 s_2'')$ by a 9-tuple $(n_1, n_2, n_3, n_4, n_5, n_6, n_7, n_8, n_9)$, where $n_i, i = (1, 2, \ldots, 9)$ are positive integers (including zero). Further, we subscript the above expression with either A or B, if the channel assigned to a particular node is of type A or B, respectively.

Example 2: A frequency channel allocated to a node for a type A call can be expressed as $(n_1 s_0 + n_2 s_1 + n_3 s_2 + n_4 s_0' + n_5 s_1' + n_6 s_2' + n_7 s_0'' + n_8 s_1'' + n_9 s_2'')_A$. Similarly, a frequency channel allocated to a node for a type B call can be expressed as $(n_1 s_0 + n_2 s_1 + n_3 s_2 + n_4 s_0' + n_5 s_1' + n_6 s_2' + n_7 s_0'' + n_8 s_1'' + n_9 s_2'')_B$.

A typical assignment order for assigning channels for type A and type B calls to the nodes $u_1, u_2, u_3, \ldots, u_p$ will be indicated by the notation

$$\begin{pmatrix} u_1 \ u_2 \ u_3 \ \cdots \ u_p \\ T \ \ T \ \ T \ \cdots \ T \end{pmatrix}$$

where T can assume a value from the set $\{A, B\}$, and each node in the sequence $(u_1, u_2, u_3, \ldots u_p)$ will appear exactly twice, once for type A assignment and once again for type B assignment.

Example 3: Let us consider the seven node subgraph of a hexagonal cellular network where each node has a single demand for each of type A and type B signals, then a typical assignment will be indicated by the notation

$$\begin{pmatrix} d \ a \ f \ e \ c \ b \ g \ a \ f \ e \ c \ b \ g \ d \\ A \ B \ B \ B \ B \ B \ B \ A \ A \ A \ A \ A \ A \ B \end{pmatrix},$$

where bold faces have been used for the assignment of the central node d in the seven-node subgraph of Fig. 9.1.

9.4 Lower Bound on Bandwidth

In this section, we derive lower bounds on the required bandwidth for assigning channels to one call of each type to each node of the seven-node subgraph of Fig. 9.1, for different relative values of the nine parameters $s_0, s_1, s_2, s_0', s_1', s_2', s_0'', s_1'', s_2''$.

To start with, we first state the following result.

Lemma 9.1 *The minimum bandwidth for allocating one band for each type of calls at every node of the seven-node subgraph must correspond to the assignment of a band containing either the lowest or the highest channel number to the central node.*

Proof: Any two nodes of the seven-node subgraph is within distance two from each other. Therefore, to avoid interference, any two frequency bands assigned to two nodes of the subgraph must be separated by at least i) s_2, when there are type A calls to each of the nodes, ii) s_2', when there is type A call in one of the nodes and type B in the other, and iii) s_2'', when there are type B calls to each of the nodes. That is, there will be an unusable minimum band gap of s_2'', before and after the frequency band assigned to a node, where no frequency channels within the gaps can be assigned to any other node of the subgraph (including the node itself). When the node is assigned the lowest or highest frequency channel, then one of these band gaps s_2'' is not required, since one of these forbidden band gaps will then lie outside the range of the usable frequency band.

Similarly, when a frequency band is assigned to the central node which is distance one from every other node of the seven-node subgraph, there must be at least a band gap of s_1'', before and after the band assigned to the central node, in order to avoid interference. That is, there would be a minimum forbidden gap of s_1'', before and after the band assigned to the central node, where no frequency channel within the gaps can be assigned to any of the nodes of the subgraph (including the node itself). When the central node is assigned the lowest or highest frequency channel, then one of these forbidden band gaps s_1'' is not required as it will lie outside the range of the usable frequency band.

Since $s_1'' \geqslant s_2''$, it follows that we can have more savings in band gap and have more number of usable frequency channels if the lowest or the highest channel number is assigned to the central node, rather than any other node of the seven-node subgraph.   □

Thus, by lemma 9.1, either the lowest or the highest numbered channel must be assigned to the central node so that we would have more number of usable channels for the remaining assignments. Further, since $BW_A \geqslant BW_B$, and hence $s_1 \geqslant s_1' \geqslant s_1''$, we get a stronger result as follows.

Lemma 9.2 *The minimum bandwidth for allocating one band for each type of calls at every node of the seven-node subgraph must correspond to the assignment of a band containing either the lowest or the highest channel number to a type A call at the central node.*

Proof: The total amount of the forbidden gaps on two sides of the band assigned for the call of type A to the central node d will be at least $2s_1'$, while that on two sides of the band assigned for the call of type B to the central node is at least $2s_1''$. Thus the total unusable gaps on both sides of these two bands assigned to the central node will be $2(s_1' + s_1'')$. If only the band for the type A call falls on one extreme (containing either the lowest or the highest channel number), then the amount of unusable gap (after allocating bands to the the central node d) reduces to $s_1' + 2s_1''$. On the other hand, if only the type B call is assigned a band on one extreme, then the total amount of unusable gap reduces to $2s_1' + s_1''$. Since $s_1' \geqslant s_1''$, the minimum bandwidth assignment must correspond to the situation where the type A call at the central node d is assigned a band containing either the lowest or the highest channel number. $\square$

We now look into the assignments of only the six peripheral node assignments, disregarding the presence of the central node for the time being, to investigate the order of assignments of these peripheral nodes amongst themselves for the minimum bandwidth. Later on, we would combine our observations on both the central node and the peripheral nodes to derive the results for the minimum bandwidth requirement.

9.4.1 *Assignment of Peripheral Nodes*

We first divide the six peripheral nodes a, b, c, e, f and g in two sets P and Q, such that any two nodes within a set is distance two apart from each other. Let these two sets be $P = \{a, f, e\}$ and $Q = \{b, c, g\}$. Then for every node $u \in P = \{a, f, e\}$, there is exactly one node in Q, which is distance two apart from u.

Example 4: Node $a \in P$ is at distance two only from the node $g \in Q$. Similarly, node $f \in P$ is at distance two only from the node $b \in Q$, and so on.

We must *visit* each node of the two sets P and Q twice for assigning channels of type A and type B signals to each of the nodes. We can start assigning channels to peripheral nodes in the following possible ways:

i) By visiting all the nodes once in a set, say P, next visiting all the nodes in set Q, followed by the second visit to all the nodes in the similar manner in the order of either PQ or QP.

Example 5: Let the order of assignments of the peripheral nodes be denoted as

$$\begin{pmatrix} a & f & e & c & b & g & a & f & e & c & b & g \\ T & T & T & T & T & T & T & T & T & T & T & T \end{pmatrix},$$

where each peripheral node has appeared twice in the above array and T assumes the value from the set $\{A,B\}$. For a particular node, once $T = A$ for type A signal and next $T = B$ for type B signal assignments. In this example, the assignment is done by first visiting all the nodes in set P (or Q), then all the nodes in Q (or P) and so on.

ii) We visit all the nodes twice in a set, say P (or Q), then visiting all the nodes twice in Q (or P) as shown below in Example 6.

Example 6:

$$\begin{pmatrix} a & f & e & a & f & e & c & b & g & c & b & g \\ T & T & T & T & T & T & T & T & T & T & T & T \end{pmatrix}$$

The assignment is done here by visiting all the nodes in set P twice in successive two rounds, then visiting all the nodes in set Q twice in next two rounds.

iii) We visit one or few nodes once in a set, say P (or Q), then visit one or few nodes in Q (or P), and so on. However, when we assign a band to a node v, say in Q immediately after assigning a node $u \in P$, then the required band gap will be minimum, if u and v are at distance two from each other.

Example 7:

$$\begin{pmatrix} a & f & b & g & c & b & g & c & e & a & f & e \\ T & T & T & T & T & T & T & T & T & T & T & T \end{pmatrix}.$$

In this example, we assign nodes b,g and c of the set Q, immediately after assigning nodes a and f of the set P. Here f is the last node assigned in the set P, and the node b of the set Q is distance two apart from the node f. We next assign bands to nodes of the set Q for the second round. Node c is the last node assigned in the set Q and node e of the set P is distance two apart from c. We next start assigning channel to node e of the set P, and so on.

We now consider below different situations arising out of the assignments of bands to these peripheral nodes where, without loss of any generality, we start with assigning the first band (containing the lowest channel number) to node a.

Case 1: $s_2 + s_2'' \leqslant 2s_2'$.

We consider two subcases as below.

Subcase a): The lowest frequency band is assigned to the type A call at node a.

Here, to reduce the total bandwidth, the next frequency band can be assigned to a type B signal of another node with the minimum gap of s_2'. After this, type B calls at other nodes may be assigned successively with a minimum band gap of s_2'', followed by assigning the type A calls at all the nodes excepting the node a. An example assignment order following this approach may be given as follows:

$$\begin{pmatrix} a\ f\ e\ c\ b\ g\ a\ f\ e\ c\ b\ g \\ A\ B\ B\ B\ B\ B\ B\ A\ A\ A\ A\ A \end{pmatrix}$$

The consecutive band gaps for the above assignment order are $s_2', s_2'', s_2'', s_2'', s_2'', s_2'', s_2', s_2, s_2, s_2, s_2$ respectively, giving rise to a total bandwidth equal to $4s_2 + 2s_2' + 5s_2'' = BW_0$, say.

As an alternative, after assigning the first band to the node a, we could also assign type A calls to a few more nodes, but not all nodes, followed by type B calls of all the nodes, followed by type A calls of the remaining nodes. A typical example situation with this scheme may be as follows:

$$\begin{pmatrix} a\ f\ e\ c\ b\ g\ a\ f\ e\ c\ b\ g \\ A\ A\ A\ B\ B\ B\ B\ B\ B\ A\ A\ A \end{pmatrix}$$

The resulting consecutive band gaps are $s_2, s_2, s_2', s_2'', s_2'', s_2'', s_2'', s_2'', s_2', s_2, s_2$ respectively, giving rise to the same total minimum bandwidth of $BW_0 = 4s_2 + 2s_2' + 5s_2''$.

However, it is interesting to note that if, after the assignment of the type A call to node a, type A calls of all other nodes are successively assigned, followed by type B calls of all the nodes, then the consecutive minimum band gaps will be $s_2, s_2, s_2, s_2, s_2, s_2', s_2'', s_2'', s_2'', s_2'', s_2''$ respectively, with the total bandwidth greater than or equal to $5s_2 + s_2' + 5s_2'' = BW_1 \geqslant BW_0$, as $s_2 \geqslant s_2'$.

We could, however, try to reduce the bandwidth by some changes in the assignment order

$$\begin{pmatrix} a\ f\ e\ c\ b\ g\ a\ f\ e\ c\ b\ g \\ A\ B\ B\ B\ B\ B\ B\ A\ A\ A\ A\ A \end{pmatrix}$$

so that a band gap of s_2 is replaced by s_2' or s_2''.

Thus, interchanging the assignment for type A and type B signals of node e in the assignment order, we get the following assignment order:

$$\begin{pmatrix} a\,f\,e\,c\,b\,g\,a\,f\,e\,c\,b\,g \\ A\,B\,A\,B\,B\,B\,B\,A\,B\,A\,A\,A \end{pmatrix}$$

As we see in the above assignment, we replaced two band gaps of s_2 and two band gaps of s_2'' by four band gaps of s_2'. Instead of the node e, if we would have interchanged the assignments for type A and type B signals of node the f, then we would have got the following assignment order with some associated changes in the node ordering so that consecutive band assignments can be done with nodes at distance two:

$$\begin{pmatrix} a\,f\,e\,f\,a\,g\,b\,c\,e\,c\,b\,g \\ A\,B\,B\,A\,B\,B\,B\,B\,A\,A\,A\,A \end{pmatrix}$$

In this arrangement, we have replaced one band gap of s_2 and one band gap of s_2'' by two band gaps of s_2'.

We thus see from the above two examples that the replacement of each band gap of s_2 by one with s_2' is associated with deletion of one band gap of s_2'' and addition of another band gap of s_2'. That is, in effect, the band gaps of a total amount $s_2 + s_2''$ is replaced by $2s_2'$. Since $s_2 + s_2'' \leqslant 2s_2'$, such a replacement leads to higher bandwidth requirement than BW_0.

Subcase b) : The lowest frequency band is assigned to the type B call at node a.

Here, we first consider one possible assignment order with type B calls of all nodes successively assigned with consecutive band gaps of s_2'', followed by type A calls of all nodes. An example assignment order may be as follows:

$$\begin{pmatrix} a\,f\,e\,c\,b\,g\,a\,f\,e\,c\,b\,g \\ B\,B\,B\,B\,B\,B\,A\,A\,A\,A\,A\,A \end{pmatrix}$$

The resulting minimum bandwidth comes out to be $5s_2 + s_2' + 5s_2'' = BW_1$.

We can also have an assignment order, where after assigning type B call to few, but not all nodes, we assign type A calls to all peripheral nodes, followed by type B calls to the remaining peripheral nodes. We get the following assignment order:

$$\begin{pmatrix} a\,f\,e\,c\,b\,g\,a\,f\,e\,c\,b\,g \\ B\,B\,A\,A\,A\,A\,A\,A\,B\,B\,B\,B \end{pmatrix}$$

The above assignment order gives rise to a total minimum bandwidth of $5s_2 + 2s_2' + 4s_2'' = BW_2$, say. Note that $BW_2 \geqslant BW_1$.

As in subcase a), we can also try to reduce the required bandwidth by suitable changes in the above assignment order, if a band gap s_2 could be replaced either by s_2' or by s_2''. However, by the same logic as given above in subcase a), such an attempt will not be beneficial since $s_2 + s_2'' \leqslant 2s_2'$.

Thus, considering both the subcases, the minimum bandwidth requirement for $s_2 + s_2'' \leqslant 2s_2'$ is given by BW_0.

Case 2: $s_2 + s_2'' \geqslant 2s_2'$.

In this case also, w.l.o.g., we can start with assigning the first band containing the lowest channel number to node a. Consider again the assignment order:

$$\begin{pmatrix} a & f & e & c & b & g & a & f & e & c & b & g \\ A & B & B & B & B & B & B & A & A & A & A & A \end{pmatrix}.$$

We can now interchange the assignment orders of the node e, so that two band gaps of s_2 are deleted (without creating any new band gap of s_2). Note that instead of the node e, if we would have chosen the node f for interchanging its orders of type A and type B assignments, then although a band gap of s_2 would have been deleted, a new band gap of s_2 would also be created in the new assignment order, with no eventual savings in the required bandwidth. So, basically we need to interchange the assignment orders of every alternate nodes in the above assignment to effect the reduction in the total bandwidth. By following this process, we can finally arrive at a situation as follows, by finally reducing the bandwidth to $10s_2' + s_2''$, when there will be no band gap of s_2 in the overall assignment:

$$\begin{pmatrix} a & f & e & c & b & g & a & f & e & c & b & g \\ A & B & A & B & A & B & B & A & B & A & B & A \end{pmatrix}$$

It is also important to note that the above assignment order contains two consecutive bands assigned to two type B calls to achieve the minimum bandwidth.

If, instead of the type A signal, we start with assigning the first band to node a for type B signal, then we can start with an assignment order as follows:

$$\begin{pmatrix} a & f & e & c & b & g & a & f & e & c & b & g \\ B & B & B & B & B & B & A & A & A & A & A & A \end{pmatrix}$$

If we now similarly try to replace a band gap of s_2 by s_2' through suitable changes in the assignment order, then we would finally arrive at the following assignment:

$$\begin{pmatrix} a & f & e & c & b & g & c & b & g & a & f & e \\ B & A & B & A & B & A & B & A & B & A & B & A \end{pmatrix}.$$

The required bandwidth in this case is $11s_2'$, which is larger than or equal to the bandwidth of $10s_2' + s_2''$ in the previous case. Following the above, it is also possible to have one more assignment order where we start with assigning the first band to a node for type B signal and end up the assignment with assigning the last band to a node for type B signal. We then have the assignment order as:

$$\begin{pmatrix} a\,f\,e\,c\,b\,g\,a\,f\,e\,c\,b\,g \\ B\,A\,B\,A\,B\,A\,A\,B\,A\,B\,A\,B \end{pmatrix}.$$

In the above scheme, there are two consecutive bands assigned to nodes for two type A calls, giving rise to a bandwidth of $10s_2' + s_2$, which is larger than or equal to the bandwidths $11s_2'$.

All these discussions lead to the following result about the assignment of peripheral nodes.

Lemma 9.3

The minimum bandwidth assignment of the peripheral nodes of a seven-node subgraph corresponds to the following situations:

i) For $s_2'' + s_2 \leqslant 2s_2'$, frequency bands are assigned first to type A calls at one or more (but not all) peripheral nodes, then to type B calls at all the six peripheral nodes, followed by type A calls at the remaining nodes, giving rise to a total bandwidth of $4s_2 + 2s_2' + 5s_2''$.

ii) For $s_2'' + s_2 \geqslant 2s_2'$, frequency bands are assigned first to type A call at some node, and then alternatively to type B and type A calls in such a way that the assignment ends with a type A call, giving rise to a total bandwidth of $10s_2' + s_2''$.

9.4.2 *Assignment of the Central Node and the Peripheral Nodes*

We now consider the entire assignment of the seven-node subgraph of Fig. 9.1. By Lemma 9.2, we must assign a band containing the lowest (or, the highest) channel number to the type A call at the central node d for keeping the bandwidth minimum. Without loss of generality, let the frequency band containing the lowest channel number be assigned to the type A call at d. Hence, the other band assigned to d must be for the type B call and it may lie anywhere in the assignment order. We consider different assignment orders for peripheral nodes along with the assignment for the central node d as given below.

Case 1: When $s_2 + s_2'' \leqslant 2s_2'$.

First, we consider the minimum bandwidth assignment orders for the peripheral nodes and try to place the band for assigning the type B call of the central node d at a suitable position in that order. To do this, consider first a typical assignment order as follows, without the assignment of type B call at the node d. We divide the assignment order into subsequences denoted as 1, 2, 3 and 4 as shown below.

$$\begin{pmatrix} d & a\,f\,e\,c\,b\,g\,a\,f\,e\,c\,b\,g \\ A & A\,A\,A\,B\,B\,B\,B\,B\,B\,A\,A\,A \end{pmatrix}$$
$$\underbrace{\quad}_{1}\;\underbrace{\quad}_{2}\;\underbrace{\quad}_{3}\;\underbrace{\quad}_{4}$$

The type B call at the central node d can now be allocated a frequency band in any of the following seven places:

i)   in between the subsequences 1 and 2,
ii) within the subsequence 2,
iii) in between the subsequences 2 and 3,
iv) within the subsequence 3,
v)   in between the subsequences 3 and 4,
vi) within the subsequence 4,
vii) on the right of the subsequence 4.

Subcase i) : If the type B call is allocated a band in between the subsequences 1 and 2, then the assignment looks as follows:

$$\begin{pmatrix} d\,d\,a\,f\,e\,c\,b\,g\,a\,f\,e\,c\,b\,g \\ A\,B\,A\,A\,A\,B\,B\,B\,B\,B\,B\,A\,A\,A \end{pmatrix}$$

The consecutive band gaps are $s'_0, s'_1, s_2, s_2, s'_2, s''_2, s''_2, s''_2, s''_2, s''_2, s'_2, s_2, s_2$, respectively with the total bandwidth of $B_1 = s'_0 + s'_1 + 4s_2 + 2s'_2 + 5s''_2$.

Subcase ii) : If the type B call is allocated a band within the subsequence 2, then the assignment looks as follows:

$$\begin{pmatrix} d\,a\,d\,f\,e\,c\,b\,g\,a\,f\,e\,c\,b\,g \\ A\,A\,B\,A\,A\,B\,B\,B\,B\,B\,B\,A\,A\,A \end{pmatrix}$$

The consecutive band gaps in this case are $s_1, s'_1, s'_1, s_2, s'_2, s''_2, s''_2, s''_2, s''_2, s''_2, s'_2, s_2, s_2$, respectively with the total bandwidth of $B_2 = s_1 + 2s'_1 + 3s_2 + 2s'_2 + 5s''_2$.

Subcase iii) : If the type B call is allocated a band in between the subsequences 2 and 3, then the assignment looks as follows:

$$\begin{pmatrix} d\,a\,f\,e\,d\,c\,b\,g\,a\,f\,e\,c\,b\,g \\ A\,A\,A\,A\,B\,B\,B\,B\,B\,B\,B\,A\,A\,A \end{pmatrix}$$

The resulting consecutive band gaps are $s_1, s_2, s_2, s'_1, s''_1, s''_2, s''_2, s''_2, s''_2, s''_2, s'_2, s_2, s_2$, respectively with the total bandwidth of $B_3 = s_1 + s'_1 + s''_1 + 4s_2 + s'_2 + 5s''_2$.

Subcase iv) : If the type B call is allocated a band within the subsequence 3, then the assignment looks as follows:

$$\begin{pmatrix} d\ a\ f\ e\ c\ b\ g\ d\ a\ f\ e\ c\ b\ g \\ A\ A\ A\ A\ B\ B\ B\ B\ B\ B\ B\ A\ A\ A \end{pmatrix}$$

In this case, the consecutive band gaps are $s_1, s_2, s_2, s_2', s_2'', s_2'', s_1'', s_1'', s_2''$, s_2'', s_2', s_2, s_2, respectively giving the total bandwidth of $B_4 = s_1 + 2s_1'' + 4s_2 + 2s_2' + 4s_2''$.

Subcase v) : If the type B call is allocated a band in between the subsequences 3 and 4, then the situation will be similar to subcase iii) above with the resulting bandwidth of $B_3 = s_1 + s_1' + s_1'' + 4s_2 + s_2' + 5s_2''$.

Subcase vi) : If the type B call is allocated a band within the subsequence 4, then the situation will be similar to subcase ii) above, with the resulting bandwidth of $B_2 = s_1 + 2s_1' + 3s_2 + 2s_2' + 5s_2''$.

Subcase vii) : If the type B call is allocated a band on the right of the subsequence 4, then the assignment looks as follows:

$$\begin{pmatrix} d\ a\ f\ e\ c\ b\ g\ a\ f\ e\ c\ b\ g\ d \\ A\ A\ A\ A\ B\ B\ B\ B\ B\ B\ A\ A\ A\ B \end{pmatrix}$$

The resulting consecutive band gaps are $s_1, s_2, s_2, s_2', s_2'', s_2'', s_2'', s_2'', s_2'', s_2', s_2$, s_2, s_1', with the total bandwidth of $B_5 = s_1 + s_1' + 4s_2 + 2s_2' + 5s_2''$.

Assignment orders and the resulting bandwidths for all these subcases corresponding to the minimum bandwidth assignments of peripheral nodes are summarized in Table 9.1.

Next, we also see whether the non-minimum bandwidth assignment orders for peripheral nodes as discussed in the earlier section, can be beneficial, when considered along with the central node assignments under the condition $s_2'' + s_2 \leqslant 2s_2'$. With the band containing the lowest channel number assigned to the type A call of the central node d, a typical assignment for this case without the type B call of the node d will look as follows.

$$\begin{pmatrix} d & a\ f\ e\ c\ b\ g\ a\ f\ e\ c\ b\ g \\ A & B\ B\ B\ B\ B\ B\ A\ A\ A\ A\ A\ A \\ \underbrace{}_{1} & \underbrace{}_{2}\ \underbrace{}_{3} \end{pmatrix}$$

We have divided the assignment order into different subsequences as marked above.

Table 9.1   Summary of complete assignments for $s_2 + s_2'' \leqslant 2s_2'$ with minimum bandwidth ordering of peripheral nodes.

No.	Assignment order	Bandwidth
1	$d\ d\ a\ f\ e\ c\ b\ g\ a\ f\ e\ c\ b\ g$ $A\ B\ A\ A\ A\ B\ B\ B\ B\ B\ B\ A\ A\ A$	$B_1 = s_0' + s_1' + 4s_2 + 2s_2' + 5s_2''$
2	$d\ a\ d\ f\ e\ c\ b\ g\ a\ f\ e\ c\ b\ g$ $A\ A\ B\ A\ A\ B\ B\ B\ B\ B\ B\ A\ A\ A$	$B_2 = s_1 + 2s_1' + 3s_2 + 2s_2' + 5s_2''$
3	$d\ a\ f\ e\ d\ c\ b\ g\ a\ f\ e\ c\ b\ g$ $A\ A\ A\ A\ B\ B\ B\ B\ B\ B\ B\ A\ A\ A$	$B_3 = s_1 + s_1' + s_1'' + 4s_2 + s_2' + 5s_2''$
4	$d\ a\ f\ e\ c\ b\ g\ d\ a\ f\ e\ c\ b\ g$ $A\ A\ A\ A\ B\ B\ B\ B\ B\ B\ B\ A\ A\ A$	$B_4 = s_1 + 2s_1'' + 4s_2 + 2s_2' + 4s_2''$
5	$d\ a\ f\ e\ c\ b\ g\ a\ f\ e\ d\ c\ b\ g$ $A\ A\ A\ A\ B\ B\ B\ B\ B\ B\ B\ A\ A\ A$	$B_3 = s_1 + s_1' + s_1'' + 4s_2 + s_2' + 5s_2''$
6	$d\ a\ f\ e\ c\ b\ g\ a\ f\ e\ c\ d\ b\ g$ $A\ A\ A\ A\ B\ B\ B\ B\ B\ B\ A\ B\ A\ A$	$B_2 = s_1 + 2s_1' + 3s_2 + 2s_2' + 5s_2''$
7	$d\ a\ f\ e\ c\ b\ g\ a\ f\ e\ c\ b\ g\ d$ $A\ A\ A\ A\ B\ B\ B\ B\ B\ B\ A\ A\ A\ B$	$B_5 = s_1 + s_1' + 4s_2 + 2s_2' + 5s_2''$

Now the second band for the type B call of the node d can lie in any of the subsequences 2 or 3, or between the subsequences 1 and 2, subsequences 2 and 3 or on the right of the subsequence 3. Different possible such assignment orders for each of the representative cases along with the required bandwidth are summarized in rows 1 to 6 of Table 9.2.

We next consider the other non-minimum bandwidth assignment order for peripheral nodes and assign the band containing the lowest channel number to a type A call at node d as follows:

$$\left(\begin{array}{c} d \quad a\ f\ e\ c\ b\ g\ a\ f\ e\ c\ b\ g \\ A \quad \underbrace{B\ B\ B}_{}\ \underbrace{A\ A\ A\ A\ A\ A}_{}\ \underbrace{B\ B\ B}_{} \\ \ _{1} \quad \ _{2} \qquad \ _{3} \qquad \ _{4} \end{array} \right).$$

We divide the assignment order into different subsequences in a similar manner where the second band for the type B call at node d can lie within any of the subsequences 2, 3 and 4, or between the subsequences 1 and 2, or subsequences 2 and 3, or subsequences 3 and 4, or on the right of the subsequence 4. We have compared different possible such assignment orders for each of the representative cases and the only assignment order corresponding to the minimum bandwidth assignment is given in row 7 of Table 9.2.

Table 9.2   Summary of complete assignments for $s_2 + s_2'' \leqslant 2s_2'$ with non-minimum bandwidth ordering of peripheral nodes.

No.	Scheme	Bandwidth
1	$d\ d\ a\ f\ e\ c\ b\ g\ a\ f\ e\ c\ b\ g$ $A\ B\ B\ B\ B\ B\ B\ B\ A\ A\ A\ A\ A\ A$	$B_1' = s_0' + s_1'' + 5s_2 + s_2' + 5s_2''$
2	$d\ a\ d\ f\ e\ c\ b\ g\ a\ f\ e\ c\ b\ g$ $A\ B\ B\ B\ B\ B\ B\ B\ A\ A\ A\ A\ A\ A$	$B_2' = s_1' + 2s_1'' + 5s_2 + s_2' + 4s_2''$
3	$d\ a\ f\ e\ c\ b\ g\ d\ a\ f\ e\ c\ b\ g$ $A\ B\ B\ B\ B\ B\ B\ B\ A\ A\ A\ A\ A\ A$	$B_3' = 2s_1' + s_1'' + 5s_2 + 5s_2''$
4	$d\ a\ f\ e\ c\ b\ g\ a\ d\ f\ e\ c\ b\ g$ $A\ B\ B\ B\ B\ B\ B\ A\ B\ A\ A\ A\ A\ A$	$B_4' = 3s_1' + 4s_2 + s_2' + 5s_2''$
5	$d\ a\ f\ e\ c\ b\ g\ a\ f\ e\ c\ b\ g\ d$ $A\ B\ B\ B\ B\ B\ B\ A\ A\ A\ A\ A\ A\ B$	$B_5' = 2s_1' + 5s_2 + s_2' + 5s_2''$
6	$d\ a\ f\ e\ c\ b\ g\ a\ f\ e\ c\ b\ g\ d$ $B\ B\ B\ B\ B\ B\ B\ A\ A\ A\ A\ A\ A\ A$	$B_6' = s_1 + s_1'' + 5s_2 + s_2' + 5s_2''$
7	$d\ a\ f\ e\ c\ b\ g\ a\ f\ e\ c\ b\ g\ d$ $A\ B\ B\ B\ A\ A\ A\ A\ A\ A\ B\ B\ B\ B$	$B_7' = s_1' + s_1'' + 4s_2'' + 2s_2' + 5s_2$

Table 9.3   Summary of complete assignments for $s_2 + s_2'' \geqslant 2s_2'$ with minimum bandwidth ordering of peripheral nodes.

No.	Scheme	Bandwidth
1	$d\ d\ a\ f\ e\ a\ f\ e\ c\ b\ g\ b\ c\ g$ $A\ B\ A\ B\ A\ B\ A\ B\ A\ B\ B\ A\ B\ A$	$B_1'' = s_0' + s_1' + 10s_2' + s_2''$
2	$d\ a\ f\ d\ e\ a\ f\ e\ c\ b\ g\ b\ c\ g$ $A\ A\ B\ B\ A\ B\ A\ B\ A\ B\ B\ A\ B\ A$	$B_2'' = s_1 + s_1' + s_1'' + 9s_2' + s_2''$
3	$d\ a\ f\ e\ a\ f\ e\ c\ b\ d\ g\ b\ c\ g$ $A\ A\ B\ A\ B\ A\ B\ A\ B\ B\ B\ A\ B\ A$	$B_3'' = s_1 + 2s_1'' + 10s_2'$
4	$d\ a\ f\ e\ a\ f\ e\ c\ b\ g\ b\ d\ c\ g$ $A\ A\ B\ A\ B\ A\ B\ A\ B\ B\ B\ A\ B\ B\ A$	$B_2'' = s_1 + s_1' + s_1'' + 9s_2' + s_2''$
5	$d\ a\ f\ e\ a\ f\ e\ c\ b\ g\ b\ c\ g\ d$ $A\ A\ B\ A\ B\ A\ B\ A\ B\ B\ B\ A\ B\ A\ B$	$B_4'' = s_1 + s_1' + 10s_2' + s_2''$

From Tables 9.1 and 9.2, we see that,

$$B_1' - B_1 = (s_2 - s_2') - (s_1' - s_1'')$$
$$B_2' - B_4 = (s_2 - s_2') - (s_1 - s_1')$$
$$B_3' - B_3 = (s_2 - s_2') - (s_1 - s_1')$$
$$B_4' - B_2 = (s_2 - s_2') - (s_1 - s_1')$$
$$B_5' - B_5 = (s_2 - s_2') - (s_1 - s_1')$$
$$B_6' - B_5 = (s_2 - s_2') - (s_1' - s_1'')$$
$$B_7' - B_7 = (s_2 - s_2'') - (s_1 - s_1'')$$

Thus, even a non-minimum assignment order for peripheral nodes can lead to a smaller bandwidth when considered with the central node assignments. For example, if $s_2 - s_2' \leqslant s_1' - s_1''$, then B_1' will be lesser than B_1, and so on.

Case 2: When $s_2 + s_2'' \geqslant 2s_2'$.

In this case also, we first consider the minimum bandwidth assignment orders for the peripheral nodes and try to place the band for assigning the type B call of the central node d at a suitable position in that order. To do this, consider first a typical assignment order as follows, without the assignment of type B call at the node d. We divide the assignment order into subsequences denoted as 1, 2 and 3 as shown below.

$$
\left(
\begin{array}{l}
d \quad a\,f\,e\,a\,f\,e\,c\,b\,g\,b\,c\,g \\
A \quad A\,B\,A\,B\,A\,B\,B\,B\,A\,B\,A\,B\,A \\
\underbrace{}_{1} \; \underbrace{}_{2} \; \underbrace{}_{3}
\end{array}
\right)
$$

The second channel with type B signal to the central node d can lie in anyone of the five places - i) in between the subsequences 1 and 2, ii) within the subsequence 2, iii) in between the subsequences 2 and 3, iv) within the subsequence 3, and v) on the right of the subsequence 3. The resulting assignment orders with the corresponding minimum bandwidths are given in Table 9.3.

We next consider cases for peripheral node assignment order for non-minimum bandwidth as well. We will now see whether the non-minimum assignment orders may be beneficial, when considered along with the central node assignments.

To do this, consider first a typical assignment order as follows, without the assignment of type B call at the node d. We divide the assignment order into subsequences denoted as 1 and 2 as shown below.

$$
\left(
\begin{array}{l}
d \quad a\,f\,e\,a\,f\,e\,c\,b\,g\,c\,b\,g \\
A \quad B\,A\,B\,A\,B\,A\,B\,A\,B\,A\,B\,A \\
\underbrace{}_{1} \; \underbrace{}_{2}
\end{array}
\right)
$$

The second channel assignment for type B signal to the node d can now lie either in between the subsequences 1 and 2, or within the subsequence 2, or on the right of the subsequence 2. The different possible assignment orders with the resulting bandwidth requirements are summarized in rows 1 to 4 of Table 9.4.

We next consider the other non-minimum bandwidth assignment order also

Table 9.4   Summary of complete assignments for $s_2 + s_2'' \geqslant 2s_2'$ with non-minimum bandwidth ordering of peripheral nodes.

No.	Assignment Order	Bandwidth
1	$d\ d\ a\ f\ e\ c\ b\ g\ c\ b\ g\ a\ f\ e$ $A\ B\ B\ A\ B\ A\ B\ A\ B\ A\ B\ A\ B\ A$	$B_1''' = s_0' + s_1'' + 11s_2'$
2	$d\ a\ f\ e\ c\ b\ g\ c\ b\ g\ d\ a\ f\ e$ $A\ B\ A\ B\ A\ B\ A\ B\ A\ B\ B\ A\ B\ A$	$B_2''' = 2s_1' + s_1'' + 10s_2'$
3	$d\ a\ f\ e\ c\ b\ g\ c\ b\ g\ a\ f\ e\ d$ $A\ B\ A\ B\ A\ B\ A\ B\ A\ B\ A\ B\ A\ B$	$B_3''' = 2s_1' + 11s_2'$
4	$d\ a\ f\ e\ c\ b\ g\ c\ b\ g\ a\ f\ e\ d$ $A\ A\ B\ A\ B\ A\ B\ A\ B\ A\ B\ A\ B\ B$	$B_4''' = s_1 + s_1'' + 11s_2'$
5	$d\ a\ f\ e\ c\ b\ g\ c\ b\ g\ a\ f\ e\ d$ $A\ B\ A\ B\ A\ B\ A\ A\ B\ A\ B\ A\ B\ B$	$B_5''' = s_1' + s_1'' + 10s_2' + s_2$

without the assignment of type B call at node d, as follows:

$$
\begin{pmatrix}
d & a\ f\ e\ a\ f\ e & c\ b\ g\ c\ b\ g \\
A & \underbrace{B\ A\ B\ A\ B\ A}_{} & \underbrace{A\ B\ A\ B\ A\ B}_{}
\end{pmatrix}
$$
$$\underbrace{}_{1}\quad\underbrace{}_{2}\quad\underbrace{}_{3}$$

The second channel assigned to type B call at node d can now lie within any of the subsequences 2 or 3, or between subsequences 1 and 2, or subsequences 2 and 3 or on the right of the subsequence 3. We have compared different possible such assignment orders for each of the representative cases and the only assignment order corresponding to the minimum bandwidth assignment is given in row 5 of Table 9.4.

9.4.3   *Approximations in Practical Situations*

Let us assume that the frequency response curves for the bands assigned to type A and type B signals are typically of trapezoidal shape, as shown in Fig. 9.2, where $BW_A \geqslant BW_B$. When we assign frequency bands for two type B calls at the same node of the seven-node subgraph of the cellular network, the required frequency separation between the center frequencies of the spectral bands allotted to them must be at least s_0'' to avoid interference, as shown in Fig. 9.3(a). It may be seen that there is some overlapping of the frequency bands below the receiver threshold as shown by the dotted line in Fig. 9.3(a), although the bands are completely separated above the receiver threshold. Let this overlapped portion of the band be x. Then, $s_0'' = BW_B - x$.

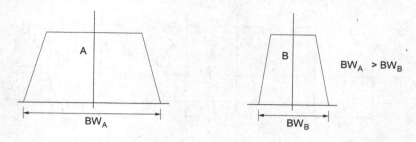

Fig. 9.2   Frequency response curves for type *A* and type *B* signals.

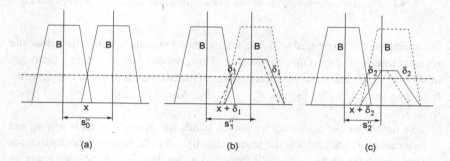

Fig. 9.3   Frequency response curves for two type *B* signals from cells at distances 0, 1 and 2.

Let us now consider two nodes that are distance one apart from each other and are assigned two consecutive frequency bands for type *B* calls. Then the type *B* signal assigned to a node must travel a distance of one cell to reach the other node for causing any interference. Thus, the signal that reaches the other node will be reduced in intensity with a response curve as shown in Fig. 9.3(b), as each frequency component in its spectrum is subject to almost the same attenuation factor due to the distance traveled. Some additional portion, say δ_1, on both sides of the attenuated spectrum falls below the receiver threshold and as a result, we can now maintain a band gap of $s_1''(\leqslant s_0'')$, between the center frequencies of the two allotted bands to avoid interference. Therefore, $s_1'' = s_0'' - \delta_1$, i.e., $s_0'' - s_1'' = \delta_1$. Also, $s_1'' = BW_B - (x + \delta_1)$.

When we assign two frequency bands to two type *B* calls to the nodes that are distance two apart, the signal from one node has to travel a longer distance to reach the other node for causing any interference. This signal would suffer more attenuation than that shown in Fig. 9.3(b) and the response curves for the two signals would appear as in Fig. 9.3(c) with further reduced height and width of the spectral band for the attenuated signal. Let δ_2 be the additional portion of

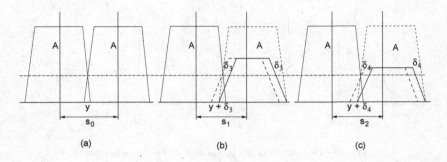

Fig. 9.4   Frequency response curves for two type A signals from cells at distances 0, 1 and 2.

the spectrum (as compared with the unattenuated signal) which falls below the receiver threshold due to this attenuation. Thus, we can now maintain a band gap of $s_2''(\leqslant s_0'')$, between the center frequencies of the two bands to avoid interference. It follows that $s_2'' = s_0'' - \delta_2$, i.e., $s_0'' - s_2'' = \delta_2$. Also, $s_2'' = BW_B - (x + \delta_2)$. Since $\delta_2 \geqslant \delta_1, s_0'' \geqslant s_1'' \geqslant s_2''$.

We next consider assigning frequency bands for two type A calls arising out in three cases: i) both calls at the same node, ii) two calls from two nodes that are distance one apart and iii) two calls from two nodes that are distance two apart, as shown in Figs. 9.4(a), (b) and (c), respectively. We derive relations for the band gaps, between the center frequencies of their allotted frequency bands for these three cases, in a similar method as above.

Case i) : In Fig. 9.4(a), the overlapped portion of the spectrum is y. Then $s_0 = BW_A - y$.

Case ii) : From Fig. 9.4(b), $s_1 = s_0 - \delta_3$, i.e., $s_0 - s_1 = \delta_3$. Hence, $s_1 = BW_A - (y + \delta_3)$.

Case iii) : From Fig. 9.4(c), $s_2 = s_0 - \delta_4$, and hence, $s_0 - s_2 = \delta_4$. Thus, $s_2 = BW_A - (y + \delta_4)$. Since $\delta_4 \geqslant \delta_3, s_0 \geqslant s_1 \geqslant s_2$.

We now consider assigning frequency bands for one type A and one type B calls at the same node of the subgraph. Then, a minimum band gap of s_0' between the center frequencies of the two allotted bands must be maintained to avoid interference, as shown in Fig. 9.5(a). Assuming that the overlapped part of their frequency spectra falling below the receiver threshold is z, the minimum band gap $s_0' = BW_A + \frac{BW_B}{2} - z$.

We now consider assignment of frequency bands to two nodes, one for type A and the other for type B call, originating from two cells i) at distance one and ii) at distance two, respectively. Then either the type B signal or the type A signal has to travel the required distance to reach the other node, for causing any interference.

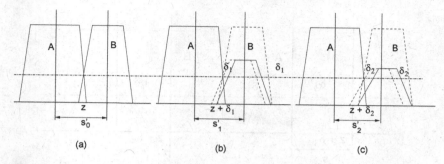

Fig. 9.5 Frequency response curves for type A signal along with a type B signal from a cell at distances 0, 1 and 2.

We consider below both these cases.

Case 1: Let the type B signal from one node reaches the other node (to which a band for type A signal is to be assigned).

Subcase a) : Distance traveled by the type B signal is that of one cell.

It is seen from Fig. 9.5(b) that the propagation characteristics of the type B signal to the distant node is similar to that of Fig. 9.3(b), and hence the band gap $s_1' = s_0' - \delta_1$, i.e., $s_0' - s_1' = \delta_1$. Also, $s_1' = BW_A + \frac{BW_B}{2} - (z + \delta_1)$.

From Fig. 9.3(b), we have $s_1'' = BW_B - (x + \delta_1)$. The overlapped portions x and z typically depend on the shape of the spectrum skirts and can be assumed approximately equal for a cellular network having similar network parameters for different types of signals. Hence, assuming $x \approx z$, we have the relation $s_1' - s_1'' = BW_A - \frac{BW_B}{2}$.

Subcase b) : Distance traveled by the type B signal is that of two cells.

Referring to Fig. 9.5(c), when the two nodes are distance two apart, the propagation of the type B signal to the distant node is similar to that of Fig. 9.3(c), and hence, the band gap $s_2' = s_0' - \delta_2$, i.e., $s_0' - s_2' = \delta_2$. Also, $s_2' = BW_A + \frac{BW_B}{2} - (z + \delta_2)$. From Fig. 9.3(c), we have $s_2'' = BW_B - (x + \delta_2)$, and assuming $x \approx z$, we have the relation, $s_2' - s_2'' = BW_A - \frac{BW_B}{2}$.

Case 2: Let the type A signal reach the other node assigned with a band for type B signal.

Subcase a) : Distance traveled by the type A signal is that of one cell.

In this subcase, referring to Fig. 9.6(a), it may be seen that the propagation characteristics of the type A signal will be similar to that of Fig. 9.4(b), and hence, the band gap $s_1' = BW_A + \frac{BW_B}{2} - (z + \delta_3)$. Note that $s_1 = BW_A - (y + \delta_3)$. y and z depend typically on the shape of the spectrum skirts and can be assumed approximately equal for a cellular network having similar network parameters for different types of signal. Hence, assuming $y \approx z$, we have the relation

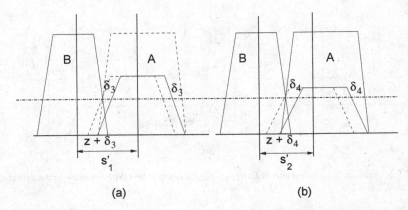

Fig. 9.6   Frequency response curves for type B signal along with a type A signal from a cell at distances 1 and 2.

$s_1 - s_1' = BW_A - \frac{BW_B}{2}$.

Subcase b) : Distance traveled by the type A signal is that of two cells.

Referring to Fig. 9.6(b), the propagation characteristics of the type A signal in this subcase is similar to that of Fig. 9.4(c), and hence, the band gap $s_2' = BW_A + \frac{BW_B}{2} - (z + \delta_4)$. Since $s_2 = BW_A - (y + \delta_4)$, assuming $y \approx z$, we have, $s_2 - s_2' = BW_A - \frac{BW_B}{2}$.

Considering all the above cases, we thus find the relationship among different minimum band gap requirements as,

$$s_1 - s_1' = s_2 - s_2' = s_1' - s_1'' = s_2' - s_2''.$$

Using these relations, we can see that multiple assignment orders indicated in the Tables 9.1-9.4 contribute to identical bandwidth requirement, resulting to a smaller number of distinct expressions for the minimum bandwidth. Thus for Table 9.1, the bandwidths of the assignment order numbers 4 and 5 become same as that of assignment order number 3, and the bandwidth of the assignment order number 6 becomes same as that of number 2, respectively. Also for Table 9.2, the bandwidth of the assignment order number 6 becomes same as that of assignment order number 7. Proceeding in this way, only four distinct expressions for minimum bandwidth follow from Tables 9.1 and 9.2, which are listed in Table 9.5 for the condition $s_2'' + s_2 \leqslant 2s_2'$.

Similarly, in Table 9.3, using the equalities $s_1 - s_1' = s_2 - s_2' = s_1' - s_1'' = s_2' - s_2''$, the bandwidths for the assignment order numbers 2 and 3 become same as that for the assignment order number 4; and in Table 9.4, the bandwidths for the assignment order number 3 becomes same as that for the assignment order

Table 9.5   Distinct possible minimum bandwidths for $s_2 + s_2'' \leqslant 2s_2'$.

No.	Bandwidth	Assignment Order Numbers
1	$s_0' + s_1' + 4s_2 + 2s_2' + 5s_2''$	#1 of Table 9.1, #1 of Table 9.2
2	$s_1 + 2s_1' + 3s_2 + 2s_2' + 5s_2''$	#2,6 of Table 9.1, #4 of Table 9.2
3	$s_1 + s_1' + s_1'' + 4s_2 + s_2' + 5s_2''$	#3, 4, 5 of Table 9.1, #2, 3 of Table 9.2
4	$s_1 + s_1' + 4s_2 + 2s_2' + 5s_2''$	#7 of Table 9.1, #5, 6 of Table 9.2

Table 9.6   Distinct possible minimum bandwidths for $s_2 + s_2'' \geqslant 2s_2'$.

No.	Bandwidth	Assignment Order Numbers
1	$s_0' + s_1' + 10s_2' + s_2''$	#1 of Table 9.3, #1 of Table 9.4
2	$s_1 + s_1' + s_1'' + 9s_2' + s_2''$	#2, 3, 4 of Table 9.3, #2 of Table 9.4
3	$s_1 + s_1' + 10s_2' + s_2''$	#5 of Table 9.3, #3, 4 of Table 9.4

Table 9.7   Distinct possible minimum bandwidths for $s_2 + s_2'' = 2s_2'$.

No.	Bandwidth	Assignment Order Numbers
1	$s_0' + s_1' + 4s_2 + 2s_2' + 5s_2''$	#1 of Table 9.5, #1 of Table 9.6
2	$s_1 + 2s_1' + 3s_2 + 2s_2' + 5s_2''$	
3	$s_1 + s_1' + s_1'' + 4s_2 + s_2' + 5s_2''$	#3 of Table 9.5, #2 of Table 9.6
4	$s_1 + s_1' + 4s_2 + 2s_2' + 5s_2''$	#4 of Table 9.5, #3 of Table 9.6

numbers 4 and 5. Proceeding similarly, for the condition $s_2'' + s_2 \geqslant 2s_2'$, we get only three distinct expressions for minimum bandwidth as shown in Table 9.6.

Also from the relationship $s_1 - s_1' = s_2 - s_2' = s_1' - s_1'' = s_2' - s_2''$, we can derive the equality $s_2'' + s_2 = 2s_2'$. As a result, the bandwidths for the assignment order numbers 1, 3 and 4 of Table 9.5 becomes same as those for the assignment order numbers 1, 2 and 3 of Table 9.6, respectively. Thus, the assignment orders of Tables 9.5 and 9.6 are further reduced to four distinct expressions for minimum bandwidth which are shown in Table 9.7.

Based on all these discussions above, we now obtain the general expressions for minimum bandwidth requirement, as stated in the following theorem.

Theorem 9.1 *Assignment of one band for type A signal and one band for type B signal to each of the nodes in a hexagonal cellular network, the required minimum bandwidth is*

$$\min\{s'_0 + s'_1 + 4s_2 + 2s'_2 + 5s''_2,$$
$$s_1 + 2s'_1 + 3s_2 + 2s'_2 + 5s''_2,$$
$$s_1 + s'_1 + s''_1 + 4s_2 + s'_2 + 5s''_2,$$
$$s_1 + s'_1 + 4s_2 + 2s'_2 + 5s''_2\}.$$

Remark: The result of Theorem 9.1 is derived by using the minimum possible band gaps between consecutively assigned nodes, without considering the feasibility of the assignment based on the relative values of the different frequency separation constraints. Also, the above lower bound is derived only for a seven node subgraph of the hexagonal cellular network and not on the whole network, disregarding the interferences that may arise from their neighboring nodes. Thus, the above lower bound is a loose one and the lower bound for a feasible assignment of the whole network may be, in general, higher than this bound.

Example 8: Let us assume that, the nine s-parameters $s_0, s'_0, s''_0, s_1, s'_1, s''_1, s_2, s'_2$ and s''_2 have values as 11, 10, 9, 7, 6, 5, 3, 2 and 1, respectively. Then the lower bound on minimum bandwidth according to Theorem 9.1 is 36, whereas the lower bound on minimum bandwidth for a feasible assignment even for the seven node subgraph will be 47.

9.5 Genetic Algorithm for Multimedia Channel Assignment Problem

We now use a genetic algorithm based technique for optimization of the channel assignment problem. We first form an algorithm for solving our channel assignment problem using the elitist model of genetic algorithm (EGA). We then show how this algorithm can be used for devising an efficient channel assignment methodology in a hexagonal cellular networks with 2-band buffering and homogeneous demand. The proposed technique basically exploits the hexagonal symmetry of the cellular network and rapidly converges to an optimal or near-optimal assignment.

9.5.1 *Multimedia Channel Assignment Problem (MMCAP) Graph*

The objective of the channel assignment problem is to assign bands for a type A and a type B calls at each of the cells of the network, satisfying all the frequency separation constraints and keeping the required bandwidth minimum. Any call to a cell is represented as a node of a graph and the nodes v_i and v_j are connected by an edge with weight c_{ij}, where $c_{ij} > 0$. We call this graph a Multimedia Channel Assignment Problem (MMCAP) graph (MMCAP graph). In our model, we assume that the bands are assigned to the nodes of the MMCAP graph in a specific order and a node will be assigned the band with a *midpoint* corresponding to the smallest integer that satisfies the frequency separation constraints with the previously assigned bands to all the nodes. It can be seen that the ordering of the nodes has strong influence and impact on the bandwidth required for whole assignment. Let, there be m band-nodes in the MMCAP graph. Then, the nodes can be ordered in $m!$ ways and for sufficiently large m, it is not feasible to find the best ordering for minimum bandwidth by an exhaustive search. We, therefore, use genetic algorithm based approach to find the optimal or near optimal solution to this problem.

9.5.2 *Problem Formulation*

We now represent the multimedia channel assignment problem by a MMCAP graph and the frequency separation constraints by the matrix $C = c_{ij}$ as described in Section 2. We assume that the MMCAP graph has n nodes. We label each node as (pq), where p is the cell number where a call is generated and q is the call number to this cell p.

Example 9: The node (21) represents the call number 1 in cell 2.

A random order of such nodes is considered as a string S or chromosome.

Example 10: A typical string can be denoted by $S = ((21), (02), (12), (00), (31), (20))$

Let M be the population size which is an even integer. Let cp be the crossover probability, which we have taken a high value, say 0.95, in our algorithm. Let q be the mutation probability and T bc the total number of iterations, having a usual value of very large positive integer. We divide the total number of iterations into five equal intervals. We start with a mutation probability of $q = 0.5$ and then vary it with the number of iterations similar to that used in.[14] The variation of the mutation probability q, in this fashion, is required due to the fact that, we have to increase the value of q for maintaining the diversity of the population and also to reduce the value of q when the optimal string is approached.

We now describe the fitness function $Fit(S)$, used in our algorithm, as follows:

function Fit(S) // S is a string.//

    $t[0] \leftarrow 0$; *// t [i] is the frequency assigned to the i-th node $node_i$ of S //*
    for $i = 1$ to $n - 1$ do
        Set t[i] to smallest integer without violating the frequency separation
        requirements specified by the matrix C with all the previously
        assigned values $t[0], t[1], \ldots, t[i-1]$.
    return max $t[0], t[1], \ldots, t[n-1]$.

9.5.3 *Algorithm GA*

Step 1: Set the iteration number $t \leftarrow 0$; Set $cp \leftarrow 0.95$; Set $M \leftarrow 20$.

Step 2: *(initial population)* For $i = 0$ to $M-1$, generate a random order of the nodes in the CAP graph and consider it as a string S_i; set $q_t \leftarrow \{S_0, S_1, \ldots, S_{M-1}\}$ as the initial population.:

Step 3: Compute $Fit(S_i)$ for each string $S_i(0 \leqslant i \leqslant M-1)$ of q_t. Find the best string S_{best1} (i.e., the string with the least fitness value) and the worst string S_{worst1} (i.e., the string with the highest fitness value) of q_t. If S_{best1} or S_{worst1} is not unique, choose one arbitrarily.

Step 4: *(Selection or reproduction)*
(a) Calculate the probability p_i of selection of $S_i(i = 0, 1, \ldots, M-1)$ as
$$p_i = \frac{\frac{1}{Fit(S_i)}}{\sum_{i=0}^{M-1} \frac{1}{Fit(S_i)}}$$
(b) Calculate the cumulative probability q_i for $S_i(i = 0, 1, \ldots, M-1)$ as $q_i = \sum_{j=0}^{i} p_j$.
(c) Generate a random number r_j from [0, 1] for $j = 0, 1, \ldots, M-1$. Now, if $r_j \leqslant q_0$, select S_0; otherwise select $S_i(1 \leqslant i \leqslant M-1)$, if $q_{i-1} < r_j \leqslant q_i$).
Note: $p_0 = q_0$ and $p_i = q_i$ - q_{i-1} for $1 \leqslant i \leqslant M-1$.

Step 5: *(Crossover)* Form $M/2$ pairs of pairing the i-th and $(M/2+i)$-th string from q_{mat} ($1 = 0, 1, \ldots, (M/2-1)$). For each pair of strings, generate a random number R from [0, 1]. If $(R \leqslant cp)$ then generate two random numbers from $\{0, n-1\}$ to define a matching section. Use this matching section to effect a cross through position-by-position exchange operation (to produce two offsprings for the next generation).

Step 6: *(Mutation)* Set $q \leftarrow m$ probability(t). For each string S_i of q_{temp1} $(0 \leqslant i \leqslant M-1)$, and for each node $node_j(0 \leqslant i \leqslant n-1)$ of string S_i, generate a random number from [0, 1], say m. If $(m \leqslant q)$ then exchange $node_j$ of S_i with any other randomly selected node $node_k$ of S_i, $(0 \leqslant k \leqslant n-1, k \neq j)$.

Step 7: Calculate $Fit(S_i)$ for each string $S_i(0 \leqslant i \leqslant M-1)$ of q_{temp2}. Find the best string S_{best2} and the worst string S_{worst2} of q_{temp2}. If S_{best2} or S_{worst2} is not unique, choose one arbitrarily.

Step 8: *(elitism)* Compare S_{best1} of q_t and S_{best2} of q_{temp2}. If $Fit(S_{best2}) > Fit(S_{best1})$, then replace S_{worst2} with S_{best1}. Rename q_{temp2} as q_t.

Step 9: $t \leftarrow t+1$. If $t < T$ then go to step 3; otherwise stop.

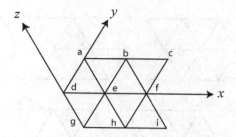

Fig. 9.7   A 9-node block of hexagonal cellular network.

9.6 Assignment Technique with Reuse of Channels

The algorithm developed above can now be applied to any MMCAP graph. However, in case of a hexagonal cellular network, we propose an elegant technique for re-using the channels in a very effective way. For this, we first consider a 9-node subgraph with nodes a, b, c, d, e, f, g, h and i, as shown in Fig. 9.7, where each node represents a cell. We refer to it as a 9-node *block*. We assign only eighteen bands to this 9-node *block*, i.e., one band for type A and one band for type B calls to each of the nine nodes, satisfying all the frequency separation constraints within the *block*, as well as, with the neighboring *blocks*. We then repeat this 9-node *block* along with the assigned eighteen bands, over the entire cellular network to complete the whole assignment.

Let $\phi_A(\alpha)$ be the band assigned to a type A call and $\phi_B(\alpha)$ to a type B call, at node α, where $\alpha \in \{a, b, c, d, e, f, g, h, i\}$. We consider three directions x, y and z on the cellular graph as shown in Fig. 9.7. In x direction, there are three different sequences of node alignments identified by their repetitive nature as type x_1: $a, b, c, a, b, c, \ldots$; type x_2: $d, e, f, d, e, f, \ldots$; and type x_3: $g, h, i, g, h, i, \ldots$. However, in cases of y and z directions, each has only one type of node sequence, identified by their repetitive natures as type y: $a, h, f, c, g, e, b, i, d, \ldots$, and type z: $d, c, i, f, b, h, e, a, g, \ldots$, respectively.

For the given problem of assigning two channels uniformly to each node of the 9-node subgraph, we have to construct a MMCAP graph with eighteen nodes, and the corresponding the frequency separation matrix with the above strategy of re-using the channels is shown in Table 9.8. Note that the frequency separations shown in Table 9.8 takes care of the adjacency of other neighboring nodes (outside the 9-node subgraph) in the network.

We next apply the *AlgorithmGA* over the 9-node *block* of Fig. 9.7, for assigning bands to one type A and one type B calls at each of the nine nodes for

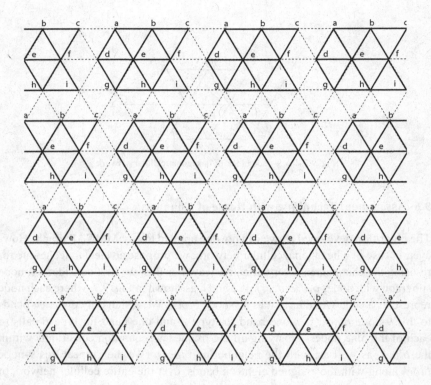

Fig. 9.8   Frequency assignment of the whole network using eighteen bands.

Fig. 9.9   Relative values of s-parameters.

minimum bandwidth, satisfying all the frequency separation requirements of Table 9.8. From the equality $s_1 - s_1' = s_2 - s_2' = s_1' - s_1'' = s_2' - s_2''$, we can plot the relative values of the above s-parameters as shown in Fig. 9.9. We can thus have five different conditions C_1, C_2, C_3, C_4 and C_5 as listed below:

1) $C_1 : 2s_2'' < s_1''$, $2s_2 < s_1$ and $s_2 + s_2' < s_1''$, 2) $C_2 : 2s_2'' < s_1''$, $2s_2 < s_1$ and $s_2 + s_2' \geqslant s_1''$, 3) $C_3 : 2s_2'' < s_1''$, $2s_2 \geqslant s_1$ and $s_2'' + s_2' < s_1''$, 4) $C_4 : 2s_2'' < s_1''$, $2s_2 \geqslant s_1$ and $s_2'' + s_2' \geqslant s_1''$, and 5) $C_5 : 2s_2'' \geqslant s_1''$.

Table 9.8 Frequency Separation matrix for 9-node block.

Nodes	a_A	a_B	b_A	b_B	c_A	c_B	d_A	d_B	e_A	e_B	f_A	f_B	g_A	g_B	h_A	h_B	i_A	i_B
a_A	s_0	s'_0	s_1	s'_1	s_1	s'_1	s_1	s'_1	s_1	s'_1	s_2	s'_2	s_1	s'_1	s_1	s'_1	s_2	s'_2
a_B	s'_0	s''_0	s'_1	s''_1	s'_1	s''_1	s'_1	s''_1	s'_1	s''_1	s'_2	s''_2	s'_1	s''_1	s'_1	s''_1	s'_2	s''_2
b_A	s_1	s'_1	s_0	s'_0	s_1	s'_1	s_2	s'_2	s_1	s'_1	s_1	s'_1	s_2	s'_2	s_1	s'_1	s_1	s'_1
b_B	s'_1	s''_1	s'_0	s''_0	s'_1	s''_1	s'_2	s''_2	s'_1	s''_1	s'_1	s''_1	s'_2	s''_2	s'_1	s''_1	s'_1	s''_1
c_A	s_1	s'_1	s_1	s'_1	s_0	s'_0	s_1	s'_1	s_2	s'_2	s_1	s'_1	s_1	s'_1	s_2	s'_2	s_1	s'_1
c_B	s'_1	s''_1	s'_1	s''_1	s'_0	s''_0	s'_1	s''_1	s'_2	s''_2	s'_1	s''_1	s'_1	s''_1	s'_2	s''_2	s'_1	s''_1
d_A	s_1	s'_1	s_2	s'_2	s_1	s'_1	s_0	s'_0	s_1	s'_1	s_1	s'_1	s_1	s'_1	s_2	s'_2	s_1	s'_1
d_B	s'_1	s''_1	s'_2	s''_2	s'_1	s''_1	s'_0	s''_0	s'_1	s''_1	s'_1	s''_1	s'_1	s''_1	s'_2	s''_2	s'_1	s''_1
e_A	s_1	s'_1	s_1	s'_1	s_2	s'_2	s_1	s'_1	s_0	s'_0	s_1	s'_1	s_1	s'_1	s_1	s'_1	s_2	s'_2
e_B	s'_1	s''_1	s'_1	s''_1	s'_2	s''_2	s'_1	s''_1	s'_0	s''_0	s'_1	s''_1	s'_1	s''_1	s'_1	s''_1	s'_2	s''_2
f_A	s_2	s'_2	s_1	s'_1	s_1	s'_1	s_1	s'_1	s_1	s'_1	s_0	s'_0	s_2	s'_2	s_1	s'_1	s_1	s'_1
f_B	s'_2	s''_2	s'_1	s''_1	s'_1	s''_1	s'_1	s''_1	s'_1	s''_1	s'_0	s''_0	s'_2	s''_2	s'_1	s''_1	s'_1	s''_1
g_A	s_1	s'_1	s_2	s'_2	s_1	s'_1	s_1	s'_1	s_1	s'_1	s_2	s'_2	s_0	s'_0	s_1	s'_1	s_1	s'_1
g_B	s'_1	s''_1	s'_2	s''_2	s'_1	s''_1	s'_1	s''_1	s'_1	s''_1	s'_2	s''_2	s'_0	s''_0	s'_1	s''_1	s'_1	s''_1
h_A	s_1	s'_1	s_1	s'_1	s_2	s'_2	s_2	s'_2	s_1	s'_1	s_1	s'_1	s_1	s'_1	s_0	s'_0	s_1	s'_1
h_B	s'_1	s''_1	s'_1	s''_1	s'_2	s''_2	s'_2	s''_2	s'_1	s''_1	s'_1	s''_1	s'_1	s''_1	s'_0	s''_0	s'_1	s''_1
i_A	s_2	s'_2	s_1	s'_1	s_1	s'_1	s_1	s'_1	s_2	s'_2	s_1	s'_1	s_1	s'_1	s_1	s'_1	s_0	s'_0
i_B	s'_2	s''_2	s'_1	s''_1	s'_1	s''_1	s'_1	s''_1	s'_2	s''_2	s'_1	s''_1	s'_1	s''_1	s'_1	s''_1	s'_0	s''_0

The GA-based assignment algorithm with the above technique of channel reuse has been run on the entire hexagonal cellular network under all these five different conditions. The resulting bandwidth requirements under different conditions are stated in the following theorem.

Theorem 9.2 *The bandwidth BW required by our proposed assignment technique under different conditions is given as follows:*

for C_1, $BW = 2s_1 + 4s'_1 + 2s''_1 + s'_2$;   for C_2, $BW = s_1 + 5s'_1 + 2s''_1 + s_2$;
for C_3, $BW = 2s'_1 + s''_1 + 6s_2 + 3s'_2$;   for C_4, $BW = s_1 + 4s'_1 + 3s_2 + 4s'_2$; and
for C_5, $BW = 5s_2 + 6s'_2 + 6s''_2$.

Proof: The proof follows from the details of channels, as shown in Table 9.9, assigned to each of the 9-nodes of the network under the above constraints for each of these five conditions. □

Table 9.9   Distinct Eighteen Bands Assigned to the Whole Network.

Band types to nodes	*Midpoints for*				
	$2s_2'' < s_1'', 2s_2 < s_1$ and $s_2 + s_2' < s_1''$	$2s_2'' < s_1'', 2s_2 < s_1$ and $s_2 + s_2' \geq s_1''$	$2s_2'' < s_1'', 2s_2 \geq s_1$ and $s_2'' + s_2' < s_1''$	$2s_2'' < s_1'', 2s_2 \geq s_1$ and $s_2'' + s_2' \geq s_1''$	$2s_2'' \geq s_1''$
$\phi_A(a)$	$s_1'' + s_2'$	$s_1'' + s_2'$	$s_1' + 2s_1'' + 4s_2 + 2s_2'$	$3s_1' + 2s_2 + 3s_2'$	$s_2 + 2s_2'$
$\phi_B(a)$	$s_1 + 4s_1' + s_1''$	$s_1 + 4s_1' + s_1''$	$s_1'' + s_2'$	s_1'	$3s_2 + 5s_2' + 4s_2''$
$\phi_A(b)$	$s_1 + s_1' + s_1''$	$s_1 + s_1' + s_1''$	$2s_1'' + s_2 + s_2'$	$4s_1' + 2s_2 + 4s_2'$	$3s_2 + 6s_2' + 6s_2''$
$\phi_B(b)$	$2s_1 + 3s_1' + 2s_1'' + s_2'$	$s_1 + 4s_1' + 2s_1'' + s_2$	$2s_1' + 2s_1'' + 5s_2 + 2s_2'$	$s_1' + s_2 + 2s_2'$	$3s_2 + 3s_2'$
$\phi_A(c)$	$s_1 + 2s_1' + s_1'' + s_2'$	$s_1 + 2s_1' + s_1'' + s_2'$	$s_1' + 2s_1'' + s_2 + 2s_2'$	$s_1 + 4s_1' + 3s_2 + 4s_2'$	0
$\phi_B(c)$	0	0	0	$2s_1' + s_2 + 3s_2'$	$3s_2 + 3s_2' + 3s_2''$
$\phi_A(d)$	$s_1 + 5s_1' + 2s_1''$	$s_1 + 5s_1' + 2s_1''$	$2s_1' + 2s_1'' + 5s_2 + 3s_2'$	$4s_1' + 3s_2 + 4s_2'$	$4s_2 + 6s_2' + 6s_2''$
$\phi_B(d)$	$s_1 + s_1' + s_1'' + s_2'$	$s_1 + s_1' + s_1'' + s_2'$	$2s_1' + s_2 + 2s_2'$	$2s_1' + s_2 + s_2'$	$3s_2 + 3s_2' + s_2''$
$\phi_A(e)$	$s_1 + 3s_1' + s_1''$	$s_1 + 3s_1' + s_1''$	$s_1' + 2s_1'' + 2s_2 + 2s_2'$	0	$3s_2 + 4s_2' + 3s_2''$
$\phi_B(e)$	s_2''	s_2''	s_2''	$3s_1' + s_2 + 2s_2'$	s_2'
$\phi_A(f)$	$s_1'' + s_1'$	$s_1'' + s_1'$	$s_1' + 2s_1'' + 5s_2 + 2s_2'$	$s_1' + s_2'$	$2s_2 + 2s_2'$
$\phi_B(f)$	$2s_1 + 3s_1' + s_1'' + s_2'$	$s_1 + 4s_1' + s_1'' + s_2$	$2s_1''$	$3s_1' + 2s_2 + 4s_2'$	$3s_2 + 5s_2' + 5s_2''$
$\phi_A(g)$	$s_1 + s_1'' + s_2'$	$s_1' + s_1'' + s_2$	$2s_1'' + s_2'$	$s_1' + s_2 + s_2'$	$3s_2 + 2s_2'$
$\phi_B(g)$	$s_1 + 4s_1' + 2s_1''$	$s_1 + 5s_1' + s_1''$	$s_1' + 2s_1'' + 5s_2 + 3s_2'$	$4s_1' + 2s_2 + 3s_2'$	$3s_2 + 5s_2' + 6s_2''$
$\phi_A(h)$	$2s_1 + 4s_1' + 2s_1'' + s_2'$	$s_1 + 5s_1' + 2s_1'' + s_2$	$2s_1' + 2s_1'' + 6s_2 + 3s_2'$	$2s_1' + s_2 + 2s_2'$	$5s_2 + 6s_2' + 6s_2''$
$\phi_B(h)$	$s_1 + 2s_1' + s_1''$	$s_1 + 2s_1' + s_1''$	$s_1' + 2s_1'' + s_2 + s_2'$	$4s_1' + 3s_2 + 5s_2'$	$3s_2 + 3s_2' + 2s_2''$
$\phi_A(i)$	$2s_1 + 2s_1' + s_1'' + s_2'$	$s_1 + 3s_1' + s_1'' + s_2'$	$s_1' + 2s_1'' + 3s_2 + 2s_2'$	$3s_1' + s_2 + 3s_2'$	$2s_2'$
$\phi_B(i)$	s_1''	s_1''	s_1''	s_2'	$3s_2 + 5s_2' + 3s_2''$

9.7 Conclusion

We have first introduced a model for the most general problem of multimedia channel assignment in a hexagonal cellular network with 2-band buffering, where interference does not extend beyond two cells. We next have considered the simplest case of multimedia cellular network with only two types of multimedia signals where each cell of the network has single demand for each type. We have derived the lower bounds on the minimum bandwidth requirement for assigning multimedia channels to a seven-node subgraph of the hexagonal cellular network. We next have estimated the lower bounds on the minimum bandwidth requirement for assigning channels in some real-life situations where the relative values of frequency separation constraints are somewhat restricted. We next have presented an algorithm for solving the multimedia channel assignment problem, in its most general form, using genetic algorithm (GA). We next have shown that this technique can effectively be applied to our network model exploiting the symmetric nature of the hexagonal cellular structure. For this purpose, we have presented a novel concept of re-using the channels by using only eighteen distinct bands on a nine node subgraph of the network. We have found our proposed technique for re-using the channels leads to a rapidly converging GA-based algorithm resulting in optimal bandwidths within a reasonable amount of computation time (less than a second on a high-end PC). Future work includes improving the lower bound on bandwidth for a feasible assignment of the whole network.

References

1. A. Sen, T. Roxborough, B. P. Sinha, On an optimal algorithm for channel assignment in cellular network, in Proc. *IEEE International Conf. Communications* (1999), 1147–1151.
2. R. Mathar and J. Mattfeldt, Channel assignment in cellular radio networks, *IEEE Transactions on Vehicular Technology* **42** (1993) 647–656.
3. K. N. Sivarajan, R. J. McEliece and J. W. Ketchum, Channel assignment in cellular radio, in Proc. *39th IEEE Vehicular Technology Conference* (1989), 846–850.
4. L. Narayanan and S. Shende, Static frequency assignment in cellular networks, in Proc. *4th International Colloquium on Structural Information and Communication Complexity*, 1997.
5. S. Kim and S. Kim, A two-phase algorithm for frequency assignment in cellular mobile systems, *IEEE Transactions on Vehicular Technology* **43** (1994) 542–548.
6. J. S. Kim, S. H. Park, P. W. Dowd, and N. M. Nasrabadi, Channel assignment in cellular radio using genetic algorithm, *Wireless Personal Commun.* **3**(3) (1996) 273–286.
7. D. Beckmann and U. Killat, A new strategy for the application of genetic algorithms to the channel assignment problem, *IEEE Transactions on Vehicular Technology* **48** (1999) 1261–1269.
8. C. Y. Ngo and V. O. K. Li, Fixed channel assignment in cellular radio networks using a modified genetic algorithm, *IEEE Transactions on Vehicular Technology* **47** (1998) 163–172.
9. C. W. Sung, and W. S. Wong, Sequential packing algorithm for channel assignment under co-channel and adjacent-channel interference constraint, *IEEE Transactions on Vehicular Technology* **46** (1997) 676–686.
10. M. R. Garey and D. S. Johnson, The complexity of near-optimal graph coloring, *Journal of the ACM* **23** (1976) 43–49.
11. W. K. Lio and G. G. Coghill, Channel assignment through evolutionary optimization, *IEEE Transactions on Vehicular Technology* **45** (1996) 91–96.
12. G. Chakraborty, An efficient heuristic algorithm for channel assignment problem in cellular radio networks, *IEEE Transactions on Vehicular Technology*, **50** (2001), 1528–1539.
13. S. C. Ghosh, B. P. Sinha and N. Das, An efficient channel assignment technique for

hexagonal cellular networks, in Proc. *6th International Symposium on Parallel Architectures, Algorithms, and Networks* (I-SPAN 2002), 361–366.

14. S. C. Ghosh, B. P. Sinha and N. Das, Channel assignment using genetic algorithm based on geometric symmetry, *IEEE Transactions on Vehicular Technology* **52** (2003) 860–875.
15. W. K. Hale, Frequency assignment: theory and application, *Proceedings of the IEEE*, **68** (1980), 1497–1514.
16. C. A. Murthy and N. Chowdhury, In search of optimal clusters using genetic algorithms, *Pattern Recognition Letters* **17** (1996) 825–832.
17. N. Funabiki and Y. Takefuji, A neural network parallel algorithm for channel assignment in cellular radio network, *IEEE Transactions on Vehicular Technology* **41** (1992) 430–437.
18. A. Gamst, Some Lower Bound for a Class of Frequency Assignment Problems, *IEEE Transactions on Vehicular Technology* **VT-35** (1986) 8–14.
19. D. Tcha, Y. Chung and T. Choi, A New Lower Bound for the Frequency Assignment Problem, *IEEE/ACM Transactions on Networking* **5** (1997) 34–39.
20. G. K. Audhya and B. P. Sinha, Lower Bound on Bandwidth for Channel Assignment in Multimedia Cellular Network with 2-Band Buffering, in Proc. *International Conference on Cmputing: Theory and Applications* (ICCTA-2007), 59–65.
21. D. Bhandari, C. A. Murthy and S. K. Pal, Genetic algorithm with elitist model and its convergence, *International Journal of Pattern Recognition and Artificial Intelligence* **10** (1996) 731–747.
22. S. Khanna and K. Kumaran, On wireless spectrum estimation and generalized graph coloring, in Proc. *17th IEEE Conference on Information and Communication* (INFOCOM 1998) **3** 1273–1283.
23. R. A. Leese, A unified approach to the assessment of radio channel on a rectangular hexagonal grid, *IEEE Transactions on Vehicular Technology* **46** (1997) 968–980.
24. A. Sen, T. Roxborough and S. R. Medidi, Upper and lower bounds of a class of channel assignment problems in cellular networks, in Proc. *IEEE Conference on Information and Communication* (INFOCOM 1998), **3**, pp. 1284–1291.
25. D. E. Goldberg, *Genetic Algorithm: Search, Optimization and Machine Learning* Addison Wesley Publishing Company, Inc., 1989.

Chapter 10

Range Assignment Problem in Wireless Network

Gautam K. Das, Sandip Das and Subhas C. Nandy

Indian Statistical Institute, Kolkata - 700 108, India

In a radio network a set of pre-placed radio stations $S = \{s_1, s_2, \ldots, s_n\}$ communicate each other by transmitting and receiving radio signals. Each radio station s_i is assigned a range $\rho(s_i)$ which is a non-negative real number. A radio station can send message to another radio station s_j if the Euclidean distance between s_i and s_j is less than or equal to $\rho(s_i)$. An exhaustive literature survey of the range assignment problem is presented in this paper. Several optimization problems that arise in assigning ranges to the radio stations in different application specific environment are discussed, and the best known algorithms for solving those problems are discussed.

Contents

10.1 Introduction

Due to the extraordinary growth of demand in mobile communication facility, design of efficient systems for providing specialized services has become an important issue in wireless mobility research. Broadly speaking, there are two major models for wireless networking: *single-hop* and *multi-hop*.[1] The single-hop model is based on the cellular network, and it provides one-hop wireless connectivity between the host and the static nodes known as base stations. Single-hop networks rely on a fixed backbone infrastructure that interconnects all the base stations by high speed wired links. On the other hand, the multi-hop model requires neither fixed wired infrastructure nor predetermined inter-connectivity. Two main examples where the multi-hop model is adopted, are ad hoc network and sensor network.[1]

Ad hoc wireless networking is a technology that enables untethered wireless networking in the environments where no wired or cellular infrastructure is available, or if available, is not adequate or cost-effective. Indeed, in an ad hoc wireless network, radio stations are already placed, and the wireless links are established based on the ranges assigned to the radio stations. This type of networking is useful in many practical applications, for example in disaster management, e-conference, etc. One of the main challenges in ad hoc wireless networks is the minimization of energy consumption. This can be achieved in several ways. The most important issues in this context are range assignment to the radio stations, and efficient routing of the packets as described below:

Range assignment: Assigning range (a non-zero real number) to the radio stations in the network. This enables each radio station to transmit packets to the other radio stations within its range. Here, the goal is to assign ranges to the radio stations such that the desired communication among the radio stations can be established, and the total power consumption of the entire network is minimized.

Routing: Transmission of packets from the source radio station to the destination radio station. Here, the power consumption of the network can be minimized by the choice of an appropriate path from source radio station to destination radio station.

On the other hand, a *wireless sensor network* (WSN) consists of large collection of co-operative small-scale nodes which can sense, perform limited computation, and can communicate over a short distance via wireless media. A WSN is self-organized in nature, and its members use short range communication to collect information from its neighborhood. It must also be capable to broadcast information to the base station in multi-hop fashion.

This article deals with the algorithmic aspects of the range assignment problem with a focus on the minimization of the total power requirement of the network maintaining its desired connectivity property. Two important sub-problems in this area are:

- The radio stations are pre-placed, and the objective is to assign ranges to the radio stations such that the network maintains some specific connectivity property. This problem is referred to as the *range assignment problem*.

- The radio stations are not pre-placed; the objective is to compute the positions and ranges of the radio stations such that the entire network maintains the desired connectivity property, and the total cost of range assignment in the entire network is minimized. This problem is referred to as the *base station placement problem*.

Specifically, we consider the range assignment problem for broadcasting, accumulation and all-to-all communication, when the radio stations are placed on a line and on a 2D plane. In the base station placement problem, we consider both the unconstrained and constrained version. In the unconstrained version, the base stations can appear anywhere inside the desired (convex) region. In the constrained version, base stations can appear only on the boundary of the desired (convex) region. In the next two sections, we discuss these two problems in detail.

10.2 Range Assignment Problem

A radio network is a finite set $S = \{s_1, s_2, \ldots, s_n\}$ of radio stations located in a geographical region. These radio stations can communicate with each other by transmitting and receiving radio signals. Each radio station $s_i \in S$ is assigned a range $\rho(s_i)$ (a non-negative real number) for communication with the other radio stations. This range assignment $\mathcal{R} = \{\rho(s_1), \rho(s_2), \ldots, \rho(s_n)\}$ defines a directed graph $G = (V, E)$, where $V = S = \{s_1, s_2, \ldots, s_n\}$, and $E = \{(s_i, s_j) | s_i, s_j \in S, d(s_i, s_j) \leqslant \rho(s_i)\}$, where $d(s_i, s_j)$ is the Euclidean distance between s_i and s_j. From now onwards, the graph G will be referred to as the *communication graph*.

A directed edge $(s_i, s_j) \in E$ indicates that $d(s_i, s_j) \leqslant \rho(s_i)$ and hence s_i can communicate (i.e., send a message) directly (i.e., in 1 hop) to any other radio station s_j. If s_i cannot communicate directly with s_j because of the insufficiency of its assigned range, then communication between them can be achieved by multi-hop transmission along a path from s_i to s_j in G; here the intermediate radio stations on the path cooperate with the source node and forward the message till its destination s_j is reached. Sometimes in a radio network, link failure occurs with some probability and all such failures occur independently. In multi-hop transmission, the probability of link failure on a transmission path increases with the number of hops. Thus, for multi-hop transmission, the reliability of communication can be ensured by bounding the number of hops for communication, considering the probability of failure of the radio stations. Several other problems related to bounded hop communication are available in the literature.[1-3] If the maximum number of hops (h) allowed is small, then communication between a pair of radio stations is established very quickly, but the power consumption of the entire radio network may become very high.[4] On the other hand, if h is large, then the power consumption decreases, but communication delay is likely to increase. The impact of tradeoff between the power consumption of the radio network and the maximum number of hops needed between a communicating pair of radio stations has been studied extensively.[5-9] The power required by a radio station s_i (denoted as $power(s_i)$) to transmit a message to another radio station s_j should satisfy

$$power(s_i) > \gamma \times (d(s_i, s_j))^\beta \tag{1}$$

where β is referred to as the distance-power gradient, and $\gamma \ (\geqslant 1)$ is the transmission quality of the message.[10] In the ideal case (i.e., free-space environment without any obstruction in the line of sight, and in the absence of reflections, scattering, diffraction caused by buildings, terrains etc.), we may assume $\beta = 2$ and $\gamma = 1$. Note that, the values of β may vary from 1 to 6 depending on various environmental factors, and the value of γ may also vary based on several other environmental factors, for example, noise, weather condition, etc. The more realistic model is to consider γ as a function of the radio station s_i. Here $\gamma(s_i)$ is referred as the weight of the radio station s_i and it depends on the positional parameters of s_i. Thus

$$power(s_i) = \gamma(s_i) \times (\rho(s_i))^2 \tag{2}$$

and the total cost of a range assignment $\mathcal{R} = \{\rho(s_i) \mid s_i \in S\}$ is written as

$$cost(\mathcal{R}) = \sum_{s_i \in S} power(s_i) = \sum_{s_i \in S} \gamma(s_i) \times (\rho(s_i))^2 \tag{3}$$

This version of the range assignment problem is referred to as the *weighted range assignment problem*. Unless otherwise specified, we assume that $\gamma(s_i) = 1$ for all $s_i \in S$, and hence the cost of the range assignment $\mathcal{R} = \{\rho(s_i) \mid s_i \in S\}$ is

$$cost(\mathcal{R}) = \sum_{s_i \in S} power(s_i) = \sum_{s_i \in S} (\rho(s_i))^2 \tag{4}$$

Note that, Equation 2 accounts for only the transmission power, i.e., the power consumed by the sender radio stations. In practice, a non-negligible amount of energy is also consumed at the receiver end to receive and decode the radio signals. Throughout this article, we consider only the energy consumed by the transmitting radio stations, since most of the existing literature do not account for the energy consumed for receiving a message.

If the radio stations are pre-placed, then the following three types of range assignment problem are considered in the literature depending on the communication criteria:

(a) Range assignment problem for broadcasting a message from a source radio station to all the target radio stations,

(b) Range assignment problem for all-to-all communication,

(c) Range assignment problem for accumulation of messages to a target radio station from all other radio stations in the network.

We assume that the radio stations are arranged on a straight line or on a 2D plane. These are referred to as 1D- and 2D-version respectively. The simple 1D model produces more accurate analysis of some typical situation arising in vehicular technology applications.[7] For an example, consider the road traffic information system where the vehicles follow roads, and messages are broadcasted along lanes.[7,11–13] Typically, the radius of curvature of a road is high in comparison to the transmission range; so we may consider the road as a straight line. Several other vehicular technology applications of this problem are available in the literature.[6,14] The 2D version of the range assignment problems are more realistic, but are often computationally hard in nature. Polynomial time approximation algorithms are available for some restricted variation of those problems.

10.2.1 *Broadcast Range Assignment Problem*

The objective of the broadcast range assignment problem is to assign transmission ranges $\rho(s_i)$ to the radio stations $s_i \in S$ so that a dedicated radio station (say $s^* \in S$) can transmit messages to all other radio stations, and the total power consumption of the entire network is minimum. The graph-theoretic formalization of the problem is as follows:

Compute a range assignment $\mathcal{R} = \{\rho(s_1), \rho(s_2), \ldots, \rho(s_n)\}$ such that there exists a directed spanning tree rooted at s^* in the communication graph G, and the total cost of the range assignment $\sum_{i=1}^{n}(\rho(s_i))^2$ is minimum.

The directed spanning tree rooted at s^* is referred to as the *broadcast tree*. In the bounded hop broadcast range assignment problem, the objective is to compute a range assignment $\mathcal{R}$ of minimum cost that realizes a broadcast tree of height bounded by a pre-specified integer h. Note that, the cost of range assignment is not the sum of weights of all the edges in the spanning tree; rather it is the sum of maximum weighted edge directed out from each node in the spanning tree. Thus, the solution of the minimum weight broadcast problem is not the same as the minimum weight spanning tree.

The hardness result of the broadcast range assignment problem depends on different parameters, namely, the distance power gradient β in the cost function (Equation 1), h the maximum number of hops allowed, the dimension of the plane where the radio stations are located, and the edge weight function. In general, we assume that the weight function $w(s_i, s_j)$ is equal to the Euclidean distance between the radio stations s_i and s_j. It is easy to show that if

$\mathcal{R} = \{\rho(s_1), \rho(s_2), \ldots, \rho(s_n)\}$ be the optimum range assignment for the broadcast problem, then $\rho(s_i) = w(s_i, s_j)$ for some $s_j \in S$.

If $0 < \beta \leqslant 1$ or $h = 1$ the problem is trivially polynomial time solvable because it suffices to set the range of the source s^* equal to the maximum weight among the edges incident on it. Or equivalently, assign the range of s^* equal to the distance of the furthest radio station from it. Clementi et al.[15] considered the general combinatorial optimization problem for the unbounded version ($h = n - 1$) of the broadcast range assignment problem, called *minimum energy consumption broadcast subgraph* (MECBS) problem, which is stated as follows:

> Given a weighted directed graph $G = (V, E)$, where $V = \{v_1, v_2, \ldots, v_n\}$, and each edge $(v_i, v_j) \in E$ is attached with a positive weight $w(v_i, v_j)$. The *transmission graph* induced by a range assignment $\mathcal{R} = \{\rho(v_1), \rho(v_2), \ldots, \rho(v_n)\}$ is a subgraph $G_{\mathcal{R}}(V, E_{\mathcal{R}})$ of G, where $E_{\mathcal{R}} = \{(v_i, v_j) | w(v_i, v_j) \leqslant \rho(v_i)\}$.
>
> The MECBS problem is then defined as follows: *given a source node $v_i \in V$, find a range assignment $\mathcal{R}$ such that $G_{\mathcal{R}}$ contains a spanning tree rooted at v_i, and $cost(\mathcal{R}) = \sum_{i=1}^{n} \rho(v_i)$ is minimum.*

The bounded hop version of MECBS problem is named as *h*-MECBS problem, and is defined as follows: *given a source node $v_i \in V$, find a range assignment $\mathcal{R}$ such that $G_{\mathcal{R}}$ contains a spanning tree of height at most h rooted at v_i, and $cost(\mathcal{R}) = \sum_{i=1}^{n} \rho(v_i)$ is minimum.*

It is proved that, both the MECBS and *h*-MECBS problems are NP-hard.[15,16] But, this does not imply that the *h*-hop broadcast range assignment problem is NP-hard. The reason is that, here the weight of each edge (v_i, v_j) in G is equal to the Euclidean distance of the radio stations in S corresponding to the nodes v_i and v_j. To our knowledge, no hardness result is available for the bounded hop broadcast range assignment problem. Fuchs[17] studied this problem in a restricted setup, called the *well-spread* instances, which is defined as follows:[5]

> Let $\Delta(S) = max\{d(u, v) \mid u, v \in S\}$ and $\delta(S) = \min\{d(u, v) \mid u, v \in S, u \neq v\}$. A set of radio stations in S are said to be *well-spread* if there exists some positive constant c such that $\delta(S) \geqslant \frac{c\Delta(S)}{\sqrt{|S|}}$.

It is shown that for any $\beta > 1$, the broadcast range assignment problem is NP-hard for a set of radio stations which are well-spread in 2D.[5]

For the 1D version of the problem, the radio stations are placed on a straight line. The range assigned to a radio station s_i ($\neq s^*$) is said to be a *bridge* if $\rho(s_i) = d(s_i, s_j)$ and s_i, s_j lie in two different sides of s^* (see Figure 10.1). The bridge is said to be functional if the removal of that bridge (edge) from the communication

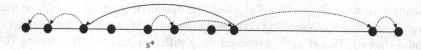

Fig. 10.1   Illustration of a bridge in the broadcast problem.

graph indicates that there exists a radio station $s_k \in S$ which is not reachable from the source s^* using a path satisfying the hop constraint.

It can be shown that the minimum cost range assignment for the h-hop broadcast in 1D may contain at most one functional bridge. Clementi et al.[14] proposed a dynamic programming based algorithm for this problem. It computes optimal solutions having (i) no functional (left/right) bridge, (ii) one functional left-bridge only, and (iii) one functional right-bridge only. Finally, the one having minimum total cost is reported. The worst case time complexity of this algorithm was $O(hn^2)$. Later, Das et al.[18] improved the time complexity to $O(n^2)$ considering the geometric properties of the problem.

Several researchers studied on developing good approximation/heuristic algorithms for the broadcast range assignment problem. The most popular heuristic for this problem is based on the minimum spanning tree (MST), and is stated below.

> Construct a weighted complete graph $G = (V, E)$, where $V = S$ (the set of radio stations), and the edge weight $w(s_i, s_j) = d(s_i, s_j)$. Compute the MST T of G. Assign the range of a radio station $s_i = \rho(s_i) = \max\{w(s_i, s_j) | s_j$ is a successor of s_i in $T\}$.

As MST is always connected, the communication graph derived from this range assignment is also connected. Wieselthier et al.[19] proposed a greedy heuristic, called *broadcast incremental power* (BIP), which is a variant of Prim's algorithm for MST, and is applicable for any arbitrary dimension d ($d > 1$). At each step, instead of adding the edge having minimum weight in the MST, a node that needs minimum extra energy is added. This formulation is obvious due to the broadcast nature of the problem, where increasing the radius of an already emitting node to reach a new node is less expensive than installing a new emitting node. Some small improvements of this method was proposed by Marks et al.[20] A different heuristic paradigm, namely *embedded wireless multicast advantage* (EWMA) is described by Cagalj et al.,[16] which is an improvement over MST based algorithm. It takes the MST as the initial feasible solution, and builds an energy efficient broadcast tree. In EWMA, every forwarding node in the initial solution is given a chance to increase its power level. This may decrease the power level of some other nodes maintaining the network connectivity. This assignment of new power level to the concerned node is acceptable if cost of the tree decreases. Each node finally chooses the power level at which the overall decrease in cost of the final tree is maximized. Assuming complete knowledge of distances for all pairs of radio stations, Das et al.[21] proposed three different integer programming (IP) formulations for the minimum energy broadcast problem.

The distributed version of this problem was studied by Wieselthier et al.[22] Although it works well for small instances, its performance degrades when the number of radio stations becomes large. The reason is that, it needs communication for exchanging data in distributed environment for constructing the global tree. Ingelrest and Simplot-Ryl[23] proposed a localized version of the BIP heuristic in the distributed set up. Here, each node applies the BIP algorithm on its 2-hop neighbors, and then includes the list of its neighbors who need to retransmit, together with the transmission ranges with the broadcast packet. It is experimentally observed that the result offered by this algorithm is very close to the one obtained by BIP with global knowledge of the network. Below, we provide a scheme for computing 1-hop and 2-hop neighbors of each node.

Let the radio stations be distributed in a 2D region. Each radio station knows its position using a location system (say GPS).[24] Each radio station broadcasts a "HELLOW" message with its own coordinate. A radio station that receives such a message can identify the sender and it notes that the sender is in its 1-hop neighborhood. Using the 1-hop neighbor information, the 2-hop neighbors of each node can be computed after the second round of exchange. After knowing the positions of 1-hop and 2-hop neighbors, each node can easily compute the distances of its 1-hop and 2-hop neighbors.

Cartigny et al.[25] proposed a distributed algorithm for the broadcast range assignment problem that is based on the *Relative Neighborhood Graph* (RNG).[26] The RNG preserves connectivity, and based on the local information, the range assignment is done as follows: for each node s_i, compute its furthest RNG neighbor s_j, excepting the one from which the message is received, and assign a range $d(s_i, s_j)$ to the node s_i. They experimentally demonstrated that their algorithm performs better than the solution obtained by the sequential version of the BIP algorithm. Cartigny et al.[27] described localized energy efficient broadcast for wireless networks with directional antennas. This is also based on RNG. Messages are sent only along RNG edges, and the produced solution requires about 50% more energy than BIP. More reviews on broadcast problem are available in.[28,29] Now we mention the state-of-the-art performance bound of the MST and BIP based algorithms for the broadcast range assignment problem.

Clementi et al.[15] first proved that the approximation factor of the MST based heuristic for the broadcast problem is $10^\beta/2 2^\beta$. Wan et al.[30] proved that if $\beta = 2$, the approximation ratio of the MST based heuristic is between 6 and 12, whereas the approximation ratio of the centralized BIP is between $\frac{13}{3}$ and 12. Unfortunately, there was a small error in the proof of Wan et al.[30] Klasing et al.[31] corrected the analysis and proved that the upper bound of the approximation ratio of the MST based algorithm is actually 12.15. Navarra[32] improved the approximation ratio to 6.33. Finally, Ambhul[33] improved the approximation factor of the MST based algorithm to 6; thus it attains the lower bound of the problem.[30]

Calamoneri et al.[34] proved an almost tight asymptotic bound on the optimal cost for the minimum energy broadcast problem on the square grid. Finally, Calinescu et al.[35] presented $(O(\log n), O(\log n))$ bicriteria approximation algorithm for *h-hop broadcast range assignment problem*. The solution produced by this algorithm needs $O(h \log n)$ number of hops, and cost is at most $O(\log n)$ times the optimum solution. They also presented an $O(\log^\beta n)$-approximation algorithm for the same problem, where the radio stations are installed in d-dimensional Euclidean spaces, and β is the distance-power gradient.

Clementi et al.[15] proved that the general MECBS problem is not approximable within a sub-logarithmic factor. The first logarithmic factor approximation algorithm for MECBS problem was proposed by Caragiannis et al.,[36] where an interesting reduction to the *node-weighted connected dominating set problem* is used. This algorithm achieves a $10.8 \ln n$ factor approximation ratio for the symmetric instances of MECBS problem. Latter, Papadimitriou and Geordiadis[37] addressed the minimum energy broadcast problem where the broadcast tree is to be constructed in such a way that different source nodes can broadcast using the same broadcast tree, and the overall cost of the range assignment is minimum. This approach differs from the most commonly used one where the determination of the broadcast tree depends on the fixed source node. It is proved that, if the same broadcast tree is used, the total power consumed is at most twice the total power consumed for creating the broadcast tree with any node as the source. It is also proved that the total power consumed for this common broadcast tree is less than $2H(n-1) * opt$, where opt denotes the minimum cost of broadcast tree with a fixed source node, and $H(n)$ is the harmonic function involving n.

Chlebikova et al.[38] and Kantor and Peleg[39] independently studied *h*-hop broadcast range assignment problem on an arbitrary edge weighted graph. By approximating edge weighted graph by paths, the authors presented an approximation algorithm for this problem as follows:

Solve the minimum linear arrangement (MLA) problem to compute a linear arrangement of the nodes in the graph $\sigma : V \rightarrow \{1, 2, \ldots, n\}$ such that $c = \Sigma_{(u,v) \in E} w(u,v)|\sigma(u) - \sigma(v)|$ is minimized. Then apply the algorithm of Das et al.[18] to compute the optimal broadcast range assignment for a linear radio network.

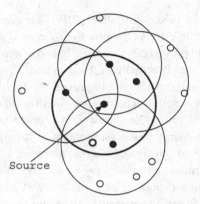

Fig. 10.2   An example of 2-hop broadcast.

The MLA problem is NP-hard.[40] A $\log n$ factor approximation algorithm is proposed in.[41] Chlibekova et al.[38] shown that the approximation factor of the solution using the above method is also $O(\log n)$, and it matches the lower bound proposed in.[42] If a graph does not contain a complete bipartite subgraph $K_{r,r}$ with $r > 2$, and $\beta \leqslant O(\frac{\log \log n}{\log \log \log n})$, then the approximation ratio can be improved to $O((\log \log n)^\beta)$,[38] where β is the distance power gradient of the cost function. The range assignment problem for h-hop broadcast can be solved optimally in $O(hn^4)$ time if the given graph is a tree.[43]

For the 2-hop broadcast in the plane, the optimum range assignment can be obtained in $O(n^7)$ time using an algorithm based on dynamic programming paradigm.[44] In the same paper, a polynomial-time approximation scheme for the h-hop broadcast range assignment problem was suggested for any $h \geqslant 1$ and $\varepsilon > 0$. The time complexity of the proposed algorithm is $O(n^\alpha)$, where $\alpha = O((8h^2/\varepsilon)^{2^h})$. In the homogeneous 2-hop broadcast problem, the range of a radio station is either a specified value ρ or 0. In Figure 10.2, an example is demonstrated for a given range value ρ. The black sites or the sites having thick boundary (except source s^*) indicate the subset of radio stations (called S_1) which are reachable from s^* in 1 hop, and the sites having thin boundary indicate the subset (called S_2) which are reachable from s^* in 2 hop. It is easy to understand that s^* must be assigned range ρ, and the members in S_1 lie inside the circle C^* having radius ρ and centered at s^*. Among the members in S_1, the black sites (denoted by S^*) are assigned range ρ for the optimum 2-hop broadcast from s^* to all the members in S, but those having thick boundary only, need not be assigned any range i.e., their range is zero. Thus, $S^* \subseteq S_1 \subset S$. All the members in S_2 lie outside C^*, and so, the range assigned to each members in S_2 is zero.

The homogeneous 2-hop broadcast problem can easily be be formulated using the traditional *set cover* problem which is known to be NP-hard, and a $\log(k)$-factor approximation algorithm is easy to get, where k is the number of radio stations inside C^*. Bronnimann and Goodrich[45] considered the *circle cover problem*, where a set S of n points is given in the plane, and a family of circles C is also given; the problem is to find a minimum number of circles in C that covers all the points in S. The circle cover problem can be easily mapped to the 2-hop broadcast problem. The proposed algorithm for the circle cover problem produces $O(1)$-approximation results in $O(n^3 \log n)$ time. The homogeneous 2-hop broadcast problem can also be formulated as follows. Consider a weighted digraph $G = (S, E)$ where $E = E_1 \cup E_2$, $E_1 = \{(s^*, s) | s \in S_1\}$ and $E_2 = \{(s, s') | s \in S_1, s' \in S_2, \delta(s, s') \leqslant r\}$. The weight of each edge in E_1 is 1, and that of each edge in E_2 is 0. The objective is to compute a minimum weight directed Steiner tree, where the terminal nodes correspond to the members in S_2, and Steiner nodes correspond to the members in S_1. The radio station s^* is a designated node such that we are searching for a Steiner tree with root at s^*. Thus, the subgraph of G with the Steiner nodes and s^* is a tree of height 2, where the Steiner nodes appear at level 1. A polynomial time $O(1)$-factor approximation algorithm for the minimum weight directed Steiner tree in graph G is proposed.[46] Calinescu et al.[47] proposed two geometric algorithms for the homogeneous 2-hop broadcast problem; the first one produces 6-approximation result in $O(n \log n)$ time and the second one produces 3-approximation result in $O(n \log^2 n)$ time. It can be shown that the time complexity of their second algorithm can easily be improved to $O(n \log n)$ using fractional cascading.[48] Thus the algorithm proposed in[47] is an improvement over both[45,46] in terms of both the time complexity results, and approximation factor. Recently, several variations of this problem are solved.[49] These are

(i) Given a range ρ, the feasibility of a 2-hop broadcast can be tested in $O(n \log n)$ time.

(ii) The problem of computing the optimum value of ρ for the homogeneous 2-hop broadcast reduces to the matrix multiplication problem, and hence it is polynomially solvable.

(iii) An $O(n^2)$ time 2-factor approximation algorithm is proposed for the problem considered in.[47]

Although the unweighted broadcast range assignment problem has been studied extensively, little is known for the case of the weighted version. Here, each radio station s_i is assigned with an weight (cost of installing a base station at s_i) $\gamma(s_i) \geqslant 0$, for $i = 1, 2, \ldots, n$. Here, the cost of range assignment is given by Equation 3. Figure 10.3 demonstrates an instance of unbounded weighted broadcast range assignment problem with five radio stations, along with the optimum solution. Here, $d(s_1, s_2) = 8$, $d(s_2, s_3) = 2$, $d(s_3, s_4) = 1$ and $d(s_4, s_5) = 4$. The weight of the radio stations $s_1, s_2, \ldots, s_5$ are 10, 1, 10000, 100 and 0.01 respectively, source is s_3 $(=s^*)$, and the cost of the optimum range assignment is 10,951.25 units.

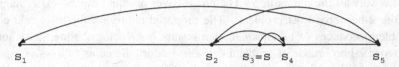

Fig. 10.3   An example of linear weighted broadcast problem.

The first work on this problem for linearly arranged n radio stations was discussed by Ambuhl et al.[50] The following variations are studied, and the algorithms are proposed using dynamic programming.

In the unbounded case (i.e., $h = n - 1$), the time and space complexities of the proposed algorithm are $O(n^3)$ and $O(n^2)$ respectively.

For h-hop broadcast, the time and space complexities of the proposed algorithms are $O(hn^4)$ and $O(hn^2)$ respectively.

For the unbounded multi-source broadcast, the time and space complexities are $O(n^6)$ and $O(n^2)$ respectively.

In higher dimension (i.e., $d > 2$) and $\beta = 1$, the problem is formulated as a shortest path problem in a graph, and the proposed algorithm produces a 3-approximation result in $O(n^3)$ time;

Das[49] considered both the unbounded and bounded version of the weighted broadcast range assignment problem in 1D. The proposed algorithm for the unbounded version of the problem outputs the optimum result in $O(n^2)$ time, and that for the bounded (h) hop broadcast problem produces the optimum solution in $O(hn^3)$ time. For further details, see.[51,52]

10.2.2 All-to-All Range Assignment Problem

The objective of the range assignment problem for h-hop all-to-all communication is to assign transmission range $\rho(s_i)$ to each radio station $s_i \in S$ so that each pair of members in S can communicate using at most h hops, and the total power consumption by the entire radio network is minimized. Typically, h can assume any value from 1 to $n-1$, where $n = |S|$. For $h = 1$, the problem is trivial. Here, for each radio station $s_i \in S$, $\rho(s_i) = Max_{s_j \in S} d(s_i, s_j)$, and the problem can be solved by computing the furthest point Voronoi diagram.[53] Basically, the hardness of the all-to-all range assignment problem depends on two parameters, namely, the distance power gradient (β in Equation 1) of the cost function and dimension (d) in which the radio stations are located. For the linear radio network ($d = 1$), the problem can be solved in polynomial time,[6] but if $d > 1$, then the problem becomes NP-hard.[5,17,54] In particular, if $\beta = 1$, then the problem can be shown to be 1.5-APX hard.[55] For $\beta > 1$, the problem is APX-hard, and so it does not admit a PTAS unless P = NP.[6] Clementi et al.[54] presented a lower and an upper bound on the minimum cost of the h-hop range assignment for a radio network in 2D. They also proved that for the well-spread instances (defined in[5]) of this problem, these two bounds remain same. In this connection it needs to be mentioned that, the homogeneous version of the problem can be solved in polynomial time. This uses a trivial result: *The common range r for all the radio stations is the distance of a pair of radio stations in S.* The algorithm proposed by Das et al.[56] works as follows:

> Compute the distance of each pair of radio stations of S in an array D. Perform a binary search among the elements in D to choose minimum i such that $d[i]$ is a feasible range (for all the members in S) for h-hop all-to-all communication. This can be tested by computing the diameter of the communication graph corresponding to range $d[i]$. Thus, the overall time complexity of this problem is $O(n^3 \log n)$.

For the linear radio network, the problem becomes relatively simple, but it results in a more accurate analysis of the situation arising in vehicular technology application.[7] In a linear radio network, several variations of the 1D range assignment problem for h-hop all-to-all communication are studied by Kirousis et al.[6] For the uniform chain case, i.e., where each pair of consecutive radio stations on the line is at a distance δ, tight upper bound on the minimum cost of range assignment is shown to be $OPT_h = \Theta(\delta^2 n^{\frac{2^{h+1}-1}{2^h-1}})$ for any fixed h. In particular, if $h = \Omega(\log n)$ in the uniform chain case, then $OPT_h = \Theta(\delta^2 \frac{n^2}{h})$. For the general problem in 1D, i.e., where the radio stations are arbitrarily placed on a line, a 2-approximation algorithm for the range assignment problem for h-hop all-to-all communication is proposed by Clementi et al.[11] The worst case running time of this algorithm is $O(hn^3)$. For the unbounded case ($h = n - 1$), a dynamic programming based $O(n^4)$ time algorithm is available[6] for generating the minimum cost. Finally, Das[57] improved the complexity result of this problem to $O(n^3)$.

Carmi and Katz[58] proved that the all-to-all range assignment problem remains NP-hard when the range of each radio station is either ρ_1 or ρ_2 with $\rho_2 > \sqrt{\frac{3}{2}}\rho_1$.

In the same paper, they also provided an $\frac{11}{6}$-approximation algorithm. Fuchs[17] studied the range assignment problem for all-to-all communication where the radio stations are well-spread[5] with $\beta > 0$. Under the assumption of symmetric connectivity as stated below, the problem is shown to be NP-hard in both 2D and 3D. It is also shown that the problem is APX-hard in 3D.

In the symmetric connectivity model, the minimum transmission power needed for a radio station s_i to reach a radio station s_j is assumed to be equal to the minimum transmission power needed for s_j to reach s_i. In other words, the symmetric connectivity means a link is established between two radio stations $s_i, s_j \in S$ only if both radio stations have transmission range at least as big as the distance between them.

Althaus et al.[59] presented an exact branch and cut algorithm based on an integer linear programming formulation for solving the unbounded (i.e., $h = n - 1$) version of the 2D all-to-all range assignment problem with symmetric connectivity assumption, and their algorithm takes 1 hour for solving instances with up to 35-40 nodes. In the same paper, a minimum spanning tree (MST) based 2-approximation algorithm has also been presented with symmetric connectivity assumption; here range of a radio station is equal to the length of the longest edge of the Euclidean MST attached with that radio station. Under the assumption of symmetric connectivity, Krumke et al.[60] presented an $(O(\log n), O(\log n))$ bicriteria approximation algorithm for h-hop all-to-all range assignment problem, i.e., their algorithm produces a solution having $O(h \log n)$ number of hops and costs at most $O(\log n)$ times the optimum solution. Latter, Calinescu et al.[35] studied the same problem independently, and provided an algorithm with same approximation result. Recently, Kucera[61] presented an algorithm for the all-to-all range assignment problem in 2D. Probabilistic analysis says that the average transmission power of the radio stations produced by this algorithm is almost surely constant if the radio stations appear in a square region of a fixed size. This algorithm can also work in any arbitrary dimension. Das et al.[56] proposed an efficient heuristic for the h-hop all-to-all range assignment problem in 2D. The experimental evidences demonstrate that it produces near optimum result in reasonable time.

Chlebikova et al.[38] studied the range assignment problem for h-hop all-to-all communication in static ad hoc networks using a graph-theoretic formulation where the edge weights can violate triangle inequality. They presented a probabilistic algorithm to approximate any edge-weighted graph by a collection of paths such that for any pair of nodes, the expected distortion of shortest path distance is at most $O(\log n)$. The paths in the collection and the corresponding probability distribution are obtained by solving a packing problem defined by Plotkin et al.,[62] and using a *minimum linear arrangement problem* solver of Robinovich and Raz[41] as an oracle. With this algorithm, they approximated a 2D static ad hoc network as a collection of paths. Then it runs the polynomial time algorithm for the minimum range assignment problem in 1D.[11] Therefore, this strategy leads to a probabilistic $O(\log n)$ factor approximation algorithm for the h-hop all-to-all range assignment problem for the static ad hoc network in 2D. A polynomial time constant factor approximation algorithm for this problem on general metrics is given by Kantor and Peleg.[39]

The weighted version of the all-to-all range assignment problem is studied only for q-spread instances in 2D.[50] The notion of q-*spread* instances in 2D is as follows:

Let s_i be a radio station in S. Consider a maximum size convex polygon containing only the radio station s_i and whose vertices are in the set $S \setminus \{s_i\}$. Let $H(s_i)$ be the vertices of this convex polygon. Now define two quantities $\Delta(H(s_i)) = \max_{s_j \in H(s_i)}(d(s_i, s_j))^\beta$ and $\delta(H(s_i)) = \min_{s_j \in H(s_i)}(d(s_i, s_j))^\beta$. Finally, choose $s_k \in S$ such that $\Delta(H(s_k)) = \max_{i=1}^n \Delta(H(s_i))$. The instance S is said to be q-spread if $\Delta(H(s_k)) \leqslant q \times \delta(H(s_k))$.

The proposed algorithm can work for any arbitrary distance power gradient $\beta > 1$, and produces a q-approximation result.

10.2.3 *Accumulation Range Assignment Problem*

The objective of *h-hop accumulation range assignment problem* is to assign transmission range $\rho(s_i)$ to the radio station $s_i \in S$ such that each radio station $s_i \in S$ can send message to a dedicated radio station $s^* \in S$ using at most h hops and the total cost $\sum_{s_i \in S}(\rho(s_i))^2$ of the network is minimized.

Clementi et al.[11] discussed h-hop accumulation range assignment problem for 1D radio network and proposed an algorithm based on dynamic programming which can produce optimum solution in $O(hn^3)$ time. This is a trivial lower bound for the cost of range assignment for the h-hop all-to-all communication. In fact, this also leads to a trivial 2-approximation algorithm for the range assignment of the h-hop all-to-all communication in a 1D radio network as follows:

Consider the communication graph based on the assigned ranges produced by the accumulation range assignment algorithm, and disregard the direction of the edges. This is an weighted undirected graph where the weight of each edge is equal to the length of that edge in Euclidean metric. Now at each node, assign range equal to the maximum weight among the edges incident to that node.

In a general graph, the h-hop accumulation range assignment problem is equivalent to finding the minimum spanning tree of height h with a designated node as the root. This problem is referred to as the h-MST problem in the literature. Experimentally tested exact super-polynomial time algorithms for the h-MST problem is already available.[63,64] Althaus et al.[65] presented a polynomial time algorithm for this problem with running time $n^{O(h)}$. The 2-MST problem can be easily reduced to the classical *uncapacitated facility location problem* (UFLP). Thus all the approximation algorithms for UFLP apply to the 2-MST as well. The best known theoretical result on UFLP is a PTAS given by Arora et al.[66] Several heuristic algorithms are available for the Euclidean 2-MST problem in 2D.[67,68] Clementi et al.[69] presented a fast and easy-to-implement heuristics for the 2D h-hop accumulation range assignment problem, and studied its behavior on the instances obtained by choosing n points at random in a square region. In the unbounded case, the accumulation range assignment problem reduces to the *arborescence problem*, stated as follows:

Given a weighted directed graph G and a root node s^*, the minimum cost arborescence problem is to find the minimum cost spanning tree in G directed out from s^*.

Given a complete digraph $G = (V, E)$ with vertex set V corresponding to the set of radio stations S, and weight of the directed edge $(s_i, s_j) \in E$ is $w_j \times (d(s_i, s_j))^2$, we can compute the optimum range assignment for the weighted version of the unbounded accumulation problem by computing the arborescence tree T directed from s^* using $O(n^2)$ time algorithm proposed by Gabow et al.[70] This algorithm works in arbitrary dimension.

10.3 Base Station Placement Problem

In this problem, the objective is to identify the locations for placing the base stations and to assign ranges to the base stations for efficient radio communication. Each mobile terminal communicates with its nearest base station, and the base stations communicate with each other over scarce wireless channels in a multi-hop fashion by receiving and transmitting radio signals. Each base station emits signal periodically and all the mobile terminals within its range can identify it as its nearest base station after receiving such radio signal. Here, the problem is to position the base stations such that a mobile terminal at any point in the entire area can communicate with at least one base station, and the total power required for all the base stations in the network is minimized. Another variation of this problem arises when there are forbidden zones for placing the base stations, but communication is to be provided over these regions. Example of such forbidden regions may include large water bodies, or stiff mountain terrains. In such cases, we need some specialized algorithms for efficiently placing the base stations on the boundary of the forbidden zone to provide services within that region. These two variations of base station placement problem are referred to as

(i) Unconstrained version of the base station placement problem, and
(ii) Constrained version of the base station placement problem.

10.3.1 *Unconstrained Base Station Placement Problem*

The base station placement problem involves placing multiple base stations within a specific deployment site, with an aim to provide an acceptable quality of service to the mobile users. Here, the formulation of the objective function depends on the hardware limitations of the specific wireless system and the particular application for which the system is to be designed.

Several authors[71-76] studied the issues of optimal base station placement in an indoor micro-cellular radio environment with an aim to optimizing several objective criteria. Most of them have primarily used local optimization strategies for optimizing the desired objective function. Stamatelos and Ephremides[73] formulated the objective function as the maximization of coverage area along with the minimization of co-channel interference under the stipulated constraint of spatial diversity. Choong and Everitt[71] investigated the role of frequency-reuse across multiple floors in a building while solving the base station placement problem to minimize co-channel interference. Howitt and Ham[77] pointed out the limitations of using local optimization algorithms for solving the base station placement problem; finally they proposed a global optimization technique based algorithm, where the objective function is modeled as a stochastic process. The authors of[72,75] indicated that the simplex method is well suited for the base station placement problem because the corresponding objective function is non-differentiable and so quasi-Newton optimization methods are not well-suited.

In the 2D version of the general base station placement problem, the objective is to place a given number of base stations in a 2D region, and assign ranges to each of these base stations such that every point in the region is covered by at least one base station, and the maximum assigned range is minimized. Since a base station with range ρ can communicate with all the mobile terminals present in the circular region of radius ρ and centered at the position where the base station is located, our problem reduces to covering a region by a given number of equal radius circles (see Figure 10.4), and the objective is to minimize the radius. For

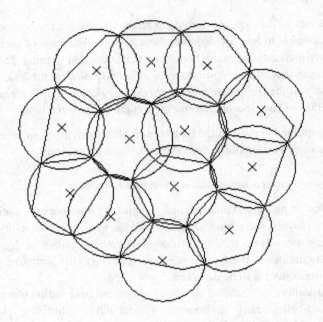

Fig. 10.4   Illustration of unconstrained base station placement problem.

simplicity, we may consider that the given region is a convex polygon, and the range of all the base stations are same, say ρ.

In the covering by circle problem, the following two variations are important: (i) find the minimum number of unit-radius circles that are necessary to cover a given polygon, and (ii) given a constant k, compute a radius ρ, such that an arrangement of k circles of radius ρ exists which can cover the entire polygon, and there does not exists any arrangement of k circles of a radius $\rho' < \rho$ which can cover the entire polygon. We also need to report the centers of these k circles (of optimum radius).

Verblunsky[78] proposed a lower bound for the first problem; it says that if m is the minimum number of unit circles required for covering a square where each side is of length σ, then $\frac{3\sqrt{3}}{2}m > \sigma^2 + c\sigma$, where $c > \frac{1}{2}$. Substantial studies have been done on the second problem. Several researcher tried to cover a unit square region with a given number (say k) of equal radius circles with the objective to minimize the radius. Tarnai and Gasper[79] proposed graph theoretic approach to obtain a locally optimal covering of a square with up to 10 equal circles. No proof for optimality was given, but later it was observed that their solution for $k = 5$ and

$k = 7$ are indeed optimal. Several results exist on covering squares and rectangles with k equal circles for small values of k (= 6, ... 10, etc.)[80,81] For a reasonably large value of k, the problem becomes more complex. Nurmela and Ostergard[82] adopted simulated annealing approach to obtain near-optimal solutions for the unit square covering problem for $k \leqslant 30$. As it is very difficult to get a good stoping criteria for a stochastic global optimization problem, they used heuristic approach to stop their program. It is mentioned that, for $k = 27$ their algorithm runs for about 2 weeks to achieve the stipulated stopping criteria. For $k \geqslant 28$, the time requirement is very high. So, they changed their stopping criteria, and presented the results. Nurmela[83] adopted the same approach for covering a equilateral triangle of unit edge length with circles of equal radius, and presented the results for different values of k less than or equal to 36. Das et al.[84] considered the covering by circles problem where the region is a convex polygon. It initially places k base stations randomly, and then uses iterative Voronoi diagram updating method where each iteration executes the following steps:

Compute the Voronoi diagram of k points, and compute the minimum enclosing circles of each cell of the Voronoi diagram. It is expected that the radius of these circles will not be the same. In order to refine (reduce) the radius of the covering circles, move each point inside a Voronoi cell at the center of its circumscribing circle.

The process continues until the difference of radii of the largest circle in two consecutive steps is less than a very small (predefined) real number. It is experimentally observed that the minimum radius obtained by this method favorably compares with the existing results. The time complexity of each iteration is $O(n \log n)$, and the entire experiment takes a fraction of a second for a reasonably large value of k ($\geqslant 100$) in a SUN Blade 1000 computing platform with 750 MHz CPU speed. In the context of practical application this method is very much acceptable since the existing methods works only for rectangle or triangle, and may even take about two weeks' time for a reasonable value of the number of circles ($\geqslant 27$).[82]

Several other variations of covering by circles problem are available in the literature. The discrete version of the covering by circle problem is the well-known k-center problem. Here the objective is to place k supply points to cover a set of n demand points on the plane such that the maximum Euclidean distance of a demand point from its nearest supply point is minimized. The simplest form of this problem is the Euclidean 1-center problem which was originally proposed by Sylvester[85] in 1857. The first algorithmic result on this problem is due to Elzinga and Hearn,[86] which gives an $O(n^2)$ time algorithm. Lee[53] proposed the furthest point Voronoi diagram, which also can be used to solve the 1-center problem in

$O(n \log n)$ time. Finally Megiddo[87] found an optimal $O(n)$ time algorithm for solving this problem using prune-and-search technique.

Jaromczyk and Kowaluk[88] studied the 2-center problem, and proposed a simple algorithm with running time $O(n^2 \log n)$. Later, Sharir[89] improved the time complexity to $O(n \log^9 n)$. The best known algorithm for this problem was proposed by Chan.[90] He suggests two algorithms. The first one is a deterministic algorithm, and it runs in $O(n \log^2 n (\log \log n)^2)$ time; the second one is a randomized algorithm that runs in $O(n \log^2 n)$ time with high probability.

A variation of this problem is the discrete 2-center problem, where the objective is to find two closed disks whose union can cover the given set P of n points, and whose centers are a pair of points in P. Kim and Shin[91] considered both the standard and discrete versions of the 2-center problem where the points to be covered are vertices of a convex polygon. Their algorithms run in $O(n \log^3 n \log \log n)$ and $O(n \log^2 n)$ time respectively.

For a given set of n demand points, the general version of both the k-center problem and the discrete k-center problem are NP-complete.[92,93] But for a fixed value of k, both the problems can be solved in $O(n^{O(\sqrt{k})})$ time.[94] Therefore, it makes sense to search for efficient approximation algorithms and heuristics for the general version. Detailed review on this topic can be found in.[95] Another variation of this problem, available in the literature, is that the center and radius of the (equal-radius) circles are fixed and the objective is to cover the points in S with minimum number of circles. Stochastic formulations of different variations of this problem are available in the literature.[96]

Apart from the base stations placement for mobile communication, the proposed problems find relevant applications in energy-aware strategic deployment of the sensor nodes in wireless sensor networks (WSN).[97,98] In particular, Boukerche et al.[98] studied the case where the sensor nodes are already placed. A distributed algorithm is proposed in that paper which can activate the sensors such that the entire area is always covered, and the total lifetime of the network is maximized. Voronoi diagram[99] is also an useful tool for dealing with the coverage problem for sensor networks, where the sensors are distributed in $\mathbb{R}^2$. Meguerdichian et al.[100] considered the problem where the objective is to find a sensor avoiding path between a pair of points s and t such that for any point p on the path, the distance of p from its closest sensor is maximized. Several other application specific covering problems related to sensor network are available in the literature.[97]

10.3.2 *Constrained Base Station Placement Problem*

In general, every point inside the desired region may not be suitable for installing a base station. Some specific situations were mentioned in the beginning of this section. A very practical constrained variation of the base station placement problem is the situation where the base stations can be erected only on the boundary of the given region. This is a variation of the constrained k-center problem. Several other constrained variations of the k-center problem exist in the literature.

First of all, let us consider the constrained versions of the Euclidean 1-center problem. Megiddo[87] studied the case where the center of the smallest enclosing circle must lie on a given straight line, and proposed a linear time algorithm. Bose and Toussaint[101] addressed another variation of 1-center problem where instead of the entire region, a given set Q of m points is to be covered by a circle whose center is constrained to lie on the boundary of a given simple polygon P of size n. They provided an output sensitive $O((n+m)\log(n+m)+\chi)$ time algorithm for this problem, where χ is the number of intersection points of the farthest-point Voronoi diagram of Q with the edges in P; this may be $O(nm)$ in the worst case. Constrained variations of the 1-center and 2-center problems are studied by Roy et al.,[102] where the target region is a convex polygon with n vertices, and the center(s) of the covering circle(s) is/are constrained to lie on a specific edge of the polygon. The time complexities of both these problems are $O(n)$. Hurtado et al.[103] used linear programming to give an $O(n+m)$ time algorithm for solving minimum enclosing circle problem for a set of points whose center satisfies m linear inequality constraints. The query version of the minimum enclosing circle problem is studied by Roy et al.,[104] where the given point set needs to be preprocessed such that given any arbitrary query line, the minimum enclosing circle with center on the query line can be reported efficiently. The preprocessing time and space of this algorithm are $O(n\log n)$ and $O(n)$ respectively, and the query time complexity is $O(\log^2 n)$. The query time can be reduced to $O(\log n)$ if $O(n^2)$ preprocessing time and space are allowed.[105]

Several other constrained variations of the k-center problem can be found in the domain of mobile communication and sensor network. Recently, Alt et al.[106] considered the problem of computing the centers of k circles on a line to cover a given set of points in 2D. The radius of the circles may not be the same. The objective is to minimize the sum of radii of all these k circles. They proposed an $O(n^2 \log n)$ time algorithm for solving this problem.

Recently, Das[49] in his Ph.D. thesis considered several constrained variations of the base station placement problem. Here the base station can be placed only on the boundary of the given convex region P. The objective is to determine the positions of k base stations (of equal range) on the boundary of P such that each point inside P is covered by at least one base station. This problem is referred to as *region-cover(k)* problem. This problem is of particular use where some portions of the target region are unsuitable for placing the base station, but the communication inside those regions need to be provided. For example, we may consider a huge water body, say lake or river, where the base stations can not be placed, but communication inside that region is needed for the fishermen. A simplified form of this problem is the *vertex-cover(k)* problem, where the objective is to establish communication among only the vertices of P instead of covering the entire polygon. This problem is also useful in some specified applications, say guards are placed on the vertices of the polygonal region P, and the communication among them is provided using these base stations.

It provides efficient algorithms for both *vertex-cover(2)* and *region-cover(2)* problems, where the base stations are to be installed on a pair of specified edges. The time complexities of these algorithms are $O(n \log n)$ and $O(n^2)$ respectively. Next, it considers the case where $k \geqslant 3$. First the restricted version of the *vertex-cover(k)* and *region-cover(k)* problems are considered, where all the k base stations are to be placed on the same edge of P. The proposed algorithm for the restricted *vertex-cover(k)* problem produces the optimum result in $O(min(n^2, nk \log n))$ time, whereas the algorithm for the restricted *region-cover(k)* problem produces an $(1+\varepsilon)$-factor approximation result in $O((n+k) \log(n+k) + n \log(\lceil \frac{1}{\varepsilon} \rceil))$ time. Finally, an efficient heuristic algorithm for the general *region-cover(k)* problem is proposed for $k \geqslant 3$.

Sohn and Jo[107] considered a different variation of the problem. It assumes that two sets of points $B = \{b_1, b_2, \ldots, b_m\}$ and $R = \{r_1, r_2, \ldots, r_n\}$, called *blue* and *red* points, are given. The objective is to cover all the red points with circles of radius ρ (given a priori) centered at minimum number of blue points. Here the blue points indicate the possible positions of base stations, and red points indicate the target locations where the message need to be communicated. A heuristic algorithm using integer linear programming is presented along with experimental results. Azad and Chockalingam[108] studied a different variation where n base stations (of same range ρ) are placed on the boundary of a square region, and m sensors are uniformly distributed inside that region. The sensors are also allowed for limited movement. The entire time span is divided into slots. At the beginning of each time slot, depending on the positions of the sensors, k base stations need to be activated. The proposed algorithm finds a feasible solution (if exists) in time $O(mn + n \log n)$ time.

10.4 Conclusion

In this article, we considered the algorithmic issues related to the range assignment of radio stations in the context of a mobile radio network. Two different variations are studied - (i) the radio stations are pre-placed, and (ii) the radio stations are to be placed. In both cases, the number of radio stations are given a priori, and the objective is to minimize the total power requirement of the entire radio network. Needless to mention that the power requirement mainly depends on the ranges assigned to the radio stations. The exact function indicating the total power requirement also involves the environmental factors where the network is to be installed.

We first studied the case where a set of pre-placed radio stations is given. The problem is to assign ranges to these radio stations so that the network satisfies some desired connectivity requirement, for example, broadcast from a designated node, all-to-all communication, etc., and the total power consumption of the entire radio network is minimized. We have considered both the 1D and 2D variations of the problems. Little is known in higher dimension, particularly in 3D, and in the polyhedral terrain. Also, the study of the range assignment problem to maintain fault tolerant connectivity with respect to all the aforesaid three problems need extensive consideration in the application domain.

We have also considered the some useful variations of the range assignment problem where the radio stations are not pre-placed, and the objective is to compute the optimal locations of the radio stations in the network such that the desired communication can be established with minimum power. Special attention is given for the homogeneous case, where the range of every radio station is same. Also, for the sake of simplicity, regions under consideration are assumed to be convex. Recently, results on more realistic non-homogeneous model have become available.[106] But, to our knowledge, no result is available where the region under consideration is of arbitrary shape.

References

1. P. Santi, Topology Control in Wireless Ad Hoc and Sensor Networks, *John Wiley & Sons*, 2005.
2. M. Haenggi, Twelve reasons not to route over many short hops, in Proc. *60th IEEE Vehicular Technology Conference* (2004), 3130–3134.
3. S. Voß, The steiner tree problem with hop constraint, *Annals of Operations Research* **86** (1999) 321–345.

4. G. Calinescu, S. Kapoor, A. Olshevsky and A. Zelikovsky, Network lifetime and power assignment in ad hoc wireless networks, in Proc. *11th Annual European Symposium* (ESA), LNCS-2832 (2003), 114–126.

5. A. E. F. Clementi, P. Penna and R. Silvestri, On the power assignment problem in radio networks, *Mobile Networks and Applications* **9** (2004) 125–140.

6. L. M. Kirousis, E. Kranakis, D. Krizanc and A. Pelc, Power consumption in packet radio networks, *Theoretical Computer Science* **243** (2000) 289–305.

7. R. Mathar and J. Mattfeldt, Optimal transmission ranges for mobile communication in linear multihop packet radio networks, *Wireless Networks* **2** (1996) 329–342.

8. M. X. Cheng, M. Cardei, J. Sun, X. Cheng, L. Wang, Y. Xu and D. -Z. Du, Topology control of ad hoc wireless networks for energy efficiency, *IEEE Trans. on Computers*, **53** (2004), 1629–1635.

9. P. Santi, Topology control in wireless ad-hoc and sensor networks, *ACM Computing Surveys* **37** (2005) 164–194.

10. K. Pahlavan and A. Levesque, *Wireless Information Networks*, John Wiley, New York, 1995.

11. A. E. F. Clementi, P. Penna, A. Ferreira, S. Perennes and R. Silvestri, The minimum range assignment problem on linear radio networks, *Algorithmica* **35** (2003) 95–110.

12. K. Diks, E. Kranakis, D. Krizanc and A. Pelc, The impact of knowledge on broadcasting time in radio networks, in Proc. *7th Annual European Symposium on Algorithms* (ESA), **LNCS-1643** (1999), 41–52.

13. E. Kranakis, D. Krizanc and A. Pelc, Fault-tolerant broadcasting in radio networks, in Proc. *6th Annual European Symposium on Algorithms* (ESA), **LNCS-1461** (1998), 283–294.

14. A. E. F. Clementi, M. D. Ianni and R. Silvestry, The minimum broadcast range assignment problem on linear multi-hop wireless networks, *Theoretical Computer Science* **299** (2003) 751–761.

15. A. E. F. Clementi, P. Crescenzi, P. Penna, G. Rossi and P. Vocca, On the complexity of computing minimum energy consumption broadcast subgraphs, in Proc. *18th Annual Symposium on Theoretical Aspects of Computer Science* (STACS), **LNCS-2010** (2001), 121–131.

16. M. Cagalj, J. P. Hubaux and C. Enz, Minimum-energy broadcast in all-wireless networks : NP-completeness and distribution issues, in Proc. *8th Annual ACM SIGMOBILE International Conference on Mobile Computing and Networking* (2002), 172–182.

17. B. Fuchs, On the hardness of range assignment problem, *Tech. Rep. TR05-113, Electronic Colloquium on Computational Complexity (ECCC)*, 2005.

18. G. K. Das, S. Das and S. C. Nandy, Range assignment for energy efficient broadcasting in linear radio networks, *Theoretical Computer Science* **352** (2006) 332–341.

19. J. E. Wieselthier, G. D. Nguyen and A. Ephremides, On the construction of energy efficient broadcast and multicast trees in wireless networks, in Proc. *19th Annual Joint Conference of the IEEE Computer and Communication Societies* (2000), 585–594.

20. R. J. Marks, A. K. Das, M. El-Sharkawi, P. Arabshahi and A. Gray, Minimum power broadcast trees for wireless networks: Optimizing using the viability lemma, in Proc. *IEEE International Symposium on Circuits and Systems* (2002), 273–276.

21. A. K. Das, R. J. Marks, M. E. -Sharkawi, P. Arabshahi and A. Gray, Minimum power broadcast trees for wireless networks: integer programming formulations, in Proc. *22nd Annual Joint Conference of the IEEE Computer and Communications Societies* (2003), 1001–1010.

22. J. E. Wieselthier, G. D. Nguyen and A. Ephremides, The energy efficiency of distributed algorithms for broadcasting in ad hoc networks, in Proc. *5th International Symposium on Wireless Personal Multimedia Communications* (2002), 499–503.

23. F. Ingelrest and D. Simplot-Ryl, Localized broadcast incremental power protocol for wireless ad hoc networks, in Proc. *10th IEEE Symposium on Computers and Communications* (2005), 28–33.

24. E. D. Kaplan, *Understanding GPS: Principles and Applications*, Second Edition, Artech House, Norwood, MA 02062, USA.

25. J. Cartigny, D. Simplot and I. Stojmenovic, Localized minimum energy broadcasting in ad hoc networks, in Proc. *22nd Annual Joint Conference of the IEEE Computer and Communications Societies* (2003), 2210–2217.

26. G. T. Toussaint, The relative neighborhood graph of a finite planar set, *Pattern Recognition* **12** (1980) 261–268.

27. J. Cartigny, D. Simplot and I. Stojmenovic, Localized energy efficient broadcast for wireless networks with directional antennas, in Proc. *1st Annual Mediterranean Workshop on Ad Hoc Networks* (2002).

28. X. -Y. Li, Algorithmic, geometric and graph issues in wireless networks, *Wireless Communications and Mobile Computing*, **3** (2003) 119–140.

29. I. Stojmenovic and J. Wu, Broadcasting and activity scheduling in ad hoc networks, *Mobile Ad Hoc Networking*, (S. Basagni, M. Conti, S. Giordano and I. Stojmenovic, eds.), IEEE/Wiley, (2004) 205–229.

30. P. -J. Wan, G. Calinescu, X. -Y. Li and O. Frieder, Minimum-energy broadcasting in static ad hoc wireless networks, *ACM/Kluwer Wireless Networks* **8** (2002) 607–617.

31. R. Klasing, A. Navarra, A. Papadopoulos and S. Perennes, Adaptive broadcast consumption(ABC), a new heuristic and new bounds for the minimum energy broadcast routing problem, in Proc. *3rd International IFIP-TC6 Networking Conference*, **LNCS-3042** (2004), 866–877.

32. A. Navarra, Tight bounds for the minimum energy broadcasting problem, in Proc. *3rd International Symposium on Modeling and Optimization in Mobile Ad Hoc and Wireless Networks* (2005), 313–322.

33. C. Ambuhl, An optimal bound for the MST algorithm to compute energy efficient broadcast tree in wireless networks, in Proc. *32th International Colloquium on Automata, Languages and Programming* (ICALP), **LNCS-3580** (2005), 1139–1150.

34. T. Calamoneri, A. E. F. Clementi, M. D. Ianni, M. Lauria, A. Monti and R. Silvestri, Minimum energy broadcast and disk cover in grid wireless networks, in Proc. *13th International Colloquium, Structural Information and Communication Complexity* (SIROCCO), **LNCS-4056** (2006), 227–239.

35. G. Calinescu, S. Kapoor and M. Sarwat, Bounded-hops power assignment in ad hoc wireless networks, *Discrete Applied Mathematics*, **154** (2006), 1358–1371.

36. I. Caragiannis, C. Kaklamanis and P. Kanellopoulos, A logarithmic approximation algorithm for the minimum energy consumption broadcast subgraph problem, *Information Prosessing Letters* **86** (2003) 149–154.

37. I. Papadimitriou and L. Geordiadis, Minimum energy broadcasting in multi-hop wireless networks using a single broadcast tree, *Mobile Networks and Applications* **11** (2006) 361–375.

38. J. Chlebikova, D. Ye and H. Zhang, Assign ranges in general ad hoc networks, *Journal of Parallel and Distributed Computing* **66** (2006) 489–498.

39. E. Kantor and D. Peleg, Approximate hierarchical facility location and applications to the shallow steiner tree and range assignment problems, in Proc. *6th Italian Conference on Algorithms and Complexity* (CIAC), **LNCS-3998** (2006), 211–222.

40. M. R. Garey and D. S. Johnson, *Computers and Intractibility: A Guide to the Theory of NP-Completeness*, W. H. freeman and Company, NY, 1979.

41. Y. Rabinovich and R. Raz, Lower bounds on the distortion of embedding finite metric spaces in graphs, *Discrete Computational Geometry* **19** (1998) 79–94.

42. G. Rossi, *The range assignment problem in static ad hoc networks*, Ph. D. Thesis, University of Siena, 2003.

43. D. Ye and H. Zhang, The range assignment problem in static ad hoc networks on metric spaces, in Proc. *11th International Colloquium on Structural Information and Communication Complexity* (SIROCCO), **LNCS 3104** (2004), 291–302.

44. C. Ambuhl, A. E. F. Clementi, M. D. Ianni, N. Lev-Tov, A. Monti, D. Peleg, G. Rossi and R. Silvestri, Efficient algorithms for low-energy bounded-hop broadcast in adhoc wireless networks, in Proc. *21st Annual Symposium on Theoretical Aspects of Computer Science* (STACS), **LNCS-2996** (2004), 418–427.

45. H. Bronnimann and M. T. Goodrich, Almost optimal set covers in finite VC-dimension, *Discrete & Computational Geometry* **14** (1995) 463–479.

46. L. Zosin and S. Khuller, On directed Steiner trees, in Proc. *13th Annual ACM-SIAM symposium on Discrete algorithms* (2002), 59–63.

47. G. Calinescu, I. I. Mandoiu, P. -J. Wan and A. Z. Zelikovsky, Selecting forwarding neighbors in wireless ad hoc networks, *Mobile Networks and Applications* **9** (2004) 101–111.

48. B. Chazelle and L. J. Guibas, Fractional cascading: I. a data structuring technique, *Algorithmica* **1** (1986) 133–162.

49. G. K. Das, *Placement and Range Assignment in Power Aware Radio Network*, Ph.D. Thesis, Indian Statistical Institute, 2007.

50. C. Ambuhl, A.E.F. Clementi, M. D. Ianni, G. Rossi, A. Monti, and R. Silvestri, The range assignment problem in non-homogeneous static ad-hoc networks, in Proc. *18th International Parallel and Distributed Processing Symposium* 2004.

51. A. E. F. Clementi, G. Huiban, P. Penna, G. Rossi and Y. C. Verhoeven, Some recent theoretical advances and open questions on energy consumption in ad-hoc wireless networks, in Proc. *3rd Workshop on Approximation and Randomization Algorithms in Communication Networks* (2002), 23–38.

52. J. Park and S. Sahni, Maximum lifetime broadcasting in wireless networks, *IEEE Transaction. on Computers* **54** (2005) 1081–1090.

53. D. T. Lee, Farthest neighbor Voronoi diagrams and applications, *Report 80-11-FC-04*, Dept. Elect. Engrg. Comput. Sci., Northwestern Univ., Evanston, USA, 1980.

54. A. E. F. Clementi, P. Penna and R. Silvestri, The power range assignment problem in radio networks on the plane, in Proc. *17th Annual Symposium on Theoretical Aspects of Computer Science* (STACS), **LNCS-1770** (2000), 651–660.

55. C. Ambuhl, A. E. F. Cleminti, P. Penna, G. Rossi and R. Silvestri, Energy consumption in radio networks: selfish agents and rewarding mechanisms, in Proc. *10th International Colloquim on Structural Information Complexity* (2003), 1–16.

56. G. K. Das, S. C. Ghosh and S. C. Nandy, An efficient heuristic algorithm for 2D *h*-hop range assignment problem, *IEEE Global Telecommunication Conference*, **2** (2004) 1051–1055.

57. G. K. Das, S. C. Ghosh and S. C. Nandy, Improved algorithm for minimum cost range assignment problem for linear radio networks, *International Journal of Foundations of Computer Science* **18** (2007) 619–635.

58. P. Carmi and M. J. Katz, Power assignment in radio networks with two power levels, *Algorithmica* **47** (2007) 183–201.

59. E. Althaus, G. Calinescu, I. I. Mandoiu, S. Prasad, N. Tchervenski, and A. Zelikovsky, Power efficient range assignment for symmetric connectivity in static ad-hoc wireless networks, *Wireless Networks* **12** (2006) 287–299.

60. S. O. Krumke, R. Liu, E. L. Lloyd, M. V. Marathe, R. Ramanathan and S. S. Ravi, Topology control problems under symmetric and asymmetric power thresholds, in Proc. *2nd International Conference on Ad-Hoc, Mobile, and Wireless Networks* (ADHOC-NOW), **LNCS-2865** (2003), 187–198.

61. L. Kucera, Low degree connectivity in ad hoc networks, in Proc. *13th Annual European Symposium on Algorithms* (ESA), **LNCS-3669** (2005), 203–214.

62. S. A. Plotkin, D. B. Shmoys and E. Tardos, Fast approximation algorithms for fractional packing and covering problems, *Mathematics of Operations Research* **20** (1995) 257–301.

63. L. Gouveia, Multicommodity flow models for spanning tree with hop constraints, *European Journal of Operational Research* **95** (1995) 178–190.

64. L. Gouveia and C. Requejo, A new Lagrangian relaxation approach for the hop-constrained minimum spanning tree problem, *European Journal of Operational Research* **132** (2001) 539–552.

65. E. Althaus, S. Funke, S. Har-Peled, J. Konemann, E. A. Ramos and M. Skutella, Approximating *k*-hop minimum spanning trees, *Operations Research Letters* **33** (2005) 115–120.

66. S. Arora, P. Raghavan and S. Rao, Approximation schemes for Euclidean *k*-medians and related problems, in Proc. *30th ACM Symposium on Theory of Computing* (1998), 106–113.

67. A. Abdalla, N. Deo and P. Gupta, Heuristics to compute a diameter-constrained MST, *Congressus Numerantium* **144** (2000) 161–182.

68. G. R. Raidl and B. A. Julstrom, Greedy heuristics and an evolutionary algorithm for the bounded-diameter minimum spanning tree problem, in Proc. *18th ACM Symposium on Applied Computing* (2003), 747–752.

69. A. E. F. Clementi, M. D. Ianni, A. Monti, G. Rossi and R. Silvestri, Experimental analysis of practically efficient algorithms for bounded-hop accumulation in ad-hoc wireless networks, in Proc. *19th International Parallel and Distributed Processing Symposium* 2005.

70. H. N. Gabow, Z. Galil, T. H. Spencer and R. E. Tarjan, Efficient algorithms for finding minimum spanning trees in undirected and directed graphs, *Combinatorica* **6** (1986) 109–122.

71. J. Choong and D. Everitt, Outage analysis and spectral efficiency of indoor mobile radio systems, in Proc. *45th IEEE Vehicular Technology Conference* (1995), 434–438.

72. S. J. Fortune, D. M. Gay, B. W. Kernighan, O. Landron, R. A. Valenzuela and M. H. Wright, WISE design of indoor wireless systems: practical computation and optimization, *IEEE Computational Science Engineering* **2** (1995) 58–68.

73. D. Stamatelos and A. Ephremides, Spectral efficiency and optimal base placement for indoor wireless networks, *IEEE Journal on Selected Areas in Communications* **14** (1996) 651–661.

74. H. D. Sherali, C. H. Pendyala and T. S. Rappaport, Optimal location of transmitters for micro-cellular radio communication system design, *IEEE Journal on Selected Areas in Communications* **14** (1996) 662–673.

75. M. H. Wright, Optimization methods for base station placement in wireless applications, in Proc. *48th IEEE Vehicular Technology Conference* (1998), 387–391.

76. S. -Y. Yang and A. Ephremides, Optimal network design : the base station placement problem, in Proc *36th Conference on Decision and Control* (1997), 2381–2386.

77. I. Howitt and S.-Y. Ham, Base station location optimization, in Proc. *50th IEEE Vehicular Technology Conference* (1999), 2067-2071.

78. S. Verblunsky, On the least number of unit circles which can cover a square, *Journal of London Mathematical Society* **24** (1949) 164–170.

79. T. Tarnai and Z. Gasper, Covering a square by equal circles, *Elementary Mathematics* **50** (1995) 167–170.

80. J. B. M. Melissen and P. C. Schuur, Improved coverings of a square with six and eight circles, *Electronic Journal on Combinatorics* **3** (1996) R–32.

81. J. B. M. Melissen and P. C. Schuur, Covering a rectangle with six and seven circles, *Discrete Applied Mathematics* **99** (2000) 149–156.

82. K. J. Nurmela and P. R. J. Ostergard, Covering a square with up to 30 equal circles, *Research Report HUT-TCS-A62*, Laboratory for Theoretical Computer Science, Helsinky University of Technology, 2000.

83. K. J. Nurmela, Conjecturally optimal coverings of an equilateral triangle with up to 36 equal circles, *Experimental Mathematics* **9** (2000) 241–250.

84. G. K. Das, S. Das, S. C. Nandy and B. P. Sinha, Efficient algorithm for placing a given number of base stations to cover a convex region, *Journal on Parallel Distributed Computing* **66** (2006) 1353–1358.

85. J. J. Sylvester, A question in the geometry of situation, *Quarterly Journal of Mathematics*, **1** (1857), p. 79.

86. J. Elzinga and D. W. Hearn, Geometrical solutions to some minimax location problems, *Transputer Science* **6** (1972) 379–394.

87. N. Megiddo, Linear-time algorithms for linear programming in R^3 and related problems, *SIAM Journal on Computing* **12** (1983) 759–776.

88. J. W. Jaromczyk and M. Kowaluk, An efficient algorithm for the Euclidean two-center problem, in Proc. *10th Annual Symposium on Computational Geometry* (1994), 303-311.

89. M. Sharir, A near-linear algorithm for the planar 2-center problem, *Discrete Computational Geometry* **18** (1997) 125–134.

90. T. M. Chan, More planar two-center algorithms, *Computational Geometry: Theory*

and Application **13** (1999) 189–198.

91. S. K. Kim and C. -S. Shin, Efficient algorithms for two-center problems for a convex polygon, in Proc. *6th Annual International Conference on Computing and Combinatorics* (COCOON), **LNCS-1858** (2000), 299–309.

92. R. J. Fowler, M. S. Paterson and S. L. Tanimoto, Optimal packing and covering in the plane are NP-complete, *Information Processing Letters* **12** (1981) 133–137.

93. A. Marchetti-Spaccamela, The p-center problem in the plane is NP-complete, in Proc. *19th Allerton Conference on Communication, Control and Computing* (1981), 31–40.

94. R. Z. Hwang, R. C. Chang and R. C. T. Lee, The searching over separators strategy to solve some NP-hard problems in subexponential time, *Algorithmica* **9** (1993) 398–423.

95. C. -S. Shin, J. -H. Kim, S. K. Kim and K. -Y. Chwa, Two-center problems for a convex polygon, in Proc. *6th Annual European Symposium on Algorithms* (ESA), **LNCS-1461** (1998), 199–210.

96. L. Booth, J. Bruck, M. Franceschetti and R. Meester, Covering algorithms, continuum percolation and the geometry of wireless networks, *The Annals of Applied Probability* **13** (2003) 722–741.

97. I. F. Akyildiz, W. Su, Y. Sankarasubramaniam and E. Cayirci, A survey on sensor networks, *IEEE Communication Magazine* **40** (2002) 102–114.

98. A. Boukerche, X. Fei and R. B. Araujo, An energy aware coverage-preserving scheme for wireless sensor networks, in Proc. *2nd ACM International Workshop on Performance Evaluation of Wireless Ad Hoc, Sensor, and Ubiquitous Networks* (2005), 205–213.

99. M. de Berg, M. van Kreveld, M. Overmars and O. Schwarzkopf, *Computational Geometry Algorithms and Applications*, Springer-Verlag, 1997.

100. S. Meguerdichian, F. Koushanfar, M. Potkonjak and M. B. Srivastava, Coverage problems in wireless ad-hoc sensor networks, in Proc. *20th Annual Joint Conference of the IEEE Computer and Communication Societies* (2001), 1380–1387.

101. P. Bose and G. T. Toussaint, Computing the constrained Euclidean geodesic and link center of a simple polygon with applications, in Proc. *14th Conference on Computer Graphics International* (1996), 102–110.

102. S. Roy, D. Bardhan and S. Das, Efficient algorithm for placing base stations by avoiding forbidden zone, *Journal of Parallel and Distributed Computing* **68** (2008) 265–273.

103. F. Hurtado, V. Sacristan and G. Toussaint, Some constrained minimax and maximin location problems, *Studies in Locational Analysis* **15** (2000) 17–35.

104. S. Roy, A. Karmakar, S. Das and S. C. Nandy, Constrained minimum enclosing circle with center on a query line segment, in Proc. *31st International Symposium on Mathematical Foundations of Computer Science* (MFCS) **LNCS-4162** (2006) 765–776.

105. A. Karmakar, S. Roy and S. Das, Fast Computation of Smallest Enclosing Circle with Center on a Query Line Segment, in Proc. *19th Annual Canadian Conference on Computational Geometry* (2007), 273-276.

106. H. Alt, E. M. Arkin, H. Bronnimann, J. Erickson, S. P. Fekete, C. Knauer, J. Lenchner, J. S. B. Mitchell, and K. Whittlesey, Minimum-cost coverage of point sets by disks, in Proc. *22nd ACM Symposium on Computational Geometry* (2006), 449–458.

107. S. Sohn and G. Jo, Optimization of base stations positioning in mobile networks, in Proc. *2006 International Conference on Computational Science and its Applications* (ICCSA) **LNCS-3981** (2006) 779–787.

108. A. P. Azad and A. Chockalingam, Mobile base stations placement and energy aware routing in wireless sensor networks, in Proc. *IEEE Wireless Communication and Networking Conference* (2006), 264–269.

Chapter 11

Privacy in the Electronic Society: Emerging Problems and Solutions*

Claudio A. Ardagna, Marco Cremonini, Ernesto Damiani,
Sabrina De Capitani di Vimercati, Pierangela Samarati

*Dipartimento di Tecnologie dell'Informazione
Università degli Studi di Milano
26013 Crema - Italy*
{*ardagna, cremonini, damiani, decapita, samarati*}*@dti.unimi.it*

As the global information infrastructure is becoming more ubiquitous, digital business transactions are increasingly performed using a variety of mobile devices and across multiple communication channels. This new service-oriented paradigm is making the protection of privacy an increasing concern, as it relies on rich context representations (e.g., of location and purpose) and requires users to provide a vast amount of information about themselves and their behavior. This information is likely to be protected by a privacy policy, but restrictions to be enforced may come from different input requirements, possibly under the control of different authorities. In addition, users retain little control over their personal information once it has been disclosed to third parties. Secondary use regulations are therefore increasingly demanding attention. In this paper, we present the emerging trends in the data protection field to address the new needs and desiderata of today's systems.

Contents

---

*A preliminary version of this paper appeared under the title "Privacy in the Electronic Society," in *Proc. of the International Conference on Information Systems Security (ICISS 2006)*, Kolkata, India, December 19-21, 2006.[1]

11.1 Introduction

Today's digital business processes increasingly rely on services accessed via a variety of mobile devices and across multiple communication channels.[2] Also, terminal devices are now equipped with sensors capable of collecting information from the environment, such as geographical positioning systems (GPS), providing a rich context representation regarding both users and the resources they access. This representation includes potentially sensitive personal information, such as the users' purpose, geographical location, and past preferences. While collecting and exploiting rich context data is indeed essential for customizing network-based processes and services, it is well known that context records can be misused well beyond the original intention of their owners. Indeed, personal information is often disclosed to third parties without the consent of legitimate data owners; also, professional services exist specializing on gathering and correlating data from heterogeneous repositories, which permit to build user profiles disclosing sensitive information not voluntarily released by their owners.

In the past few years, increasing awareness of the privacy risks of unauthorized user profiling has led to stricter regulations on personal data storage and sharing. It is now widely acknowledged that business processes requiring large-scale information sharing will become widespread only if their users have some convincing assurance that, while they release the information needed to access a service, disclosure of really sensitive data is not a risk. Unfortunately, some of the emerging technological and organizational requirements for preserving users' privacy are still not completely understood; as a consequence, personal data are often poorly managed and sometimes abused. Protecting privacy requires the investigation of different aspects, including:

- *data protection requirements composition* to take into consideration requirements coming from the data owner, the data holder, and possible privacy law. These multiple authorities scenario should be supported from the administration point of view providing solutions for modular, large-scale, scalable policy composition and interaction;[3,4]

- *security and privacy specifications and secondary usage control* to identify under which conditions a party can trust others for their security and privacy. Trust models are one of the techniques be evaluated.[5,6] In particular, *digital certificates* (statements certified by given entities) can be used to establish properties of their holder (such as identity, accreditation, or authorizations).[7-11] Moreover, since users often have no idea on how their personal information may be used subsequently, it must also be given a mechanism to specify whether or not to consent to the future use of that information in secondary applications;[12]

- *inference and linking attacks protection* that is often impossible, if not at the price of not disclosing any information at all. Among the techniques used to protect the released data, *k*-anonymity promises to be a successful solution towards increasing privacy;

- *context information (including location) protection* to avoid unauthorized leaks that may cause loss of privacy, for example, on the user's whereabouts.

These issues pose several new challenges to the design and implementation of privacy-aware systems. As far as mobile devices systems are concerned, a major concern is on-board memory and storage limitations. Lightweight terminals require usage logs to be held by the infrastructure, making inference and linking attacks more likely. On the other hand, usage logs need to contain enough information to enable analysis for detection of violations to the privacy policies in place. Another challenge relates to the fact that clients and servers alike will not be under the control of trustworthy authorities, so they cannot be assumed to be trusted. Each device and operating system must provide measures to protect the integrity and confidentiality of sensitive personal data and of the privacy control policies. Finally, the lack of resources available on portable devices such as cell phones and laptops may pose some constraints on the effectiveness of purely cryptographic approaches to privacy solutions. Therefore, adversaries trying to access personal data, could have much more computational resources at their disposal than legitimate clients. In this paper, we discuss these problems and illustrate some current approaches and ongoing research. The remainder of the paper is organized as follows. Section 11.2 addresses the problem of combining authorization specifications that may be independently stated. We describe the characteristics that a policy composition framework should have and illustrate some current approaches and open issues. Section 11.3 addresses the problem of defining policies in open environments such as the Internet. We then describe current approaches and open issues. Section 11.4 addresses the problem of protecting released data against inference and linking attacks. We describe the *k*-anonymity concept and illustrate some related current approaches and open issues. Section 11.5 discusses the problem of protecting privacy of location information in pervasive environments. We describe the *location privacy* concept and illustrate some current approaches and open issues. Finally, Section 11.6 concludes the paper.

11.2 Policy Composition

Traditionally, authorization policies have been expressed and managed in a centralized manner: one party administers and enforces the access control requirements. In many cases however, access control needs to combine restrictions independently stated that should be enforced as one, while retaining their independence and administrative autonomy. For instance, the global policy of a large organization can be the combination of the policies of its independent and geographically distributed departments. Each of these departments is responsible for defining access control rules to protect resources and each brings its own set of constraints. To address these issues, a *policy composition framework* by which different component policies can be integrated while retaining their independence should be designed. The framework should be flexible to support different kinds of composition, yet remain simple so to keep control over complex compound policies. It should be based on a solid formal framework and a clear semantics to avoid ambiguities and enable correctness proofs.

Some of the main requirements that a policy composition framework should have can be summarized as follows.[3]

- *Heterogeneous policy support.* The composition framework should be able to combine policies expressed in arbitrary languages and possibly enforced by different mechanisms. For instance, a datawarehouse may collect data from different data sources where the security restrictions autonomously stated by the sources and associated with the data are stated with different specification languages, or refer to different paradigms (e.g., open vs closed policy).
- *Support of unknown policies.* It should be possible to account for policies that may be not completely known or even be specified and enforced in external systems. These policies are like "black-boxes" for which no (complete) specification is provided, but that can be queried at access control time. Think, for instance, of a situation where given accesses are subject, in addition to other policies, to a policy P enforcing "central administration approval". Neither the description of P, nor the specific accesses that it allows might be available; whereas P can respond yes or no to each specific request. Run-time evaluation is therefore the only possible option for P. In the context of a more complex and complete policy including P as a component, the specification

could be partially compiled, leaving only P (and its possible consequences) to be evaluated at run time.

- *Controlled interference.* Policies cannot always be combined by simply merging their specifications (even if they are formulated in the same language), as this could have undesired side effects. The accesses granted/denied might not correctly reflect the specifications anymore. As a simple example, consider the combination of two systems P_{closed}, which applies a closed policy, based on rules of the form *"grant access if $(s, o, +a)$"*, and P_{open} which applies an open policy, based on rules of the form *"grant access if $\neg(s, o, -a)$"*. Merging the two specifications would cause the latter decision rule to derive all authorizations not blocked by P_{open}, regardless of the contents of P_{closed}. Similar problems may arise from uncontrolled interaction of the derivation rules of the two specifications. Besides, if the adopted language is a logic language with negation, the merged program might not be stratified (which may lead to ambiguous or undefined semantics).

- *Expressiveness.* The language should be able to conveniently express a wide range of combinations (spanning from minimum privileges to maximum privileges, encompassing priority levels, overriding, confinement, refinement etc.) in a uniform language. The different kinds of combinations must be expressed without changing the input specifications (as it would be necessary even in most recent and flexible approaches) and without ad-hoc extensions to authorizations (like those introduced to support priorities). For instance, consider a policy P_1 regulating access to given documents and the central administration policy P_2. Assume that access to administrative documents can be granted only if authorized by both P_1 and P_2. This requisite can be expressed in existing approaches only by explicitly extending all the rules possibly referred to administrative documents to include the additional conditions specified by P_2. Among the drawbacks of this approach is the rule explosion that it would cause and the complex structure and loss of controls of two specifications; which, in particular, cannot be maintained and managed autonomously anymore.

- *Support of different abstraction levels.* The composition language should highlight the different components and their interplay at different levels of abstraction. This is important to: *i)* facilitate specification analysis and design; *ii)* facilitate cooperative administration and agreement on global policies; *iii)* support incremental specification by refinement.

- *Support for dynamic expressions and controlled modifications.* Mobile policies that follow (*stick with*) the data and can be enriched, subject to constraints, as the data move.

- *Formal semantics.* The composition language should be declarative, implementation independent, and based on a solid formal framework. The need
  of an underlying formal framework is widely recognized and in particular it
  is important to *i)* ensure non-ambiguous behavior, and *ii)* reason about and
  prove specifications properties and correctness.[13] In our framework this is
  particularly important in the presence of *incomplete* specifications.

11.2.1 *Overview of Ongoing Work*

Various models have been proposed to reason about security policies.[14–17] In[14,16]
the authors focused on the secure behavior of program modules. McLean[17] proposed a formal approach including combination operators: he introduced an algebra of security which enables to reason about the problem of policy conflict that
can arise when different policies are combined. However, even though this approach permits to detect conflicts between policies, it did not propose a method
to resolve the conflicts and to construct a security policy from inconsistent sub-
policies. Hosmer[15] introduced the notion of meta-policies (i.e., policies about
policies), an informal framework for combining security policies. Subsequently,
Bell[18] formalized the combination of two policies with a function, called *policy
combiner*, and introduced the notion of *policy attenuation* to allow the composition of conflicting security policies. Other approaches are targeted to the development of a uniform framework to express possibly heterogeneous policies.[19–21]
Recently, Bonatti et al.[3] proposed an algebra for combining security policies together with its formal semantics. Following Bonatti et al.'s work, Jajodia et al.[4]
presented a propositional algebra for policies with a syntax consisting of abstract
symbols for atomic policy expressions and composition operators. The basic idea
of these proposals is to define a set of policy operators used for combining different policies. In particular, in[3] a policy is defined as a set of triples of the form
(s, o, a), where s is a constant in (or a variable over) the set of subjects S, o is a
constant in (or a variable over) the set of objects O, and a is a constant in (or a
variable over) the set of actions A. Here, complex policies can then be obtained by
combining policy identifiers, denoted P_i, through the following *algebra operators*.

- *Addition* ($+$) merges two policies by returning their set union. For instance,
  in an organization composed of different divisions, access to the main gate
  can be authorized by any of the administrator of the divisions (each of them
  knows users who needs the access to get to their division). The totality of the
  accesses through the main gate to be authorized would then be the union of
  the statements of each single division. Intuitively, additions can be applied
  in any situation where accesses can be authorized if allowed by any of the

component (operand) policies.

- *Conjunction* (&) merges two policies by returning their intersection. For instance, consider an organization in which divisions share certain documents (e.g., clinical folders of patients). Access to the documents is to be allowed only if all the authorities that have a say on the document agree on it. Intuitively, while addition enforces maximum privilege, conjunction enforces minimum privilege.

- *Subtraction* (−) restricts a policy by eliminating all the accesses in the second policy. Intuitively, subtraction specifies exceptions to statements made by a policy and it encompasses the functionality of negative authorizations in existing approaches, while probably providing a clearer view of the combination of positive and negative statements. The advantages of subtraction over explicit denials include a simplification of the conflict resolution policies and a clearer semantics. In particular, the scoping of a difference operation allows to clearly and unambiguously express the two different uses of negative authorizations, namely *exceptions to positive statements* and *explicit prohibitions*, which are often confused in the models or require explicit ad-hoc extension to the authorization form. The use of subtraction provides extensible as the policy can be enriched to include different overriding/conflict resolution criteria as needed in each specific context, without affecting the form of the authorizations.

- *Closure* (∗) closes a policy under a set of inference (derivation) rules. Intuitively, derivation rules can be thought of as logic rules whose head is the authorization to be derived and whose body is the condition under which the authorization can be derived. Examples of derivation rules can be found in essentially all logic based authorization languages proposed in the literature, where derivation rules are used, for example, to enforce propagation of authorizations along hierarchies in the data system, or to enforce more general forms of implication, related to the presence or absence of other authorizations, or depending on properties of the authorizations.[19]

- *Scoping restriction* (ˆ) restricts the application of a policy to a given set of subjects, objects, and actions. Scoping is particularly useful to "limit" the statements that can be established by a policy and, in some way, enforcing authority confinement. Intuitively, all authorizations in the policy which do not satisfy the scoping restriction are ignored, and therefore ineffective. For instance, the global policy of an organization can identify several component policies which need to be merged together; each component policy may be restricted in terms of properties of the subjects, objects and actions occurring in its authorizations.[a]

- *Overriding* (*o*) replaces part of a policy with a corresponding fragment of the second policy. The portion to be replaced is specified by means of a third policy. For instance, consider the case where users of a library who have passed the due date for returning a book cannot borrow the same book anymore *unless* the responsible librarian vouchers for (authorizes) the loan. While the accesses otherwise granted by the library are stated as a policy P_{lib}, black-list of accesses, meaning triples (user, book, loan) are stated as a policy P_{block}. In the absence of the *unless* portion of the policy, the accesses to be allowed would simply be $P_{lib} - P_{block}$. By allowing the librarian discretion for "overriding" the black list, calling P_{vouch} the triples authorized by the librarians, we can express the overall policy as $o(P_{lib}, P_{vouch}, P_{block})$.

- *Template* (τ) defines a partially specified policy that can be completed by supplying the parameters. Templates are useful for representing partially specified policies, where some component X are to be specified at a later stage. For instance, X might be the result of further policy refinement, or it might be specified by a different authority.

To fix ideas and make concrete examples, consider a drug-effects warehouse that might draw information from many hospitals. We assume that the warehouse receives information from three hospitals, denoted h_1, h_2, and h_3, respectively. These hospitals are responsible for granting access to information under their (possibly overlapping) authority domains, where domains are specified by a scoping function. The statements made by the hospitals are then unioned meaning that an access is authorized if any of the hospital policy states so. In term of the algebra, the warehouse policy can be represented as an expression of the form $P_1\char94[o \leqslant O_{h_1}] + P_2\char94[o \leqslant O_{h_2}] + P_3\char94[o \leqslant O_{h_3}]$, where P_i denotes the policy defined by hospital h_i, and the scope restriction $\char94[o \leqslant O_{h_i}]$ selects the authorizations referred to objects released by hospital h_i.[b] Each policy P_i can then be further refined. For instance, consider policy P_1. Suppose that hospital h_1 defines a policy P_{drug} regulating the access to drug-effects information. Assume also that the drug-effects information can be released only if the hospital's researchers obtain a patient's consent; $P_{consents}$ reports accesses to drug-effects information that the patients agree to release. We can then express P_1 as $P_{drug}\&P_{consents}$.

---

[a] A simple example of scoping constraint is the limitation of authorizations that can be stated by a policy to a specific portion of the data system hierarchy.[19]

[b] We assume that the information collected from the hospitals can be organized in abstractions defining groups of objects that can be collectively referred to with a given name. Objects and groups thereof define a partial order that naturally introduces a hierarchy, where O_{h_i} contains objects obtained from hospital h_i.

11.2.2 *Open Issues*

We briefly describe some open issues that need to be taken into consideration in the future development of a policy composition framework.

- Investigate different *algebra operators and formal languages* for enforcing the algebra and proving properties. The proposed policy composition frameworks can be enriched by adding new operators. Also, the influence of different rule languages on the expressiveness of the algebra has to be investigated.
- *Administrative policies and language* with support for multiple authorities. The proposed approaches could be enriched by adding administrative policies that define who can specify authorizations/rules (i.e., who can define a component policy) governing access control.
- *Policy enforcement.* The resolution of the algebraic expression defining a policy P determines a set of ground authorization terms, which define exactly the accesses to be granted according to P. Different strategies can be used to evaluate the algebraic expression for enforcing access control: materialization, run-time evaluation, and partial evaluation. The first one allows a one-time compilation of the policy against which all accesses can be efficiently evaluated and which will then need to be updated only if the policy changes. The second strategy consists in enforcing a run-time evaluation of each request (access triple) against the policy expression to determine whether the access should be allowed. Between these two extremes, possibly combining the advantages of them, there are partial evaluation approaches, which can enforce different degrees of computation/materialization.
- Incremental approaches to enforce *changes to component policies*. When a materialization approach is used to evaluate the algebraic expression for enforcing access control, incremental approaches[22] can be applied to minimize the recomputation of the policy.
- *Mobile policies.* Intuitively, a *mobile policy* is the policy associated with an object and that follows the object when it is passed to another site. Because different and possibly independent authorities can define different parts of the mobile policy in different time instants, the policy can be expressed as a policy expression. In such a context, there is the problem on how ensure the obedience of policies when the associated objects move around. Within the context of mobile policies we can also classify the problem of providing support for handling "sticky" policies,[23] that is, policies that remain attached to data as they move between entities and are needed to enforce secondary use constraints (see Section 11.3). Mobile policies encompass also the problem of digital right management (DRM) as they also require constraints of the owner to remain attached to the data.

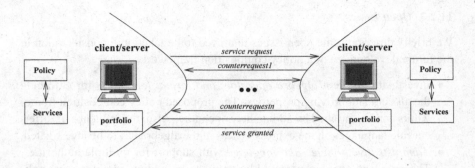

Fig. 11.1   Client/server interaction.

11.3 Access Control in Open Systems

Open environments are characterized by a number of systems offering different resources/services. In such a scenario, interoperability is a very important issue and traditional assumptions for establishing and enforcing policies do not hold anymore. A server may receive requests not just from the local community of users, but also from remote, previously unknown users. The server may not be able to authenticate these users or to specify authorizations for them (with respect to their identity). Early approaches that attempt to solve these issues, PolicyMaker[6] and KeyNote,[5] basically use credentials to describe specific delegation of trusts among keys and to bind public keys to authorizations. Although early trust management systems do provide an interesting framework for reasoning about trust between unknown parties, assigning authorizations to keys may result limiting and make authorization specifications difficult to manage.

A promising direction to overcome such a disadvantage is represented by *digital certificates*. A digital certificate is basically the on-line counterparts of paper credentials (e.g., drivers licenses). Digital certificates can be used to determine whether or not a party may execute an access on the basis properties that the requesting party may have. These properties can be proven by presenting one or more certificates.[8-11] The development and effective use of credential-based models require tackling several problems related to credential management and disclosure strategies, delegation and revocation of credentials, and establishment of credential chains.[24-30]

Figure 11.1 depicts the basic scenario we consider, where there are different parties that interact with each other to offer services. A party can act both as a server and a client and each party has *i)* a set of services it provides and *ii)* a *portfolio* of properties (attributes) that the party enjoys. Access restrictions to the services are expressed by policies that specified the properties that a requester should enjoy to gain access to the services. The services are meant to offer certain functionalities that depend on the input parameters supplied by its users. Often input parameters must fulfill certain conditions to assure correct behavior of a service. We identified the following requirements for specifying credential-based access control.

- *Attribute interchange.* A server should be able to communicate to the client the requirements it need to satisfy to get access. Also, a client should be able to prove its eligibility for a service. This communication interchange could be performed in different ways (e.g., the involved parties can apply different strategies with respect to which properties are submitted).
- *Support for fine-grained reference to attributes within a credential.* The system should allow the selective disclosure of credentials which is a requirement that is not usually supported because users attributes are defined according to functional needs, making it easier to collect all credentials in a row instead of iteratively asking for the ones strictly necessary for a given service only.
- *Support for hierarchical relationships and abstractions on services and portfolio.* Attribute-based access control policies should be able to specify accesses to collection of services based upon collection of attributes processed by the requester.
- *Expressiveness and flexibility.* The system must support the specification of complex access control requirements. For instance, consider a service that offers telephone contracts and requires that the customer is at least 18 years of age. The telephone selling service has two input parameters, namely `homeAddress` and `noticePeriod`. The `homeAddress` must be a valid address in Italy and `noticePeriod` must be either one or three months. Further, the service's access control policy requires that contracts with one month notice period and home address outside a particular geographical region are closed only with users who can prove their AAA membership. Hence, we see that the access control requirements of a service may require more than one interaction between a client and a server.
- *Purpose specific permission.* The permission to release data should relate to the purpose for which data are being used or distributed. The model should prevent information collected for one purpose from being used for other purposes.
- *Support for meta-policies.* The system should provide meta-policies for protecting the policy when communication requisites. This happens when a list of alternatives (policies) that must be fulfilled to gain the access to the

data/service is returned to the counterpart. For instance, suppose that the policy returned by the system is "citizenship=EU". The party can decide to return to the client either the policy as it is or a modified policy simply requesting the user to prove its nationality (then protecting the information that access is restricted to EU citizens).

- *Support for secondary use specifications and control.* The information owner should be able to control further dissemination and use of personal information. This represents a novel feature that is no simply concerned with authorizing the access to data and resources but also with defining and enforcing the way data and resources are subsequently managed.

11.3.1 *Overview of Ongoing Work*

The first proposals investigating the application of credential-based access control regulating access to a server were made by Winslett et al.[26,29] Here, access control rules are expressed in a logic language and rules applicable to an access can be communicated by the server to clients. In[30,31] the authors investigated trust negotiation issues and strategies that a party can apply to select credentials to submit to the opponent party in a negotiation. In[7] the authors proposed a uniform framework for regulating service access and information disclosure in an open, distributed network system like the Web. As in previous proposals, access regulations are specified as logical rules, where some predicates are explicitly identified. Certificates are modeled as *credential expressions* of the form "credential_name(attribute_list)", where credential_name is the credential name and attribute_list is a possibly empty list of elements of the form "attribute_name=value_term", where value_term is either a ground value or a variable. Besides credentials, the proposal also allows to reason about declarations (i.e., unsigned statements) and user-profiles that the server can maintain and exploit for taking the access decision. Communication of requisites to be satisfied by the requester is based on a filtering and renaming process applied on the server's policy, which exploits partial evaluation techniques in logic programs. Yu et al.[11,30,32] developed a service negotiation framework for requesters and providers to gradually expose their attributes. In[30] the *PRUdent NEgotiation Strategy* (PRUNES) has been presented. This strategy ensures that the client communicates its credentials to the server only if the access will be granted and the set of certificates communicated to the server is the minimal necessary for granting it. Each party defines a set of *credential policies* that regulates how and under what conditions the party releases its credentials. The negotiation consists of a series of requests for credentials and counter-requests on the basis of the parties' credential policies.

The credential policies established can be graphically represented through a tree, called *negotiation search tree*, composed of two kinds of nodes: *credential nodes*, representing the need for a specific credential, and *disjunctive nodes*, representing the logic operators connecting the conditions for credential release. The root of a tree node is a service (i.e., the resource the client wants to access). The negotiation can therefore be seen as a backtracking operation on the tree. The backtracking can be executed according to different strategies. For instance, a *brute-force* backtracking is complete and correct, but is too expensive to be used in a real scenario. The authors therefore proposed the PRUNES method that prunes the search tree without compromising completeness or correctness of the negotiation process. The basic idea is that if a credential C has just been evaluated and the state of the system is not changed too much, then it is useless to evaluate again the same credential, as the result will be exactly as the result previously computed. The same research group proposed also a method for allowing parties adopting different negotiation strategies to interoperate through the definition of a *Disclosure Tree Strategy* (DTS) family.[32] The authors show that if two parties use different strategies from the DST family, they are able to establish a negotiation process. The DTS family is a closed set, that is, if a negotiation strategy can interoperate with any DST strategy, it must also be a member of the DST family.

In[33] a *Unified Schema for Resource Protection* (UniPro) has been proposed. This mechanism is used to protect the information in policies. UniPro gives (opaque) names to policies and allows any named policy P_1 to have its own policy P_2 meaning that the contents of P_1 can only be disclosed to parties who have shown that they satisfy P_2. Another approach for implementing access control based on credentials is the *Adaptive Trust Negotiation and Access Control* (ATNAC).[34] This method grants or denies access to a resource on the basis of a *suspicion level* associated with subjects. The suspicion level is not fixed but may vary on the basis of the probability that the user has malicious intents. In[35] the authors proposed to apply the automated trust negotiation technology for enabling secure transactions between portable devices that have no pre-existing relationship. In[11] the authors presented a negotiation architecture, called *TrustBuilder*, that is independent from the language used for policy definition and from the strategies adopted by the two parties for policy enforcement. Other logic-based access control languages based on credentials have been introduced. For instance, D1LP and RT,[36,37] the SD3 language,[38] and Binder.[39] In[19,21] logic languages are adopted to specify access restrictions in a certificate-based access control model.

Few proposals have instead addressed the problem of how to regulate the use of personal information in secondary applications. In[40] the authors proposed an XML-based privacy preference expression language, called *PReference Expres-*

sion for Privacy (PREP), for storing the user's privacy preferences with Liberty Alliance. PREP allows users to specify, for each attribute, a *privacy label* that is characterized by a purpose, type of access, recipient, data retention, remedies, and disputes. The *Platform for Privacy Preferences Project* (P3P)[41] is another XML-based language that allows service providers and users to reach an agreement on the release of personal data. Basically, a service provider can define a P3P policy, which is an XML document, where it is possible to define the recipient of the data, desired data, consequence of data release, purpose of data collection, data retention policy, and dispute resolution mechanisms. Users specify their privacy preferences in term of a policy language, called APPEL,[42] and enforce privacy protection through a user agent: the user agent compares the user privacy preferences with the service provider P3P policy and checks whether the P3P policy conforms to the user privacy preferences. Although P3P is a good starting point, it is not widely adopted by the service providers and presents some limitations on the user side.[43] The main limitation is that the definition of simple privacy preferences is a complex task and writing APPEL preferences is error prone. For this reason, Agrawal et al.[43] proposed a new language, called XPref, for user preferences. However, both APPEL and XPref are not sufficiently expressive because, for example, they do not support negotiation and contextual information, and they do not allow the definition of attribute-based conditions. Another important disadvantage of these approaches is that users have a passive role: a service provider defines a privacy policy that users can only accept or reject. In[12] a new type of privacy policy, called *data handling policy*, that regulates the secondary use of a user's personal data has been discussed. A data handling policy regulates how Personal Identifiable Information (PII) will be used (e.g., information collected through a service will be combined with information collected from other services and used in aggregation for market research purposes), how long PII will be retained (e.g., information will be retained as long as necessary to perform the service), and so on. Users can therefore use these policies to define how their information will be used and processed by the counterpart.

11.3.2 *Open Issues*

Although current approaches supporting attribute-based policies are technically mature enough to be used in practical scenarios, there are still some issues that need to be investigated in more detail to enable more complex applications. We summarize these issues as follows.[7]

- *Ontologies.* Due to the openness of the scenario and the richness and variety of security requirements and attributes that may need to be considered, it is important to provide parties with a means to understand each other with respect to the properties they enjoy (or request the counterpart to enjoy). Therefore, common languages, dictionaries, and ontologies must be developed.

- *Access control evaluation and outcome.* Users may be occasional and they may not know under what conditions a service can be accessed. Therefore, to make a service "usable", access control mechanisms cannot simply return "yes" or "no" answers. It may be necessary to explain why authorizations are denied, or - better - how to obtain the desired permissions. Therefore, the system can return an undefined response meaning that current information is insufficient to determine whether the request can be granted or denied. For instance, suppose that a user can use a particular service only if she is at least eighteen and provides a credit card. According to this policy, two cases can occur: *i)* the system knows that the user is not yet eighteen and therefore returns a negative response; *ii)* the user has proved that she is eighteen and the system returns an undefined response together with the request to provide the information of a credit card.

- *Privacy-enhanced policy communication.* Since access control does not return only a "yes" or "no" access decision, but it returns the information about which conditions need to be satisfied for the access to be granted ("undefined" decision), the problem of communicating such conditions to the counterpart arises. To fix the ideas, let us see the problem from the point of view of the server (the client's point of view is symmetrical). A naive solution consists in giving the client a list with all the possible sets of credentials that would enable the service. This solution is however not feasible due to the large number of possible alternatives. Also, the communication process should not disclose "too much" of the underlying security policy, which might also be regarded as sensitive information.

- *Negotiation strategy.* Credentials grant parties different choices with respect to what release (or ask) the counterpart and when to do it, thus allowing for multiple trust negotiation strategies.[32] For instance, an *eager* strategy, requires parties to turn over all their credentials if the release policy for them is satisfied, without waiting for the credentials to be requested. By contrast, a *parsimonious* strategy requires that parties only release credentials upon explicit request by the server (avoiding unnecessary releases).

- *Composite services.* In case of a composite service (i.e., a service that is decomposable into other services called component services) there must be a semi-automatic mechanism to calculate the policy of a composite service from the policies of its component services.

- *Semantics-aware rules.* Although attribute-based policies allow the specifications of restrictions based on generic attributes or properties of the requestor and the resources, they do not fully exploit the semantic power and reasoning capabilities of emerging web applications. It is therefore important to be able to specify access control rules about subjects accessing the information and about resources to be accessed in terms of rich ontology-based metadata (e.g., Semantic Web-style ones) increasingly available in advanced e-services applications.[44]

11.4 Privacy Issues in Data Collection and Disclosure

Internet provides novel opportunities for the collection and sharing of privacy-sensitive information from and about users. Information about users is collected every day, as they join associations or groups, shop for groceries, or execute most of their common daily activities. Consequently, users have very strong concerns about the privacy of their personal information and they fear that their personal information can be misused. Protecting privacy requires therefore the investigation of many different issues including the problem of *protecting released information against inference and linking attacks*, which are becoming easier and easier because of the increased information availability and ease of access as well as the increased computational power provided by today's technology. In fact, released data too often open up privacy vulnerabilities through, for example, data mining techniques and record linkage. Indeed, the restricted access to information and its expensive processing, which represented a form of protection in the past, do not hold anymore. In addition, while in the past data were principally released in tabular form (macrodata) and through statistical databases, many situations require today that the specific stored data themselves, called microdata, be released. The advantage of releasing microdata instead of specific pre-computed statistics is an increased flexibility and availability of information for the users. At the same time however microdata, releasing more specific information, are subject to a greater risk of privacy breaches. To this purpose, the main requirements that must be taken into account are the following.

- *Identity disclosure protection.* Identity disclosure occurs whenever it is possible to identify a subject, called *respondent*, from the released data. It should therefore be adopted techniques for limiting the possibility of identifying respondents.
- *Attribute disclosure protection.* Identity disclosure protection alone does not guarantee privacy of sensitive information because all the respondents in a

group could have the same sensitive information. To overcome this issue, mechanisms that protect sensitive information about respondents should be adopted.

- *Inference channel.* Given the possibly enormous amount of data to be considered, and the possible inter-relationships between data, it is important that the security specification and enforcement mechanisms provide automatic support for complex security requirements, such as those due to inference and data association channels.

To protect the anonymity of the respondents to whom the released data refer, data holders often remove, encrypt, or code identity information. Identity information removed or encoded to produce anonymous data includes names, telephone numbers, and Social Security Numbers. Although apparently anonymous, however, the de-identified data may contain other quasi-identifying attributes such as race, date of birth, sex, and geographical location. By linking such attributes to publicly available databases associating them with the individual's identity, data recipients can determine to which individual each piece of released data belongs, or restrict their uncertainty to a specific subset of individuals. This problem has raised particular concerns in the medical and financial fields, where microdata, which are increasingly released for circulation or research, can be or have been subject to abuses, compromising the privacy of individuals.

To better illustrate the problem, consider the microdata table in Figure 11.2(a) and the non de-identified public available table in Figure 11.2(b). In the microdata table, which we refer to as private table (PT), data have been de-identified by suppressing names and Social Security Numbers (SSNs) so not to explicitly disclose the identities of respondents. However, the released attributes Race, Date of birth, Sex, ZIP, and Marital status can be linked to the public tuples in Figure 11.2(b) and reveal information on Name, Address, and City. In the private table, for example, there is only one single female (F) born on 71/07/05 and living in the 20222 area. This combination, if unique in the external world as well, uniquely identifies the corresponding tuple as pertaining to "Susan Doe, 20222 Eye Street, Washington DC", thus revealing that she has reported hypertension. While this example demonstrates an exact match, in some cases, linking allows one to detect a restricted set of individuals among whom there is the actual data respondent.

Among the microdata protection techniques used to protect de-identified microdata from linking attacks, there are the commonly used approaches like sampling, swapping values, and adding noise to the data while maintaining some overall statistical properties of the resulting table.[45] However, many uses require the

SSN	Name	Race	Date of birth	Sex	ZIP	Marital status	Disease
		asian	71/07/05	F	20222	Single	hypertension
		asian	74/04/13	F	20223	Divorced	Flu
		asian	74/04/15	F	20239	Married	chest pain
		asian	73/03/13	M	20239	Married	Obesity
		asian	73/03/18	M	20239	Married	hypertension
		black	74/11/22	F	20238	Single	short breath
		black	74/11/22	F	20239	Single	Obesity
		white	74/11/22	F	20239	Single	Flu
		white	74/11/22	F	20223	Widow	chest pain

(a)

Name	Address	City	ZIP	DOB	Sex	Status
....	....	....	......	......	......	....
Susan Doe	Eye street	Washington DC	20222	71/07/05	F	single
....	....	....	......	......	......	....

(b)

Fig. 11.2　An example of private table PT (a) and non de-identified public available table (b).

release and explicit management of microdata while needing truthful information within each tuple. This "data quality" requirement makes inappropriate those techniques that disturb data and therefore, although preserving statistical properties, compromise the correctness of single tuples.[45] *k*-anonymity, together with its enforcement via *generalization* and *suppression*, has been proposed as an approach to protect respondents' identities while releasing truthful information.[46]

The concept of *k*-anonymity tries to capture, on the private table to be released, one of the main requirements that has been followed by the statistical community and by agencies releasing the data, and according to which the *released data should be indistinguishably related to no less than a certain number of respondents*.

The set of attributes included in the private table, also externally available and therefore exploitable for linking, is called *quasi-identifier*. The requirement above-mentioned is then translated in the *k*-anonymity requirement:[46] *each release of data must be such that every combination of values of quasi-identifiers can be indistinctly matched to at least k respondents*. Since it seems impossible, or highly impractical and limiting, to make assumptions on the datasets available for linking to external attackers or curious data recipients, essentially *k*-anonymity takes a safe approach requiring that, in the released table itself, the respondents be indistinguishable (within a given set) with respect to a set of attributes. To guarantee the k-anonymity requirement, k-anonymity requires each quasi-identifier value in the released table to have at least k occurrences. This is clearly a sufficient condition for the k-anonymity requirement: if a set of attributes of external tables appears in the quasi-identifier associated with the private table PT, and the table satisfies this condition, the combination of the released data with the external data will never allow the recipient to associate each released tuple with less than k respondents. For instance, with respect to the microdata table in Figure 11.2 and the quasi-identifier Race, Date of birth, Sex, ZIP, Marital status, the table satisfies k-anonymity with $k = 1$ only, since there are single occurrences of values over the quasi-identifier (e.g., "asian, 71/07/05, F, 20222, single").

11.4.1 *Overview of Ongoing Work*

As above-mentioned, k-anonymity proposals focus on *generalization* and *suppression* techniques. Generalization consists in representing the values of a given attribute by using more general values. This technique is based on the definition of a *generalization hierarchy*, where the most general value is at the root of the hierarchy and the leaves correspond to the most specific values. Formally, the notion of *domain* (i.e., the set of values that an attribute can assume) is extended by assuming the existence of a set of *generalized domains*. The set of original domains together with their generalizations are referred to as Dom. Each generalized domain contains generalized values and there exists a mapping between each domain and its generalizations. This mapping is stated by means of a *generalization relationship* $\leq_D$. Given two domains D_i and $D_j \in$ Dom, $D_i \leq_D D_j$ states that values in domain D_j are generalizations of values in D_i. The generalization relationship $\leq_D$ defines a partial order on the set Dom of domains, where each D_i has at most *one* direct generalization domain D_j, and all values in each domain can always be generalized to a single value. The definition of a generalization relationship implies the existence, for each domain $D \in$ Dom, of a totally ordered hierarchy, called *domain generalization hierarchy*, denoted DGH$_D$. As an example, consider attribute ZIP code and suppose that a step in the corresponding generalization hierarchy consists in suppressing the least significant digit in the ZIP code. Figure 11.3 illustrates the corresponding domain generalization hierarchy. In this case, for example, if we choose to apply one generalization step, values 20222, 20223, 20238, and 20239 are generalized to 2022* and 2023*. A generalization process therefore proceeds by replacing the values represented by the leaf nodes with one of their ancestor nodes at a higher level. Different gener-

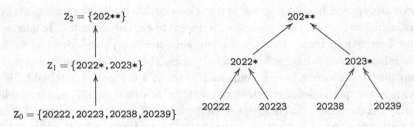

Fig. 11.3   An example of domain generalization hierarchy for attribute ZIP.

alized microdata tables can be built, depending on the amount of generalization applied on the considered attribute.

Suppression is a well-known technique that consists in protecting sensitive information by removing it. The introduction of suppression can reduce the amount of generalization necessary to satisfy the k-anonymity constraint.

Generalization and suppression can be applied at different levels of granularity. Generalization can be applied at the level of single column (i.e., a generalization step generalizes all the values in the column) or single cell (i.e., for a specific column, the table may contain values at different generalization levels). Suppression can be applied at the level of row (i.e., a suppression operation removes a whole tuple), column (i.e., a suppression operation obscures all the values of a column), or single cells (i.e., a k-anonymized table may wipe out only certain cells of a given tuple/attribute). The possible combinations of the different choices for generalization and suppression (including also the choice of not applying one of the two techniques) result in different k-anonymity proposals and different algorithms for k-anonymity.

Note that the algorithms for solving k-anonymity aim at finding a k-minimal table, that is, one that does not generalize (or suppress) more than it is needed to reach the threshold k. As an example, consider the microdata table in Figure 11.4 and suppose that the quasi-identifier is {Race, Date of birth, Sex, ZIP}.

Figure 11.4 illustrates an example of 2-anonymous table obtained by applying the algorithm described in,[46] where generalization is applied at the column level and suppression is applied at the row level. Note that the first tuple in the original table has been suppressed, attribute Date of birth has been generalized by removing the day, and attribute ZIP has been generalized by applying two generalization steps along the domain generalization hierarchy in Figure 11.3.

In[47] we defined a possible taxonomy for k-anonymity and discussed the main proposals existing in the literature for solving the k-anonymity problems. Basically, the algorithms for enforcing k-anonymity can be partitioned into three main

SSN	Name	Race	Date of birth	Sex	ZIP	Marital status	Disease
		asian	74/04/	F	202**	divorced	Flu
		asian	74/04/	F	202**	married	chest pain
		asian	73/03/	M	202**	married	obesity
		asian	73/03/	M	202**	married	hypertension
		black	74/11/	F	202**	single	short breath
		black	74/11/	F	202**	single	obesity
		white	74/11/	F	202**	single	flu
		white	74/11/	F	202**	Widow	chest pain

Fig. 11.4   An example of a 2-anonymized table for the private table PT in Figure 11.2(a).

classes: *exact*, *heuristic*, and *approximation* algorithms, respectively. While exact and heuristic algorithms produce *k*-anonymous tables by applying attribute generalization and tuple suppression and are exponential in the size of the quasi-identifier,[46,48–53] approximation algorithms produce *k*-anonymous tables by applying cell suppression without generalization or cell generalization without suppression.[54–56] In these case, exact algorithms are not applicable because the computational time could be exponential in the number of tuples in the table.

Samarati[46] presented an algorithm that exploits a binary search on the domain generalization hierarchy to avoid an exhaustive visit of the whole generalization space. Since the *k*-anonymity definition is based on a quasi-identifier, the algorithm works only on this set of attributes and on tables with more than *k* tuples (this last constraint being clearly a necessary condition for a table to satisfy *k*-anonymity). Bayardo and Agrawal[48] presented an optimal algorithm, called *k-Optimize*, which starts from a fully generalized table (with all tuples equal) and specializes the dataset in a minimal *k*-anonymous table, exploiting ad-hoc pruning techniques. LeFevre, DeWitt, and Ramakrishnan[51] described an algorithm that uses a bottom-up technique and a priori computation.

Iyengar[53] presented genetic heuristic algorithms and solves the *k*-anonymity problem using an incomplete stochastic search method. The method does not assure the quality of the solution proposed, but experimental results show the validity of the approach. Winkler[50] proposed a method based on simulated annealing for finding locally minimal solutions, which requires high computational time and does not assure the quality of the solution. Fung, Wang and Yu[49] presented a top-down heuristic to make a table to be released *k*-anonymous. The algorithm starts from the most general solution, and iteratively specializes some values of the current solution until the *k*-anonymity requirement is violated. Each step of specialization increases the information and decreases the anonymity.

Meyerson and Williams[56] presented an algorithm for *k*-anonymity, which guarantees a $O(k\log(k))$-approximation. Aggarwal et al.[54,55] illustrated two approximation algorithms that guarantee a $O(k)$-approximation solution. Note that although both heuristics and approximation algorithms do not guarantee the minimality of their solution, and we cannot perform any evaluation on the result of a heuristic, an approximation algorithm guarantees near-optimum solutions.

k-anonymity is also currently the subject of many interesting studies. In particular, these studies aim at: studying efficient algorithms for k-anonymity enforcement; using k-anonymity as a measure on information disclosure due to a set of views;[57] extending its definition to protect the released data against attribute, in contrast to identity, disclosure (ℓ-diversity);[58] supporting fine-grained application of generalization and suppression; and investigating additional techniques for k-anonymity enforcement.[59]

11.4.2 *Open Issues*

We now summarize the main open issues in developing a k-anonymity solution.

- *Extensions and enrichment of the definition.* k-anonymity captures only the defence against identity disclosure attacks, while remaining exposed to attribute disclosure attacks.[46] Some researchers have just started proposing extensions to k-anonymity[58] to capture also attribute disclosure, however research is still to be done.
- *Protection against utility measures.* As we can imagine the more the protection, the less precise or complete the data will be. Research is needed to develop measures to allow users to assess, besides the protection offered by the data, the utility of the released data. Clearly, utility may be different depending on the data recipients and the use intended for the information. Approaches should be therefore devised that maximize information utility with respect to intended uses, while properly guaranteeing privacy.
- *Efficient algorithms.* Computing a table that satisfies k-anonymity guaranteeing minimality (i.e., minimal information loss or, in other words, maximal utility) is an NP-hard problem and therefore computationally expensive. Efficient heuristic algorithms have been designed, but still research is needed to improve the performance. Indexing techniques could be exploited in this respect.
- *New techniques.* The original k-anonymity proposal assumed the use of generalization as suppression since, unlike others, they preserve truthfulness of the data. The k-anonymity property is however not tied to a specific technique and alternative techniques could be investigated.

- *Merging of different tables and views.* The original *k*-anonymity proposal as well as most subsequent works assume the existence of a single table to be released with the further constraints that the table contains at most one tuple for each respondent. Work is needed to release these two constraints. In particular, the problem of releasing different tables providing anonymity even in presence of join that can allow inferring new information needs to be investigated.

- *External knowledge.* *k*-anonymity assumes the data recipient has access to external database linking identities with quasi identifiers; it did not however model external knowledge that can be further exploited for inference and expose the data to identity or attribute disclosure. Work is needed to allow modeling external knowledge and taking it into account in the process of computing the table to be released.

11.5 Location Privacy Issues

The pervasive diffusion of mobile communication devices and technical improvements of location technologies are fostering the development of a new wave of applications that use the physical position of individuals to offer location-based services for business, social, or informational purposes.[60] Location awareness supports an extended context of interaction for each user and resource in the environment, eventually modeling a number of spatial-temporal relationships among users and resources. In a location-aware environment, context is not the static situation of a predefined environment; rather, it is a dynamic part of the process of interacting with a changing environment, composed of mobile users and resources.[61]

Location-related information can be classified as follows.

- *Punctual location,* absolute longitude-and-latitude geographical location provided by systems like GPS (*Global Positioning System*). In outdoor and rural environments GPS, when at least three satellites are visible, delivers position information with an acceptable accuracy[c]. Today, GPS chipsets are integrated into most mainstream cell phones and PDAs; when it is not available, the cellular network itself can be used as a basic geo-location service.[62]

- *Logical* or *local* location, composed of location assertions with different

---

[c]In dense urban areas or inside buildings, localization with GPS may becomes critical because the satellites are not visible from the mobile terminal. By 2008 the European Union will deploy Galileo, a next-generation GPS system that promises greater accuracy and operation covering both indoors and out, due to stronger radio signals that should penetrate most buildings.

levels of precision, for example, specifying that the user in a specific country, city, building or room.

Obviously, given the necessary background information (e.g., in the form of a map with geographical coordinates) there may be a function that maps punctual locations to logical ones. Recent research has proposed many location techniques producing a user's logical location, punctual location or both, depending on application requirements. Location techniques have been proposed for local wireless networking: for example, Microsoft Research's RADAR system requires an initial calibration in which 802.11 readings are made on a 1 meter (1m) grid. Then, this grid is used for positioning 802.11 access points (APs). If the APs are positioned correctly, knowing the readings of a device is sufficient for estimating its location. The Place Lab project[63] does not rely on the availability of a grid of previous readings; rather, it predicts location via the positions of the APs, read from a database cached on each device.

Today, the public database wigle.net contains the position of more than 2 million APs in the US and in Europe, providing quick-and-dirty location in some key urban areas.

In this scenario, it comes with no surprise that personal privacy, which is already the center of many concerns for the risks posed by current on-line services,[64,65] is considered seriously threatened by location-based services. In addition, the publicity gained by recent security incidents that have targeted individuals privacy, revealed faulty data management practices and unauthorized trading of users personal information (including, ID thefts and unauthorized profiling). For instance, some legal cases have been reported, where rental companies used GPS technology to track their cars and charge users for agreement infringements,[66] or where an organization used a "Friend finder" service to track its own employees.[67] Research on privacy issue has also gained a relevant boost since providers of online and mobile services, often, largely exceeded in collecting personal information as a requirement for service provision.

In such a worrisome scenario, the concept of *location privacy* can be defined as the right of individuals to decide how, when, and for which purposes their location information could be released to other parties. The lack of location privacy protection could result in severe consequences that make users the target of fraudulent attacks:[68]

- *unsolicited advertising,* the location of the user could be exploited, without her consent, to provide advertisements of products and services available nearby the user position;
- *physical attacks or harassment,* the location of the user could be used to carry

physical assaults to individuals;

- *users profiling,* the location of the user, which intrinsically carries personal information, could be used to infer other sensitive information such as state of health, personal habits, professional duties, and the like;
- *denial of service,* the location of the user could be used to deny accesses to services under some circumstances.

Situations in which sensing technologies have been used for stalking users locations and harassing individuals have been already reported.[67,69]

In this context, location privacy can assume several meanings and pursue different objectives depending on the scenario in which the users are moving and on the services with which the users are interacting with. Location privacy protection could be aimed either at preserving: the privacy of the user identity, the single user location measurement, or the location movement of the user monitored in a certain period of time. The following categories of location privacy can then be identified.[60]

- *Identity privacy.* The main goal is to protect the identities of the users associated with or inferable from location information. For instance, many online services provide a person with the ability to establish a relationship with some other entities without her personal identity being disclosed to those entities. In this case, the best possible location measurement can be provided to the others entities but the identity of the users must be preserved.
- *Position privacy.* The main goal is to perturb locations of the users to protect the positions of individual users. In particular, this type of location privacy is suitable for environments where users identities are required for a successful service provisioning. An example of a technique that most solutions either explicitly or implicitly exploit, consists in scaling a location to a coarser granularity (e.g., from meters to hundreds of meters, from a city block to the whole town, and so on).
- *Path privacy.* The main goal is to protect the privacy of the users that are monitored during a certain period of time. The location-based services will no longer receive a single location measurement, but they will gather many samples allowing them to track users. In particular, path privacy can be guaranteed by adapting the techniques used for identity and position privacy to preserve the privacy of a user that is continuously monitored.

These categories of location privacy pose different requirements that are guaranteed by different privacy technologies, which we will analyze in the following Section. Note that no technique is able to provide a general solution satisfying all the privacy requirements.

11.5.1 *Overview of Ongoing Work*

Accordingly to the categories of location privacy previously described, three different classes of location privacy techniques can be introduced: anonymity-based, obfuscation-based, and policy-based. These classes are partially overlapped in scope and could be potentially suitable to cover requirements coming from one or more of the categories of location privacy. Anonymity-based and obfuscation-based techniques can be usually regarded as dual categories. While anonymity-based techniques have been primarily defined to protect identity privacy and are less suitable for protecting position privacy, obfuscation-based techniques are well suited for position protection and less appropriate for identity protection. Anonymity-based and obfuscation-based techniques are well-suited for protecting path privacy. Nevertheless, more studies and proposals have been focused on anonymity-based rather than on obfuscation-based techniques. Policy-based techniques are in general suitable for all the location privacy categories. However, they can be difficult to understand and manage for end users.

Anonymity-based solutions. An important line of research in location privacy protection relies on the notion of *anonymity*.[70–74] Anonymity typically refers to an individual, and it means that the personal identity, or personally identifiable information of that person is not known.

Mix zones is the method developed by Beresford and Stajano[70,75] to enhance privacy in location-based services by means of an anonymity service based on an infrastructure that delays and reorders messages from subscribers within predefined zones. In particular, Mix zone model is managed by a trusted middleware that lies between the positioning systems and the third party applications and is responsible for limiting the information collected by applications. The Mix zone model is based on the concepts of *application zone* and *mix zones*. The former represents homogeneous application interests in a specific geographic area, while the latter represents areas in which a user cannot be tracked. In particular, within mix zones, a user is anonymous in the sense that the identities of all users coexisting in the same zone are mixed and become indiscernible. Furthermore, the infrastructure makes a user entering the mix zone unlinkable from other users leaving it. The authors also provide an analysis of an attacker behavior by defining and calculating the level of anonymity assured to the users.[70] In particular, the success of an attack aimed at recovering users identities is inversely proportional to the anonymity level. To conclude, the Mix zones model is aimed at protecting long-term user movements still allowing the interaction with many location-based services.

Bettini et al.[71] proposed a framework able to evaluate the risk of sensitive location-based information dissemination, and introduces a technique aimed at supporting k-anonymity.[46] In particular, the authors put forward the idea that the geo-localized history of the requests submitted by a user can be considered as a *quasi-identifier* to access sensitive information about that individual. For instance, a user tracked during working days is likely to commute from her house to the workplace in a specific time frame in the morning and come back in another specific time frame in the evening. This information can be used to easily re-identify the user. The privacy preservation framework based on the concepts of quasi-identifier and k-anonymity is designed for such scenario. In particular, the service provider gathering both users requests and personal histories of locations should never be able to link a subset of requests to a single user. To make this possible, there must exist k-1 users having a personal history of locations compatible with the requests that have been issued.

Gruteser and Grunwald[73] defined k-anonymity in the context of location obfuscation. The paper proposes a middleware architecture and an adaptive algorithm to adjust location information resolution, in spatial or temporal dimensions, to comply with specific anonymity requirements. The authors proposed the concepts of *spatial* and *temporal cloaking* used to transform a user's location to comply with the requested k level of anonymity. In particular, spatial cloaking guarantees the k-anonymity required by the users by enlarging the area in which a user is located until enough indistinguishable individuals are contained. The same reasoning could be done for the temporal cloaking, which is an orthogonal process with respect to the spatial one. Whereas this method could provide spatial coordinates with higher accuracy, it reduces the accuracy in time.

Gedik and Liu[72] described another k-anonymity model aimed at protecting location privacy against various privacy threats, and provided a framework supporting location k-anonymity. Each user is able to define the minimum level of anonymity and the maximum acceptable temporal and spatial resolution for her location measurement. Then, the focus of the paper is on the definition of a message perturbation engine responsible for providing location anonymization of user's request messages through identity removal and spatio-temporal obfuscation of location information.

Mokbel et al.[74] presented a framework, named Casper, aimed at changing traditional location-based servers and query processors to provide the users with anonymous services. Users can define their privacy preferences through a k, which is the number of users to be indistinguishable, and A_{min} representing the minimal area that the user is willing to release. Casper framework is composed by a *location anonymizer,* responsible for perturbing the users location to achieve the privacy preferences of users, and by a *privacy-aware query processor,* responsible for the management of anonymous queries and cloaked spatial areas.

To conclude, another line of research that relies on the concept of anonymity is aimed at protecting the path privacy of the users.[76–78] This research area is particularly relevant since in the near past many location tracking applications have been designed and developed also for devices with limited capabilities (e.g. cellular phones). Nowadays, in fact, data about users moving in a particular area are collected by external services, such as navigation systems, that use them to provide their services effectively. In such a scenario the need for privacy techniques aimed at protecting the privacy of the path becomes urgent.

Obfuscation-based solution.   Another line of research in location privacy protection consists in the adoption of obfuscation techniques. Obfuscation is the process of degrading the accuracy of the information, to provide privacy protection. Differently from anonymity-based techniques the major goal of obfuscation techniques is to perturb the location information still maintaining a binding with the identity of users. Several location-based services in fact requires a user to present her identity to access the requested service.

Duckham and Kulik[79] analyzed obfuscation techniques for protecting the location privacy of users. The paper sets out a formal framework that provides a mechanism for balancing individuals needs for high-quality information services and for location privacy. The technique is based on the imprecision concept, which means the lack of specificity of location information. The authors proposed to degrade location information quality and to provide obfuscation features by adding *n* points at the same probability to the real user position. The algorithm assumes a graph-based representation of the environment. In[80] the defined obfuscation methods are validated and evaluated through a set of simulations. The results show that obfuscation can provide at the same time both high quality of service and high privacy level.

In addition, today, some commercial location platforms include a gateway that mediates between location providers and location-based applications. In those architectures, such as Openwave,[81] the location gateway obtains users location information from multiple sources and delivers them, possibly modified, according to privacy requirements. Openwave assumes that users specify their privacy preferences in terms of a minimum distance representing the maximum accuracy they are willing to provide.

Bellavista et al.[82] studied a solution based on a middleware that balances between the proper level of user privacy and the needs of location-based services precision. The location data are then exposed at the proper level of granularity depending on privacy/efficiency requirements negotiated by the parties. Hence, instead of exact client positions, a downscaled location information (with lower precision and lower geographical granularity) is returned.

Finally, some proposals[83-85] presented several obfuscation-based techniques for location privacy protection that are particularly suitable for location-based services. These techniques are based on a simple and intuitive mechanism for the definition of the privacy preferences, and on a formal estimator, named *relevance*, of both privacy and accuracy of location. In summary, these techniques provide a degree of privacy to the users by degrading the location accuracy of each measurement and offer a measurable accuracy to service providers.

Policies-based solution. Other works studied the possibility of protecting users privacy through the definition of complex rule-based policies.

Hauser and Kabatnik[86] addressed this problem in a privacy-aware architecture for a global location service, which allows users to define rules that will be evaluated to manage access to location information. Hengartner and Steenkiste[87] described a method of using digital certificates combined with rule-based policies to protect location information. The IETF Geopriv working group[88] addressed privacy and security issues related to the disclosure of location information over the Internet. The main goal is to define an environment (i.e., an architecture, protocols, and policies) supporting both location information and policy data. The Geopriv infrastructure relies on both *authorization policies,* posing restrictions on location management and access, and *privacy rules* associated with the location information, defining restrictions on how the released information can be managed by the counterparts.

Some proposals used the Platform for Privacy Preferences (P3P)[41] to encode users privacy preferences. In particular, Hong et al.[89] provided an extension to P3P for representing user privacy preferences for context-aware applications, while Langheinrich[90] proposed the *pawS* system that provides a privacy enabling technology for end-users.

11.5.2 *Open Issues*

We briefly describe some open issues that need to be taken into consideration in the future development of location privacy techniques.

- *Privacy preference definition.* A key aspect for the success of location privacy techniques is the definition of a mechanism for privacy preferences specification that balances between the two traditionally conflicting requirements of *usability* and *expressiveness*. Despite its importance for the effectiveness of a privacy solution, this issue has received little attention in previous works on location privacy.

- *Balancing location privacy and accuracy.* Location privacy solutions should be able to balance the need of privacy protection required by users and the need of accuracy required by service providers. Location privacy techniques, which are focused on users needs, could make the service provisioning impossible in practice due to the excessively degradation of location measurement accuracy. A possible direction to avoid excessive degradation is the definition of an estimator of the accuracy of location information, abstracting from any physical attribute of sensing technology, which permits to quantitatively evaluate both the degree of privacy introduced into a location measurement and the location accuracy requested by a service provider. Both quality of online services and location privacy could then be adjusted, negotiated, or specified as contractual terms. A quantitative estimation of the provided privacy level makes simpler the integration of privacy solutions into a full fledged location-based application scenario.[91,92]

- *Composition of privacy techniques.* Usually, all location privacy solutions implement a single privacy technique. This is clear in the case of obfuscation-based techniques, where most of the solutions rely on traditional obfuscation by scaling the location area. An important requirement for next generation solutions is to provide more techniques and combine them to increase their robustness with respect to possible de-obfuscation attempts performed by adversaries.

- *Degree of privacy protection.* Although some works[83–85] provide an estimation of the degree of privacy introduced by location privacy techniques, the *real* degree of privacy is not estimated yet. The real degree of privacy must be calculated by analyzing the possibilities of an adversary to reduce the effects of the privacy techniques. As an example, consider a traditional obfuscation-based technique by scaling the location area. Let assume that the location of a user walking in an urban area has been obfuscated by just increasing the radius to return an area that covers the whole city, rather than an area with radius of some hundreds of meters. It would be reasonable for an adversary to infer that the original area covers just few neighborhoods rather than the whole city. Whereas such trivial de-obfuscation does not produce exactly the original measure, it provides the adversary with a better approximation of the original measurement than the obfuscated area, hence, reducing the user's location privacy.

11.6 Conclusions

This paper discussed aspects related to the protection of information in today's globally networked society. We investigated recent proposals and ongoing work addressing different privacy issues in emerging applications and new scenarios focussing on: the combination of security policies and their interchange and enforcement in open scenarios, the protection of personal data undergoing public or semi-public release, and on the protection of location information in location-based services. For all these areas, we have briefly illustrated the challenges to be addressed, current research, and open issues to be investigated.

11.7 Acknowledgements

This work was partially supported by the European Union within the PRIME Project in the FP6/IST Programme under contract IST-2002-507591, by the Italian Ministry of Research Fund for Basic Research (FIRB) under project RBNE05FKZ2 and by the Italian MIUR under project MAPS.

References

1. S. De Capitani di Vimercati and P. Samarati, Privacy in the electronic society. In *Proc. of the International Conference on Information Systems Security (ICISS 2006)*, Kolkata, India (December, 2006).
2. P. Liu, P. Mitra, C. Pan, and V. Atluri, Privacy-preserving semantic interoperation and access control of heterogeneous databases. In *ACM Symposium on InformAtion, Computer and Communications Security*, Taipei, Taiwan (March, 2006).
3. P. Bonatti, S. De Capitani di Vimercati, and P. Samarati, An algebra for composing access control policies, *ACM Transactions on Information and System Security* 5(1), 1–35 (February, 2002).
4. D. Wijesekera and S. Jajodia, A propositional policy algebra for access control, *ACM Transactions on Information and System Security* 6(2), 286–325 (May, 2003).
5. M. Blaze, J. Feigenbaum, J. Ioannidis, and A. Keromytis, *The KeyNote Trust Management System (Version 2)*, internet rfc 2704 edition, (1999).
6. M. Blaze, J. Feigenbaum, and J. Lacy, Decentralized trust management. In *Proc. of the 17th Symposium on Security and Privacy*, Oakland, California, USA (May, 1996).
7. P. Bonatti and P. Samarati, A unified framework for regulating access and information release on the web, *Journal of Computer Security* 10(3), 241–272, (2002).
8. K. Irwin and T. Yu, Preventing attribute information leakage in automated trust negotiation. In *Proc. of the 12th ACM Conference on Computer and Communications Security*, Alexandria, VA, USA (November, 2005).
9. N. Li, J. Mitchell, and W. Winsborough, Beyond proof-of-compliance: Security analysis in trust management, *Journal of the ACM* 52(3), 474–514 (May, 2005).

10. J. Ni, N. Li, and W. Winsborough, Automated trust negotiation using cryptographic credentials. In *Proc. of the 12th ACM Conference on Computer and Communications Security*, Alexandria, VA, USA (November, 2005).
11. T. Yu, M. Winslett, and K. Seamons, Supporting structured credentials and sensitive policies trough interoperable strategies for automated trust, *ACM Transactions on Information and System Security (TISSEC)* 6(1), 1–42 (February, 2003).
12. C. Ardagna, S. De Capitani di Vimercati, and P. Samarati, Enhancing user privacy through data handling policies. In *Proc. of the 20th Annual IFIP WG 11.3 Working Conference on Data and Applications Security*, Sophia Antipolis, France (August, 2006).
13. C. Landwehr, Formal models for computer security, *Computing Surveys* 13(3), 247–278 (September, 1981).
14. M. Abadi and L. Lamport, Composing specifications, *ACM Transactions on Programming Languages* 14(4), 1–60 (October, 1992).
15. H. Hosmer, Metapolicies ii. In *Proc. of the 15th NIST-NSA National Computer Security Conference*, Baltimore, Maryland (October, 1992).
16. T. Jaeger, Access control in configurable systems, *Lecture Notes in Computer Science* 1603, 289–316, (2001).
17. J. McLean, The algebra of security. In *Proc. of the 1988 IEEE Computer Society Symposium on Security and Privacy*, Oakland, CA, USA (April, 1988).
18. D. Bell, Modeling the multipolicy machine. In *Proc. of the New Security Paradigm Workshop* (August, 1994).
19. S. Jajodia, P. Samarati, M. Sapino, and V. Subrahmanian, Flexible support for multiple access control policies, *ACM Transactions on Database Systems* 26(2), 18–28 (June, 2001).
20. N. Li, J. Feigenbaum, and B. Grosof, A logic-based knowledge representation for authorization with delegation. In *Proc. of the 12th IEEE Computer Security Foundations Workshop*, pp. 162–174 (July, 1999).
21. T. Woo and S. Lam, Authorizations in distributed systems: A new approach, *Journal of Computer Security* 2(2,3), 107–136, (1993).
22. V. Subrahmanian, S. Adali, A. Brink, J. Lu, A. Rajput, T. Rogers, R. Ross, and C. Ward, *Hermes: heterogeneous reasoning and mediator system*. http://www.cs.umd.edu/projects/hermes.
23. M. Casassa Mont, S. Pearson, and P. Bramhall, Towards accountable management of identity and privacy: Sticky policies and enforceable tracing services. In *Proc. of the 14th International Workshop on Database and Expert Systems Applications*, Prague, Czech (September, 2003).
24. C. Ellison, B. Frantz, B. Lampson, R. Rivest, B. Thomas, and T. Ylonen, SPKI certificate theory. RFC2693 (September, 1999).
25. N. Li, W. Winsborough, and J. Mitchell, Distributed credential chain discovery in trust management, *Journal of Computer Security* 11(1), 35–86 (February, 2003).
26. K. E. Seamons, W. Winsborough, and M. Winslett, Internet credential acceptance policies. In *Proc. of the Workshop on Logic Programming for Internet Applications*, Leuven, Belgium (July, 1997).
27. K. E. Seamons, M. Winslett, T. Yu, B. Smith, E. Child, J. Jacobson, H. Mills, and L. Yu, Requirements for policy languages for trust negotiation. In *Proc. of the 3rd*

International Workshop on Policies for Distributed Systems and Networks (POLICY 2002), Monterey, CA (June, 2002).

28. L. Wang, D. Wijesekera, and S. Jajodia, A logic-based framework for attribute based access control. In *Proc. of the 2004 ACM Workshop on Formal Methods in Security Engineering*, Washington DC, USA (October, 2004).

29. M. Winslett, N. Ching, V. Jones, and I. Slepchin, Assuring security and privacy for digital library transactions on the web: Client and server security policies. In *Proc. of the ADL '97 — Forum on Research and Tech. Advances in Digital Libraries*, Washington, DC (May, 1997).

30. T. Yu, M. Winslett, and K. Seamons, Prunes: An efficient and complete strategy for automated trust negotiation over the internet. In *Proc. of the 7th ACM Conference on Computer and Communications Security*, Athens, Greece (November, 2000).

31. W. Winsborough, K. E. Seamons, and V. Jones, Automated trust negotiation. In *Proc. of the DARPA Information Survivability Conf. & Exposition*, Hilton Head Island, SC, USA (January, 2000).

32. T. Yu, M. Winslett, and K. Seamons, Interoperable strategies in automated trust negotiation. In *Proc. of the 8th ACM Conference on Computer and Communications Security*, Philadelphia, PA, USA (November, 2001).

33. T. Yu and M. Winslett, A unified scheme for resource protection in automated trust negotiation. In *Proc. of the IEEE Symposium on Security and Privacy*, Berkeley, California (May, 2003).

34. T. Ryutov, L. Zhou, C. Neuman, T. Leithead, and K. Seamons, Adaptive trust negotiation and access control. In *Proc. of the 10th ACM Symposium on Access Control Models and Technologies*, Stockholm, Sweden (June, 2005).

35. T. van der Horst, T. Sundelin, K. Seamons, and C. Knutson, Mobile trust negotiation: Authentication and authorization in dynamic mobile networks. In *Proc. of the Eighth IFIP Conference on Communications and Multimedia Security*, Lake Windermere, England (September, 2004).

36. N. Li, B. Grosof, and Feigenbaum, Delegation logic: A logic-based approach to distributed authorization, *ACM Transactions on Information and System Security* 6(1), 128–171 (February, 2003).

37. N. Li and J. Mitchell, Datalog with constraints: A foundation for trust-management languages. In *Proc. of the Fifth International Symposium on Practical Aspects of Declarative Languages (PADL 2003)*, New Orleans, LA, USA (January, 2003).

38. T. Jim, SD3: A trust management system with certified evaluation. In *Proc. of the 2001 IEEE Symposium on Security and Privacy*, Oakland, CA, USA (May, 2001).

39. J. DeTreville, Binder, a logic-based security language. In *Proc. of the 2001 IEEE Symposium on Security and Privacy*, Oakland, CA, USA (May, 2002).

40. G.-J. Ahn and J. Lam, Managing privacy preferences in federated identity management. In *Proc. of the ACM Workshop on Digital Identity Management (In conjuction with 12th ACM Conference on Computer and Communications Security)*, Fairfax, VA, USA (November, 2005).

41. The Platform for Privacy Preferences 1.0 (P3P1.0) Specification. *The Platform for Privacy Preferences 1.1 (P3P1.1) Specification*. World Wide Web Consortium (July, 2005). http://www.w3.org/TR/2005/WD-P3P11-20050701.

42. A P3P Preference Exchange Language 1.0 (APPEL1.0). *A P3P Preference Ex-*

change Language 1.0 (APPEL1.0). World Wide Web Consortium (April, 2002). http://www.w3.org/TR/P3P-preferences/.

43. R. Agrawal, J. Kiernan, R. Srikant, and Y. Xu, An xpath based preference language for P3P. In *Proc. of the 12th International World Wide Web Conference*, Budapest, Hungary (May, 2003).

44. E. Damiani, S. De Capitani di Vimercati, C. Fugazza, and P. Samarati, Extending policy languages to the semantic web. In *Proc. of the International Conference on Web Engineering*, Munich, Germany (July, 2004).

45. V. Ciriani, S. De Capitani di Vimercati, S. Foresti, and P. Samarati, Microdata protection. In *Security in Decentralized Data Management*. Springer, (2007).

46. P. Samarati, Protecting respondents' identities in microdata release, *IEEE Transactions on Knowledge and Data Engineering* 13(6), 1010–1027 (November, 2001).

47. V. Ciriani, S. De Capitani di Vimercati, S. Foresti, and P. Samarati, *k*-anonymity. In *Security in Decentralized Data Management*. Springer, (2007).

48. R. Bayardo and R. Agrawal, Data privacy through optimal *k*-anonymization. In *Proc. of the 21st International Conference on Data Engineering (ICDE'05)*, pp. 217–228, Tokyo, Japan (April, 2005).

49. B. Fung, K. Wang, and P. Yu, Top-down specialization for information and privacy preservation. In *Proc. of the 21st International Conference on Data Engineering (ICDE'05)*, Tokyo, Japan (April, 2005).

50. W. Winkler, Masking and re-identification methods for public-use microdata: Overview and research problems. In ed. J. Domingo-Ferrer, *Privacy in Statistical Databases 2004*. Springer, New York, (2004).

51. K. LeFevre, D. DeWitt., and R. Ramakrishnan, Incognito: Efficient full-domain *k*-anonymity. In *Proc. of the 24th ACM SIGMOD International Conference on Management of Data*, pp. 49–60, Baltimore, Maryland, USA (June, 2005).

52. L. Sweeney, Achieving *k*-anonynity privacy protection using generalization and suppression, *International Journal on Uncertainty, Fuzziness and Knowledge-based Systems* 10(5), 571–588, (2002).

53. V. Iyengar, Transforming data to satisfy privacy constraints. In *Proc. of the Eigth ACM SIGKDD International Conference on Knowledge Discovery and Data Mining*, pp. 279–288, Edmonton, Alberta, Canada (July, 2002).

54. G. Aggarwal, T. Feder, K. Kenthapadi, R. Motwani, R. Panigrahy, D. Thomas, and A. Zhu, Anonymizing tables. In *Proc. of the 10th International Conference on Database Theory (ICDT'05)*, pp. 246–258, Edinburgh, Scotland (January, 2005).

55. G. Aggarwal, T. Feder, K. Kenthapadi, R. Motwani, R. Panigrahy, D. Thomas, and A. Zhu, Approximation algorithms for *k*-anonymity, *Journal of Privacy Technology*, (2005).

56. A. Meyerson and R. Williams, On the complexity of optimal *k*-anonymity. In *Proc. of the 23rd ACM-SIGMOD-SIGACT-SIGART Symposium on the Principles of Database Systems*, pp. 223–228, Paris, France (2004).

57. C. Yao, X. Wang, and S. Jajodia, Checking for *k*-anonymity violation by views. In *Proc. of the 31st International Conference on Very Large Data Bases (VLDB'05)*, Trondheim, Norway (August, 2005).

58. A. Machanavajjhala, J. Gehrke, and D. Kifer, ℓ-diversity: Privacy beyond *k*-anonymity. In *Proc. of the International Conference on Data Engineering (ICDE'06)*,

Atlanta, GA, USA (April, 2006).

59. J. Domingo-Ferrer and J. Mateo-Sanz, Practical data-oriented microaggregation for statistical disclosure control, *IEEE Transactions on Knowledge and Data Engineering* **14**(1), 189–201, (2002).

60. C. Ardagna, M. Cremonini, E. Damiani, S. De Capitani di Vimercati, and P. Samarati, Privacy-enhanced location services. In *Digital Privacy: Theory, Technologies and Practices*, Taylor and Francis (to appear), (2007).

61. J. Coutaz, J. Crowley, S. Dobson, and D. Garlan, Context is key, *Comm. of the ACM* **48**(3) (March, 2005).

62. M. Anisetti, V.Bellandi, E. Damiani, and S. Reale, Localization and tracking of mobile antennas in urban environment. In *Proc. of the International Symposium on Telecommunications (IST 2005)*, Shiraz, Iran (September, 2005).

63. B. Schlit, Ubiquitous location-aware computing and the place lab initiative. In *Proc. of the First ACM International Workshop on Wireless Mobile Applications and Services on WLANHotspots*, San Diego, CA (September, 2003).

64. L. Barkhuus and A. Dey, Location-based services for mobile telephony: a study of user's privacy concerns. In *Proc. of the 9th IFIP TC13 International Conference on Human-Computer Interaction (INTERACT 2003)*, pp. 709–712, Zurich, Switzerland (September, 2003).

65. A Chronology of Data Breaches, *A Chronology of Data Breaches*. Privacy Rights Clearinghouse/UCAN (2006).
http://www.privacyrights.org/ar/ChronDataBreaches.htm.

66. *Rental firm uses GPS in speeding fine*, Chicago Tribune, July 2nd, p9. Associated Press: Chicago, IL, 2001.

67. J.-W. Lee, *Location-tracing sparks privacy concerns*, Korea Times.
http://times.hankooki.com, 16 November 2004. Accessed 22 December 2006.

68. M. Duckham and L. Kulik, *Dynamic & Mobile GIS: Investigating Change in Space and Time*, chapter Location privacy and location-aware computing, pp. 34–51. Taylor & Francis, (2006).

69. Fox News, *Man Accused of Stalking Ex-Girlfriend With GPS*.
http://www.fox-news.com/story/0,2933,131487,00.html,
04 September 2004. Accessed 22 March 2007.

70. A. Beresford and F. Stajano, Mix zones: User privacy in location-aware services. In *Proc. of the 2nd IEEE Annual Conference on Pervasive Computing and Communications Workshops (PERCOMW04)*, (2004).

71. C. Bettini, S. Jajodia, X. S. Wang, and D. Wijesekera, Provisions and obligations in policy management and security applications. In *Proc. 28th Conference Very Large Data Bases (VLDB'02)* (August, 2002).

72. B. Gedik and L. Liu, Location privacy in mobile systems: A personalized anonymization model. In *Proc. of the 25th International Conference on Distributed Computing Systems (IEEE ICDCS 2005)*, Columbus, Ohio (June, 2005).

73. M. Gruteser and D. Grunwald, Anonymous usage of location-based services through spatial and temporal cloaking. In *Proc. of the 1st International Conference on Mobile Systems, Applications, and Services (MobiSys 2003)* (May, 2003).

74. M. Mokbel, C.-Y. Chow, and W. Aref, The new casper: Query processing for location services without compromising privacy. In *Proc. of the 32nd International Conference*

on Very Large Data Bases (VLDB 2006), Korea (September, 2006).

75. A. Beresford and F. Stajano, Location privacy in pervasive computing, *IEEE Pervasive Computing* **2**(1), 46–55, (2003). ISSN 1536-1268.

76. M. Gruteser, J. Bredin, and D. Grunwald, Path privacy in location-aware computing. In *Proc. of the Second International Conference on Mobile Systems, Application and Services (MobiSys2004)*, Boston, Massachussetts, USA (June, 2004).

77. M. Gruteser and X. Liu, Protecting privacy in continuous location-tracking applications, *IEEE Security & Privacy Magazine* **2**(2), 28–34 (March-April, 2004).

78. B. Ho and M. Gruteser, Protecting location privacy through path confusion. In *Proc. of IEEE/CreateNet International Conference on Security and Privacy for Emerging Areas in Communication Networks (SecureComm)*, Athens, Greece (September, 2005).

79. M. Duckham and L. Kulik, A formal model of obfuscation and negotiation for location privacy. In *Proc. of the 3rd International Conference PERVASIVE 2005*, Munich, Germany (May, 2005).

80. M. Duckham and L. Kulik, Simulation of obfuscation and negotiation for location privacy. In *Proc. of Conference On Spatial Information Theory (COSIT 2005)*, (September, 2005).

81. Openwave Location Manager, *Openwave Location Manager*. Openwave, (2006). http://www.openwave.com/.

82. P. Bellavista, A. Corradi, and C. Giannelli, Efficiently managing location information with privacy requirements in wi-fi networks: a middleware approach. In *Proc. of the International Symposium on Wireless Communication Systems (ISWCS'05)*, Siena, Italy (September, 2005).

83. C. A. Ardagna, M. Cremonini, E. Damiani, S. De Capitani di Vimercati, and P. Samarati, A middleware architecture for integrating privacy preferences and location accuracy. In *Proc. of the 22nd IFIP TC-11 International Information Security Conference (SEC 2007)*, Sandton, South Africa (May, 2007).

84. C. A. Ardagna, M. Cremonini, E. Damiani, S. De Capitani di Vimercati, and P. Samarati, Location privacy protection through obfuscation-based techniques. In *Proc. of the 21st Annual IFIP WG 11.3 Working Conference on Data and Applications Security*, Redondo Beach, CA, USA (July, 2007).

85. C. Ardagna, M. Cremonini, E. Damiani, S. D. C. di Vimercati, and P. Samarati, Managing privacy in LBAC systems. In *Proc. of the Second IEEE International Symposium on Pervasive Computing and Ad Hoc Communications (PCAC-07)*, Niagara Falls, Canada (May, 2007).

86. C. Hauser and M. Kabatnik, Towards Privacy Support in a Global Location Service. In *Proc. of the IFIP Workshop on IP and ATM Traffic Management (WATM/EUNICE 2001)*, Paris, France (March, 2001).

87. U. Hengartner and P. Steenkiste, Protecting access to people location information, *Security in Pervasive Computing* (March, 2003).

88. Geographic Location/Privacy (geopriv), *Geographic Location/Privacy (geopriv)* (September, 2006). http://www.ietf.org/html.charters/geopriv-charter.html.

89. D. Hong, M. Yuan, and V. Y. Shen, Dynamic privacy management: a plug-in service for the middleware in pervasive computing. In *Proc. of the 7th International Conference on Human Computer Interaction with Mobile Devices & Services (Mobile-*

HCI'05), Salzburg, Austria (2005).

90. M. Langheinrich, A privacy awareness system for ubiquitous computing environments. In eds. G. Borriello and L. E. Holmquist, *Proc. of the 4th International Conference on Ubiquitous Computing (Ubicomp 2002)*, (September, 2002).

91. C. Ardagna, M. Cremonini, E. Damiani, S. De Capitani di Vimercati, and P. Samarati, Supporting location-based conditions in access control policies. In *Proc. of the ACM Symposium on Information, Computer and Communications Security (ASIACCS'06)*, Taipei, Taiwan (March, 2006).

92. V. Atluri and H. Shin, Efficient enforcement of security policies based on tracking of mobile users. In *Proc. of the 20th Annual IFIP WG 11.3 Working Conference on Data and Applications Security*, Sophia Antipolis, France (July-August, 2006).

Chapter 12

Risk-Based Access Control for Personal Data Services

Soon Ae Chun* and Vijayalakshmi Atluri†

*College of Staten Island, City University of New York
Staten Island, NY 10314
chun@mail.csi.cuny.edu

†MSIS Department and CIMIC, Rutgers University
Newark, NJ 07102
atluri@rutgers.edu

In the context of ubiquitous computing, small mobile devices are used as a platform for consuming various services. Specifically, personal data and information of a person (called data owner) are distributed over many third party organizations (data/service provider), and are being made available as Web services, such as monthly financial statements, personal medical data services (e.g., X-ray results), etc. Very often, the personal information Web services are not just used by the data owner, but by third party data consumers who work on the cases on behalf of the data owner, such as financial advisers or doctors. In this environment, the data consumers often are not the same as the data owner. Access control is enforced to prevent confidential personal information from falling into the hands of "unauthorized" users. However, in many critical situations, such as emergencies, relevant information may need to be released even if users are not explicitly authorized. In this paper, we present the notion of situational role and propose a risk-based access control model that makes the decisions by assessing the risk in releasing data in the situation at hand. Specifically, it employs the "access first and verify later" strategy so that needed personal information is released without delaying access for a decision making third-party, and yet providing an adequate mechanism for appropriate release of personal information by a third party provider. Our approach employs the notion of situation role and uses semantics in building situation role hierarchies. It computes the semantic distance between the credential attributes required by the situational role and the actual role of a user requesting access, which essentially is used in assessing the risk.

Contents

12.1 Introduction

In today's networked environment, often much of the personal information and contents are created, stored and are in custody of private third parties. For example, medical records of one person are distributed over many different doctors (general doctors, specialists, etc.), financial information is in custody of financial institutions (banks, investment companies, credit card companies), legal documents in various institutions (e.g. lawyer's office, courts, police stations), photo services and so on. With the ease of using Internet and Web technologies, the data custodians can make the data readily accessible through Web services.

Very often, the personal data stored by the service providers needs to be accessed by another third party consumer who works on one's case for decision making or evaluation. For instance, the doctor who is diagnosing a patient's condition may need X-ray results from the outpatient center as well as lab results from a laboratory. Traditional solutions may impose access control on the service provider side to ensure proper access by the third party's (e.g. doctor's) role and credentials. However, this solution may not be the best solution or a feasible one, since the third party data requestor can be an ad-hoc one time user. Pre-defining and considering all possible data requestor roles on the data provider side is not feasible. The data owner is separate from the data provider and data consumer (requestor). In this largely distributed and ad-hoc environment, personal data is distributed over many different providers, and it is accessed not only by the data owners, but also by third party data consumers. A security challenge is to provide a mechanism to control access to the Web service usages, without a pre-defined trusted entity.

Another security challenge is that users do not have much control over their data transfer by the legitimate third party data consumer to another party, who may be partnered or affiliated with the data consumer. Web service aggregators for value-added services may need to use data services that are provided by multiple providers. The value-added services or decision making requires multiple data services. An inappropriate data transfer from one organization to another without appropriate consent is not desirable. At least a notification of the unexpected data transfer, may alert the improper data flow. With the Web services environment that allows automated agents to exchange data freely, this kind of warnings of data transfers may be useful to prevent or to remedy the damage that can be caused with further propagation.

These challenges are also applicable in the organizational level. Most organizations now use several outsourcing or consulting services where the company information, such as payroll and employee information, is served by third party data providers, and the data requestors/consumers may be another third party organization which evaluates the company's financial health, or standing.

However, in time critical situations such as a medical emergency, disaster situation, or incident involving critical information needs, the personal data is being accessed and released without much concern. However, after the mitigation and responses to the situations are handled, it is not clear whether the private data was accessed unnecessarily or not. Our goal in this paper is that even in those emergency situations some access control can be enforced based on the assessment of risks in releasing private, confidential data, and there should be traces to track back and audit the data access in case needs arise.

Example 12.1 We present an example that illustrates the need for obtaining data from third party entities. John just had a car accident, and his injury requires immediate medical attention. The police needs to call an ambulance, his car needs to be towed, and the paramedic transported him to a medical emergency room for immediate treatment. The nurses need to have his medical records and health insurance data. John's blood type information and his medical history records reside distributed in several doctor's offices. The nurse can quickly authenticate to the medical group network, and finds the services of doctors who provide the medical records for their patients. Using the car accident report ID sent over by the police laptop as an authentication credential, the nurse on behalf of John will be able to access the appropriate blood type from the lab, and diabetic conditions and drug/allergic conditions from general and specialist doctor's. For proper diagnosis, the X-ray of the chest is also required, and the images are retrieved from another X-ray lab. The blood type identified is used for blood transfusion, and the ER doctor will use drug reaction conditions to prevent certain drug uses. The nurse also finds out the insurance information to alert the insurance carrier.

Normally these medical records are not released to a third party, but in an emergency situation as in the above example, the availability of these records for the third party is necessary and desirable to handle the case without delaying or without duplicating the existing records. In this case the nurse uses the police report ID as an informal guarantor for her access rights. However, the nurse should not be allowed to view John's salary or employment related records, or financial investment information in this context. It also should not let the nurse propagate/leak John's sensitive information to another individual or group, such as another specialist doctor who may be interested in expanding his own patient pool, or an insurance company that may affect John's future insurance costs or benefits.

In this paper, we propose a risk-based distributed data access control that utilizes a risk-reasoner that calculates the risk level in releasing data. The data is first released even though the data disclosure policies of each data provider do not exactly match with the credentials of the data requestor, but the risk level is acceptable. The risk reasoner calculates the risk-level based on the semantic similarity between the disclosure policy statement and the given credentials. An exact or approximate matching will result in a low risk level, while no matching or a low matching level will result in a high risk level. The risk reasoner considers the semantic and situational relationships between the policy statements and the credentials provided by the requestors. The risk reasoner also considers the risk entailed by having several data items requested at the same time, implying that there may be a higher risk when the data items are combined together than when each data item stands by itself. The data requests are modeled with a connected graph, consisting of a common set of data for a particular situation.

To ensure the accountability of released data under a certain level of risk, we have a monitoring and auditing component that monitors and records the traces of released data, and notifies the data owner about the released items with a capability to "explain" the circumstances under which the release decision was made, if needed.

This paper is organized as follows. In section 12.2, we briefly discuss the previous related work on trust-based access control and Web service authentication and authorization. In Section 12.3, the situation role model and risk-based access control model is introduced followed by section 12.5 that shows the risk-based access control evaluation. In section 12.6 we present a system architecture for proposed risked-based access control and implementation issues. In section 12.7 we conclude and present future research directions.

12.2 Related Work

In a typical federated Web service environment, there is a central federation organization with several partner organizations as service providers. In this federated environment, these business partner organizations establish some or all of business agreements, cryptographic trust, and user identifiers or attributes across security and policy domains to enable more seamless cross-domain business interactions. Federation is the dominant movement in identity management today. It decouples or loosely couples at the identity management layer, insulating each domain from the details of the others' authentication and authorization infrastructure. Key to this loose coupling at the identity management layer are standardized mechanisms and formats for the communication of identity information between the domains.

A standard language SAML (Security Assertion Markup Language) developed by the OASIS XML-Based Security Services Technical Committee (SSTC) is an XML-based framework that provides a common language for exchanging the security information and sharing security services between partner organizations engaged in B2B and B2C business transactions. SAML allows business entities to make assertions regarding the identity, attributes, and entitlements of a subject (an entity that is often a human user). This approach allows a single sign on (SSO) for federation partners. However, this approach assumes the prior-agreement between participating organizations, that is, all the participants are trusted partners.

To address the ad-hoc access requests in many open systems, such as the Web environment, where no prior known entities are involved, there have been several approaches, such as trust negotiations,[1] content-based trust generation[2] among many, in open systems, and resource services access[3] in pervasive computing environments. These approaches concern establishing trust among the unknown Web interaction parties, e.g. clients and servers, when there are no pre-existing trust relationships.

For trust negotiations, the access policy for a resource is specified in declarative credentials required to gain access to the resource. The credentials required for accessing resource access may be also sensitive, and needs to be released only when the server is trustworthy, and the data resources are released only when the client meets the necessary credentials.[1,2] The trust building involves bilateral and iterative negotiation between parties, to establish trust in one another.

In the study,[2] the trust-building credentials are dynamically generated from the contents of messages or resources exchanged. The contents are filtered and if they are classified to be of a sensitive nature, then the appropriate credentials are requested for checking whether the message is allowed to be sent out to the recipient or to be received from the sender. There are four types of message security monitoring: client sending message monitor, client receiving message monitor, server sending message monitor and server receiving message monitor. These monitors capture the messages coming in or going out and classify the messages using similarity match with pre-defined sets of sensitive data queries.

In the work,[3] the access to distributed resources in a pervasive computing environment is allowed even though the requesting client is not a user with pre-established trust with the security agent who checks the credentials. The client first obtains the access right delegation from a trusted party (i.e. authorized party by the security agent). With the delegation of a trusted party, the client unknown to the system decides grants to access. This mechanism provides dynamic access rights according to the trust delegation from a trusted party. It presents a distributed model in which security agents are hierarchically arranged and security agents manage security and trust, and X.509 authentication certificates identify users and services. Authorized users can make delegations and revocations in the form of signed assertions. These signed assertions with delegation information are evaluated by security agents, and matched with the appropriate security policies for providing access control. The study assumes that it is easy to identify the trusted party as a "Certificate Authority" for obtaining authorization delegation for the access requestor. However, in a fully distributed environment, identifying the right trusted party for obtaining delegation rights is in itself a daunting task. Also the delegated assertions only consider the trusted party's authentication credentials that are statically stored with the security agent.

The Benefit And Risk Access Control (BARAC)[4] is proposed where the access control is based on the benefits of information sharing and risks of information disclosure. Typically, there are allowed transactions (represented as AT graph) and information flow paths and BARAC accessiblity graph (AC graph) that describes the objects accessible by subjects, but in some circumstances, the allowed paths are not possible due to different circumstances. Thus, it needs to modify AT graph and AC graph. The modifications of these BARAC configurations may result in adding risks or subtracting benefits. The access control system maintains the total benefits and risks of these modified AT and AC graphs with allocated budgets.

In the study,[5] different permissions are associated with different levels of risks, and there exist security risk ordering relations among these tasks. The role hierarchy relations are used for selecting a role for delegation of tasks that would yield least risks. This is similar to our approach in role related risks, but it does not consider semantic distance, but strict seniority of the role hierarchy. It also assumes a thorough study on risk relationship orderings of role and permission relations within an organization, e.g. a hospital. This is not feasible in an open system as in the Internet-based information access.

The approach[6] is a risk-adaptive access control model where the estimated quantified risk is based on the probability of damage times value of damage. These estimated risks belong to some band in the risk scale that represents the range of quantified risk estimates that is further divided into multiple bands of risk. Each band on the risk scale is associated with an access control decision. If the estimated risk falls in a lower band, then the access is allowed. On the other hand, if it falls in the top band, the access is denied.

12.3 Risk-Based Access Control Model

In this section, we present a risk model for roles and objects, followed by a risk-based policy representation.

12.3.1 *Situational Role Model*

The risks involved with roles typically arise when the access control system needs to grant an access, even though the requester's role deviates from a role speci-fied in the authorization rules, or the credentials of the requester do not exactly match with the authorized role. Thus, even though the authorization rule states that "doctors in general hospitals are allowed to read a patient record" but the request is made by "a nurse in a general hospital" or "a doctor in a medical uni-versity," the system tries to grant access to the requester due to a circumstance such as an emergency or urgent nature of data needs. In these cases, the usual exact specification of the role in an access control rule or an exact matching of the rule may need to be relaxed and the role specification needs to be enriched.

Our approach is to model more enriched roles, using a role in a situation called a *situational role*. A situation is modeled with a workflow which denotes a process with a set of coordinated tasks and typical roles (default or prototypical roles) that are assigned to execute these tasks.

We assume that each user is associated with one or more *credentials*. Creden-tials are assigned when a user is created and are updated accordingly. We assume the policy states that a user should provide a set of *credentials* to assume a role. Thus, a role is defined in terms of a bundle of credentials with a unique type ID called *credential type*. Let $CT = \{ct_1, ct_2, ...\}$ be the set of credential types.

Definition 12.1 (Credential-type) *A credential-type ct is a pair (ct_id, CA), where $ct\_id \in CT$ is a unique identifier and CA is the set of attributes belonging to ct_id. Each $ca_i \in CA$ has an attribute name and $CA(ct)$ is the set of attributes belonging to ct.*

Credential types are organized in a subtype relationship hierarchy, called *Credential Type Hierarchy*. We use C_u to denote the set of credentials associated with a user u.

Definition 12.2 (Credential) *A credential c, an instance of a credential-type ct, is a triple (ct_id, c_id, C_u), where $ct\_id \in CT$, c_id is a unique credential identifier, and $C_u = (a_1 : v_1, \ldots, a_n : v_n)$, where $\{a_1, \ldots a_n\} \subseteq CA(ct)$.*

Example 12.2 An example of a credential for a credential type "doctor" is as follows:
(doctor, D123, (medical license: Board of New York Medical Association, license Number: 12412, specialty:surgery, affiliation: St. Barnabas Hospital)).

A *situation* is modeled with a typical process that contains a set of interrelated tasks performed by a set of authorized roles. For instance, in a medical emergency situation, there are typically a doctor, nurse, residents, medical aids, ambulance drivers, etc., who interact in a process of providing patient care, such as patient record retrieval by a nurse, critical health indicator comparison by a nurse, medical treatment information lookup by a resident, look up for medical diagnostic information, and prescription alternatives by a doctor etc. In the following, we define a situation.

Definition 12.3 (Situation) *A situation s is represented as a pair (sid, W) where sid denotes the unique name of a situation, and $W = (T, D)$, is a workflow (process), where $T = \{t_1, t_2, \ldots\}$ is a set of tasks and D is the set of edges denoting inter-task dependencies among tasks.*

Dependencies are typically of the form $t_i \xrightarrow{x} t_j$ where x the dependency expression. It essentially means that t_i follows t_j only if x is true. Several categories of dependencies are possible, including control-flow, value and external dependencies. Additionally, the set of tasks are not always sequential but may be of the form, AND/OR join and AND/OR split, etc.[7,8]

An example of a situation is (car-accident, W_1) that shows a car accident situation that is typically associated with a workflow, W_1="Process-Accident-scene", where W_1 consists of tasks T= {call-police, look-up-license, generate-police-report, call-ambulance, lookup-blood-type, lookup-medical-records, call-insurance, call-criminal-records, call-accident-history...}, where each task has a typical role that executes it, e.g. the policeman performs generate-police-report. The dependencies among the tasks, for example, include call-criminal-records follows look-up-license only if look-up-license has resulted in the discovery of an expired license.

Situations, expressed as sid's, are nothing but their descriptors. We use the notations $W(s), T(s), D(s), t_i(s)$ to refer to the workflow, its task set, its dependency set, and a particular task t_i in $W(s)$ of a situation s, respectively.

We assume that situations are organized hierarchically based on the semantic relationships among one another. Specifically speaking, a situation is at a higher level in the hierarchy, if it pertains to a more general concept. We define a situation hierarchy as follows:

Definition 12.4 (Situation Ontology) *A situation hierarchy H_s is defined as a set of situations $\{s_1, s_2, ...\}$ with partial ordering relationships among all s_i and s_j in H_s.*

An example of the situation hierarchy is a disaster situation having subtypes such as "natural disaster" and "man-made disaster" situations, and natural disaster in turn can have several subtype situations such as "weather disaster," "earthquake."

Each task $t \in T$ in the situation workflow W has a specified role(s) that is allowed to execute the task. We denote this role as the *situational role*. Given a task $t \in W$, for a situation, we denote the situational role as $r_s(t)$. We assume that for every situational role $r_s(t)$, there exists a required set of credential attributes associated with it, which is defined as follows.

Definition 12.5 (Situational Role) *A situational role $r_s(t)$ is represented as a triple $(s, t_i(s), C)$ where $C = \{a_1 : v_1 ... a_n : v_n\}$ is the required credential set associated with r, denoted as $C(r)$.*

$C(r)$ is a set of attribute and value pairs $(a_1 : v_1, ..., a_n : v_n)$.

Based on the semantics associated with the roles, we assume that situation roles can be partially ordered, which we denote as a *situation role hierarchy*, which captures the partial ordering relationships among them. Note that this may not be the same as the role hierarchy as traditionally assumed. We assume that the situation role hierarchy always has a root and the level of a situation role is higher if it is closer to the root. A role r_i is said to be at a higher level than the other if its situation is more general than that of the other.

12.3.2 *Objects*

We assume objects in the system comprise of properties/attributes and their values. Policy rules over objects are specified using the above attributes and logical combinations of these attribute-value relationships. The requested objects should have matching properties with those specified in the policy, in order to grant an access to the specified object. The privacy and confidentiality level of the object determines the risk of allowing access to that object. For instance, exposure of an object with personally identifiable attributes such as social security number or name, may cause more risk than when objects with other non-identifying attributes are accessed, since those may violate a person's privacy (e.g. medical records with identifying information), or may risk identify thefts.

In this paper, we assume that the data objects can differ in the risk level depending on three types of sensitive attributes: (1) anonymous attributes, (2) personally non-identifying attributes, such as age, date of birth, gender, occupation, education, income, ZIP Code with no address, interest and hobbies; and (3) personally identifying information that refers to information that can be used to identify or locate an individual. The exposure of the object may cause different levels of risk depending on whether personally identifying attributes are included in the request or not, since they can pose more risks to the privacy of an individual.

We assume that there exists a set of data providers $P = \{p_1, p_2, ...\}$ that may generate, manage and disseminate data objects, and assume that each data provider p_i has a set of data objects $O_i = \{o_{i1}, o_{i2}, ...\}$ residing at p_i. Each data object has a set of attributes $o_i = \langle at_1, at_2, ... \rangle$. Thus we have multiple data sets $OBJ = \{O_1, O_2, ...\}$ distributed over different providers in P. At present, we do not consider that data are replicated and maintained by multiple data providers.

Each data set O_i is provided by a service provider p_i through a set of Web services $WS_i = \{ws_1, ws_2, ...\}$. Each Web service ws_i takes Input IN and Output OUT. The input and output behavior of each Web service is described and available in a directory service UDDI (Universal Description Discovery and Integration). UDDI also contains all the Web services provided by all the service providers $WS = \{WS_1, WS_2, ...\}$.

Each attribute can be associated with a risk level inherent to data items such as data attributes that can identify the personal identity or reveal any confidential information when it is revealed, so called linkable attributes. These risks can be at different levels, and we define these as risk indices. Thus a data attribute that has a potential of leaking sensitive information has a higher risk index than those that do not reveal private or secure information. For instance, by default the attributes that can identify individuals uniquely as in the unique primary attribute, or the attributes that have greater linkability will be assigned as high risk. However, each individual may have different levels of concerns in privacy. In such a case, he or she can express her privacy preferences using privacy coefficients to denote the personal privacy policy expressing how one values the level of sensitivity or privacy of data attributes.

Definition 12.6 (Privacy Coefficient) *The privacy coefficient pv_i of an attribute at_i of an object o is a natural number that specifies its level of privacy. For an object o, $PV(o) = \{pv_1, pv_2, ...\}$ represents the privacy coefficients of all the attributes of o.*

These privacy coefficients will be used to calculate the risk levels, where the value of privacy pv_i is directly translated into the risk level of the attribute at_i. Thus the higher the privacy coefficient, the higher the risk level. The personal privacy policy may not list all the attributes' privacy levels. For the attributes not specified in the policy, the attribute is not considered as private or sensitive. Based on the level of privacy, one may define the risk associated with releasing an object attribute. Obviously, risk is some function of the privacy.

Definition 12.7 (Object Risk Vector) *Given object $o \in O$ with attribute vector $A(o) = \{at_1, at_2, ...\}$ and the privacy coefficient $PV(o)$ associated with the attributes, a risk vector of o is defined as $RK(o)=(rk_1, rk_2, ...)$ where $rk_i = f(pv(at_i))$.*

$RK(o)$ represents a risk index of each attribute in $A(o)$. A higher risk index rk_i represents a higher risk of revealing sensitive information about o or the data owner of o. For instance, the social security information for an employee salary record or a patient medical record may have a higher risk index than the transaction amount or the blood pressure information.

12.4 Situation Role-Based Access Control

When a user logs in, a new session is activated and he/she presents credential certificates and a situation identifier. Upon providing user credentials, namely a set of attribute and value pairs, the system checks with the credentials for potential roles. When the credentials provided by the user match with the credentials associated with a situation role, then it is activated.

For instance, if a user is logged on as a doctor with appropriate credentials for a doctor role, and the situation is medical emergency, then the situation role, emergency doctor = (emergency, operate, surgeon), is activated. As stated in definition 12.3, a situation workflow consists of tasks and the typical roles associated with each task in the workflow. Thus, for the medical emergency, the "Operate" workflow may have roles such as {nurse, doctor, anesthetist, clerk, ...}. Among these, the situation workflows that have overlapping roles with those verified with the roles according to the certificate credentials are activated. Thus role activation is a mapping of a user session in a situation to situation roles.

Definition 12.8 (Situation Role-based Access Control)

- $U, SROLES, T$, and OBJ are the set of users, situation roles, tasks, and operations.
- $UA \subseteq U \times SROLES$, a many-to-many mapping user-to-role assignment relation.
- $assigned\_users(r) \rightarrow 2^{SROLES}$, the mapping of r onto a set of users. Formally: $assigned\_users(r) = \{u \in U | (u,r) \in UA\}$, the mapping of role r onto a set of users.
- $PRMS = 2^{(T \times OBJ)}$ (the set of permissions) Formally: $p \subseteq \{(t,o) | t \in T \wedge o \in OBJ\}$
- $PA \subseteq PRMS \times SROLES$, a many-to-many mapping of permission-to-role assignments.
- $assigned\_permissions(r) \rightarrow 2^{PERMS}$, the mapping of role r onto a set of permissions. Formally, $assigned\_permissions(r) = \{p \in PRMS | (p,r) \in PA\}$.
- $t \rightarrow \{P \in PRMS\}$, the task-to-permission mapping, which gives the set of permissions associated with a task t.
- SES, the set of sessions.
- $user\_sessions(u : USERS) \rightarrow 2^{SES}$, the mapping of user u onto a set of sessions.
- $session\_roles(ses : SESSIONS) \rightarrow 2^{SROLES}$, the mapping of session ses onto a set of roles. Formally: $session\_roles(ses_i) \subseteq \{r \in SROLES | (session\_users(ses_i), r) \in UA\}$.
- S, the set of situations.
- $situation(s : S) \rightarrow 2^{W}$, the mapping of situation s onto a set of workflows.

Since objects are stored and delivered by the data providers, the requests for data objects are done through data Web services. The Web services access and manipulate the data items. Web services include not only traditional simple access functions such as "read/retrieve/view" and "write/update," but also include other types of data record manipulations using different types of Web services such as "anonymize" which limits access to only the non-identifying attributes of data objects, or "select" where a subset of the data records are retrieved. In addition, the operations can include not only these atomic Web services, but also composite Web services which combine several Web services in sequence to arrive at the desired value-added data. Thus, each Web service data provider can maintain the situation roles and the set of required credentials for that role and also evaluate the eligibility of the consumers of the Web services.

Definition 12.9 (Web services) *Given a set of Web services WS, a Web service*

$ws \in WS$ is defined as an operation on data objects $ws(O_{IN}, O_{OUT})$ where $O_{IN} \subset OBJ$ and $O_{OUT} \subset OBJ$.

Definition 12.10 (Permission Assignment Rules) *Permission assignment is a relationship that maps a situation role to a set of Web services, denoted as* $R \rightarrow 2^{WS}$.

Definition 12.11 (Access Request) *An access request is defined as a tuple* $< u, s, ws, o_{in}, o_{out} >$, *where a user in a session, u, in a situation s requests service ws, given input object o_{in} to receive output o_{out}.*

John Smith handles a medical emergency in the ER unit of a hospital H, and he needs to access the patient's previous hospitalization history records from another hospital H' before any medical treatment. He can look up the Web service repository, and finds a Web service that returns the hospital records given the patient's ID as an input. He directly communicates with H' and proceeds with the authentication process. Once H' authenticates the user and the situation he is in, H' evaluates if a doctor in a medical emergency can access patient's hospital records. The doctor role is activated with the credentials, and the permissions given to doctors in emergency operations are retrieved. The tasks in situation-related workflows contain the permitted activities, i.e. provide patient hospitalization records; then the role is granted to access the objects.

12.5 Risk-Based Access Control Reasoning and Auditing

The basic model for the situational role based access control presented in the previous section assumes that the user's roles exactly match with the permission assignment rules. In an emergency or crisis management situation, the ability for forming a dynamic coalition of organizations and people available at the time is a key for successful situation handling. In this kind of situations, resources may have to be released to users who may not fit to the roles specified for the permissions, because the right people may not be available at the time, and the situation needs to be handled. On the other hand, there are risks in releasing information to the wrong party. There should be a mechanism to assess the risks involved in a structured manner, and also leave enough auditable information in case the needs arise to trace back for accountability.

Our approach provides a reasoning mechanism (Risk Reasoner) to measure risks based on role structure, information sensitivity levels, and process dependency sequences. This section will present the risk measurements and risk-based reasoning for access control decisions.

12.5.1 Risks Based on Role Structure

In this case, we consider the case where the credentials associated with the user do not match exactly with those of the situational roles required for the workflow. Recall that each situation is associated with a workflow W, and the workflow consists of a set of tasks $T = \{t_1, t_2, ...\}$, and each task $t \in T$ is associated with situational roles $R(T) = \{r(t_1), r(t_2), ...\}$ that are granted for executing tasks in each workflow W. However, the set of credentials associated with the user may not match with any situational roles associated with the tasks within the workflow pertaining to the situation. In this case, the system tries to see if the authorized role $(AR \in R(T))$ for a task $t_i \in T$ in W and the user's roles (UR) have any semantic relationships, such as UR is a child of AR, or sibling of AR in a situational role hierarchy. Using the situational role hierarchy and the required credential attributes, we can measure the risk based on the semantic and structural distances between AR and UR. The distance from AR to UR is the risk measure. We look at the attribute dimension as well as the child-parent (specific-general) relationships between these two roles in the role hierarchy.

The semantic distance (SD) of two roles is measured with the semantic attribute distance (AD) that is measured with the cosine distance with multidimensional attributes of two roles, and the relationship distance (RD) is measured by the level distance of the nodes in the role hierarchy.

$$AD(r,r') = \sqrt{((v_1 - v_1')^2 + (v_2 - v_2')^2 + ...)} \tag{1}$$

where v_i, v_i' are values for an attribute a_i, and $a_i \in C(r)$.

$$RD(r,r') = |level(r) - level(r')| \tag{2}$$

where *level* refers to the height in the situation role hierarchy.

$$Risk(r) = SD(r,r') = AD + RD \tag{3}$$

The risk involved in role deviation, i.e. using one role (r') different from the required role (r) in the policy, is greater when the semantic distance between these roles is greater. For example, in the medical emergency situation, the medical office clerk normally does not access the injured person's confidential medical records, but the doctor or nurses may. In the case of a medical office clerk, the access may be granted due to its semantic relationships to the doctor or nurses. The semantic distance of the medical clerk from the granted roles, namely doctors or nurses, is calculated and when the risk level is within the risk threshold, the access is granted. On the other hand, if a lawyer in the situation tries to access the medical records, the risk level calculated from the semantic distance between the granted role (doctors or nurses) and the lawyer may be higher than a threshold to grant access.

12.5.2 *Risks Based on Information Sensitivity*

When the policy specifies the roles that are allowed to use a permitted task (Web services) to access an output, or allowed to execute tasks in a situational workflow W, the risk is not an issue. However, when the roles that are deviating from the authorized role try to execute a task that returns an output object, the risk of revealing the object to this role should be considered. The privacy coefficient of object attributes is used to calculate the risks.

Given privacy coefficients $PV = \{pv_1, pv_2, ...\}$ for some attribute sets of a data object, the risk vector of the requested output object o_{out} is defined as the privacy coefficient of attributes in the requested output. Thus, $RK(o_{out}) = \{pv_i | at_i \in o_{out}\}$. Then the risk level of the requested data object is calculated as the sum of the privacy coefficients in the risk vector for the output of a task (i.e. Web services).

$$Risk(o) = \sum_{i=1}^{n} rk_i \qquad (4)$$

for all attributes in o_{out}.

12.5.3 *Risks Based on Process Dependency*

Another risk is when the typical situational workflow has a sequence of tasks to be performed. Each task may have certain preconditions to be met before it can be performed by a role. However, in some cases, these preconditions may not be met and the task needs to be executed. In other words, there is a risk associated with the task execution since its preconditions are not fully met and pending to be satisfied. For instance, the medical emergency situation may contain the doctor's task to retrieve the medical history of a patient's mental health may depend on a police record of some incident. With the absence of the police record, the task may proceed with some risk of revealing the mental health records of a patient. In some cases, a task may require several preconditions to be met, but not all the preconditions are satisfied before the task is executed.

To calculate the risk for a task according to a dependency (or pre-condition) satisfaction, the following risk is used:

Given a dependency between two tasks, $t' \xrightarrow{d} t$, the risk of task t

$$rk(t) = \begin{cases} 1 & \text{if d(t) is not satisfied;} \\ 0 & \text{otherwise.} \end{cases} \qquad (5)$$

For calculating the risk associated for all task $t \in T$, we use $Risk(t)$ to denote the sum of $rk(t)$ for all t where each t has dependencies, $t' \xrightarrow{d} t$, as shown in equation (6).

$$Risk(t) = \sum_{\forall t' | t' \xrightarrow{d} t} rk(t) \qquad (6)$$

Thus given a situation s, the overall risk of a request for accessing an object through a task (Web service) by a role in a situation s can be expressed as:

$$Risk(r, t(o)) = Risk(r) + Risk(o) + Risk(t) \qquad (7)$$

The risk-based access control algorithm enforces the access control based on the risk values. The algorithm *Risk-based Access Control* shows a pseudo-code for the the risk reasoning and access control:

Risk-based Access Control (rq):

Input: rq= $< us, s, ws, o_{in}, o_{out} >$, where us, a user in a session in a situation
    s, requests service ws given input object o_{in} to receive output o_{out}.
Uses: Repositories of Situation Workflows, Role-credential Hierarchy,
    Situation hierarchy, Web services, Object Risk Vector;
Output: Permissions for grant access, empty for denying access
1. $R(us) \leftarrow$ Validate user's credentials to determine roles.
2. $W(s) \leftarrow$ Determine situation workflows
3. $SR \leftarrow$ Identify Situation Roles $(R(us), ws \in W(s))$
4. $AR \leftarrow$ Identify Permitted Roles $(ws))$
5. if $SR \cap AR \neq \emptyset$
6.   then Return $(SR \cap AR)$
7.   else /* if no permitted role exists */ {
8. $Risk(r) =$ Semantic Distance (SR, AR)
9.   Risk $(o) =$ Sensitivity Risk $(RK(o))$
10. $Risk(t) =$ Dependency Risk for all t' where $t' \xrightarrow{d} t$
11.   Risk $=$ Risk$(r) +$ Risk$(o) +$ Risk$(t) >$
12.   if $(Risk > \delta)$
13.   then Return$(\emptyset)$
14.   else { RecordTrace $(us, R(us), WS(o_{in}, o_{out}), Risk, timestamp)$
15.   Return(WS, δ)}}

12.6 Architecture of Prototype and Implementation Issues

This section describes the component architecture of the Risk-based Access Control to provide an overall approach. We consider a decentralized architecture where a user may send a request for an object to a particular service provider. Figure 12.1 shows the data service providers and the users. Each provider maintains a proxy server that can run the situation-based and risk-based access control with risk reasoning to evaluate a proper access and to maintain the audit records. The proxy server also authenticates the user for a session, using the certificate and its credentials provided by a certificate authority (CA). In addition, this architecture is based on the third party services, on the situation related workflow information that contains the typical processes associated with a situation and its typical roles, and on Web service registry services such as UDDI to locate the service profiles. This architecture is scalable as the ad-hoc users may be authenticated and authorized for the requested object, and it allows each service provider to maintain its own access policies and privacy policies, i.e. privacy coefficients for objects, and enable them to be more flexible without losing control over their own company's rules and without unnecessarily revealing the company policies.

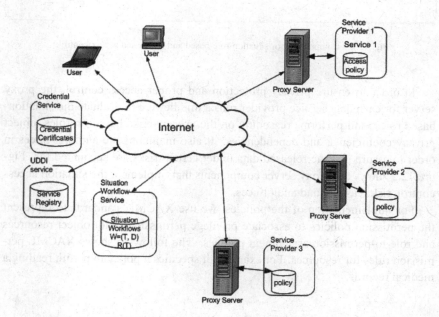

Fig. 12.1 Decentralized system architecture for risk-based access control.

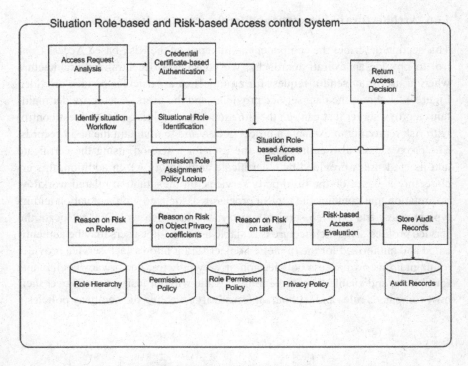

Fig. 12.2   Components for situation role-based and risk-based access control.

In order to ensure privacy protection and proper access control, the proxy server for each data service provider runs a middleware to evaluate the situation-based roles, and performs reasoning on the privacy risks based on roles, object privacy coefficients, and dependencies. It also maintains the auditing traces in order to assure that the released data under certain risks are accounted for. Figure 12.2 shows the proxy server components that implement the situation access control, risk reasoner and audit traces.

For implementation of the policies, we use XACML standards to represent the permission policies to associate privilege permissions with object resources and role-to-permission assignment policies. The following shows XACML permission rules for resources. For example, it specifies a policy to permit reading a medical records.

Table 1.1 XACML specification of permission rules.

```
<PolicySet xmlns="urn:oasis:names:tc:xacml:2.0:policy:schema:os" PolicySetId="PPS:doctor:role"
PolicyCombiningAlgId="&policy-combine;permit-overrides">
  <!-- Permissions specifically for the doctor role -->
  <Policy PolicyId="Permissions:specifically:for:the:doctor:role"
    RuleCombiningAlgId="&rule-combine;permit-overrides">
  <!-- Permission to access medical records-->
  <Rule RuleId="Permission:to:access:medical:records" Effect="Permit">
   <Target>
   <Resources>
    <Resource>
    <ResourceMatch MatchId="&function;string-equal">
      <AttributeValue DataType="&xml;string">medical records</AttributeValue>
      <ResourceAttributeDesignator AttributeId="&resource;resource-id" DataType="&xml;string"/>
    </ResourceMatch>
    </Resource>
   </Resources>
   <Actions>
    <Action>
    <ActionMatch MatchId="&function;string-equal">
    <AttributeValue DataType="&xml;string">read</AttributeValue>
      <ActionAttributeDesignator AttributeId="&action;action-id" DataType="&xml;string"/>
      </ActionMatch>
     </Action>
   </Actions>
   </Target>
  </Rule>
</Policy>
</PolicySet>
```

The following shows a role to permission assignment. It says a subject who has the attribute value of "doctor" has a permission that is specified in the permission policy set "PPS:doctor:role."

Table 1.2 An example of a role to permission assignment.

```
<PolicySet xmlns="urn:oasis:names:tc:xacml:2.0:policy:schema:os" PolicySetId="RPS:doctor:role"
PolicyCombiningAlgId="&policy-combine;permit-overrides">
<Target>
<Subjects>
  <Subject>
  <SubjectMatch MatchId="&function;anyURI-equal">
  <AttributeValue DataType="&xml;anyURI">&roles;doctor</AttributeValue>
  <SubjectAttributeDesignator AttributeId="&role;" DataType="&xml;anyURI"/>
  </SubjectMatch>
  </Subject>
</Subjects>
</Target>
<!-- Use permissions associated with the doctor role -->
<PolicySetIdReference>PPS:doctor:role</PolicySetIdReference>
</PolicySet>
```

In addition, a policy to assign a role to a situation workflow is similarly specified. Each workflow is specified in Business Process Execution Language (BPEL)
which has been a popular process specification for Web services, and the policy
of associating a BPEL with a specific situation is defined in the situation policy.
One situation can be associated with multiple BPEL specifications. The following shows the situationr-to-workflow BPEL assignment policies. We do not show
the BPEL specification for the medical emergency process, which is a set of Web
services (tasks) with dependencies.

Table 1.3 An example of a Situation-to-Workflow assignment.

```
<policyset xmlns="urn:oasis:names:tc:xacml:2.0:policy:schema:os"
PolicySetId="SWS:situation:process" PolicyCombiningAlgId="&policy-combine;permit-overrides">
<Target>
<Situations>
  <Situation>
  <SubjectMatch MatchId="&function;anyURI-equal">
  <AttributeValue DataType="&xml;anyURI">&situation;medical-emergency</AttributeValue>
  <SubjectAttributeDesignator AttributeId="&situation;" DataType="&xml;anyURI"/>
  </SituationMatch>
  </Situation>
</Situations>
</Target>
<!-- Use process specification associated with the medical-emergency-situation -->
<PolicySetIdReference>BPS:medical-emergency:process</PolicySetIdReference>
</PolicySet>
```

12.7 Conclusion and Future Work

In this paper, we have presented a situational role-based access control model
and risk-based access control reasoning approach. We presented an approach that
allows a situation to be a factor in evaluating the access control decision. Often
the situations, such as emergency or time-critical situations, determine the typical
processes of responding. An access control in the emergency or crisis situations
is either not provided, ignoring most of policies, or not adequately addressed.
Even in these situations, the information, especially privacy-related data, should
be disseminated in a controlled manner and in an accountable fashion, such that
there is no gross privacy violation. We introduced a model of a situation as a
set of typical workflows (response processes, or typical activities) that consists of
sequenced tasks (or services) that are executed by typical roles. This is called a
situational role. The situational role-based access control is a method to match
the situational role activation with permitted roles. In case of a user whose roles
do not match with these typical roles, a set of risks is computed based on the

role mismatch, the object privacy coefficient that represents the level of damage if the object is exposed, and the task dependencies that may not be completely satisfied although they are typically required in order for a task (data release) to be performed. When risk-based access control is used, the system records the audit information to report back in case needs arise.

A decentralized architecture for access control is proposed for data or service providers such that the access reasoning and its required policies can be locally managed by the service providers. The situation related information, such as typical processes and roles, can be provided by third party service providers. An implementation within the Web service framework is proposed with the decentralized policy specifications using XACML standards.

Future work includes more detailed work on the audits and accountabilities and privacy object provenance. An implementation of the prototype system is under way. In this paper, the events or situations are not "authenticated." One research direction is how to verify the situation as it is claimed by the user to exist. Most of user authentication resorts to a fixed set of credentials, except some contextual information. The certification of a situation should be dynamic and it may not resort to a static trusted third party, but the authentication of the emerging situation may depend on other sources, and the types of credentials used to verify the situation may be quite different from the conventional user identity related credentials.

References

1. A. J. Lee, M. Winslett, J. Basney, and V. Welch. Traust: a trust negotiation-based authorization service for open systems. In *SACMAT*, pp. 39–48, (2006).
2. A. Hess. Content-triggered trust negotiation. Master's thesis, Department of Computer Science, Bringham Young University, (2003). Unpublished.
3. L. Kagal, T. W. Finin, and A. Joshi. A policy language for a pervasive computing environment. In *POLICY*, pp. 63–, (2003).
4. L. Zhang, A. Brodsky, and S. Jajodia. Toward information sharing: Benefit and risk access control (barac). In *POLICY*, pp. 45–53, (2006).
5. N. Nissanke and E. J. Khayat. Risk based security analysis of permissions in rbac. In *WOSIS*, pp. 332–341, (2004).
6. P.-C. Cheng, P. Rohatgi, C. Keser, P. A. Karger, G. M. Wagner, and A. S. Reninger. Fuzzy multi-level security: An experiment on quantified risk-adaptive access control. In *IEEE Symposium on Security and Privacy*, pp. 222–230, (2007).
7. V. Atluri and S. A. Chun, A geotemporal role-based authorization system, *International Journal of Information and Computer Security*. 1(1/2), 143–168, (2007).
8. V. Atluri, S. A. Chun, and P. Mazzoleni, Chinese wall security for decentralized workflow management systems., *Journal of Computer Security*. 12(6), 799–840, (2004).

Chapter 13

Topological Vulnerability Analysis: A Powerful New Approach For Network Attack Prevention, Detection, and Response

Sushil Jajodia and Steven Noel

Center for Secure Information Systems
George Mason University
Fairfax, VA 22030-4444, USA
{jajodia,snoel}@gmu.edu

This chapter examines issues and methods for survivability of systems under malicious penetrating attacks. To protect from such attacks, it is necessary to take steps to prevent them from succeeding. At the same time, it is important to recognize that not all attacks can be averted at the outset; those that are partially successful may be unavoidable, and comprehensive support is required for identifying and responding to such attacks. We describe our Topological Vulnerability Analysis (TVA) system, which analyzes vulnerability to multi-step network penetration. At the core of the TVA system are graphs that represent known exploit sequences that attackers can use to penetrate computer networks. We show how TVA attack graphs can be used to compute actual sets of hardening measures that guarantee the safety of given critical resources. TVA can also correlate received alerts, hypothesize missing alerts, and predict future alerts. Thus, TVA offers a promising solution for administrators to monitor and predict the progress of an intrusion, and take quick appropriate countermeasures.

Contents

13.1 Introduction

Computer networks are inherently difficult to secure against attack. They are often connected to the Internet, for which security was not an original design goal. Default configurations for many software components are insecure, and these configurations often remain unchanged by the user. There is generally little economic incentive to develop secure software, so vulnerabilities are commonplace.

Moreover, network security concerns are highly interdependent, so that a machine's susceptibility to attack can depend on vulnerabilities across the network. Attackers can combine vulnerabilities in unexpected ways, allowing them to incrementally penetrate a network and compromise critical systems. We can reduce the impact of attacks by knowing the paths of vulnerability through our networks. To do so, we need to transform raw security data into topological maps that let us prepare for attacks, manage risks, and have real-time situational awareness.

Traditional tools for network vulnerability assessment simply scan individual machines on a network and report their known vulnerabilities. Security conscious organizations may then employ Red Teams of network penetration testers, who attempt to combine vulnerabilities in ways that real attackers might. But penetration-testing experts are expensive, changes to the network configuration render the test results obsolete, and the only attack paths reported are those found within the allotted test time.

The processes for tracking network vulnerabilities are labor-intensive, require a great deal of expertise, and are error prone because of the complexity, volume, and frequent changes in security data and network configurations. But through automated simulation of possible attack paths, we can understand our overall security posture in the face of actual attacks.

Our innovative approach to network attack survivability is termed *Topological Vulnerability Analysis* (TVA).[1] TVA simulates incremental network penetration, building complete maps of multi-step attacks showing all possible paths into a network. It maintains models of the network configuration and potential threats. From these models, it discovers attack graphs that convey the impact of combined vulnerabilities on overall security. TVA technology includes recursive attack graph aggregation with interactive drill down of scenarios in the cyber domain. It incorporates a variety of types of network scan data, providing the ability to easily model and analyze even large networks.

Currently available tools generally give few clues as to how attackers might exploit combinations of vulnerabilities among multiple hosts to advance an attack on a network. The security analyst is left with just a set of known vulnerabilities. It can be difficult even for experienced analysts to recognize how an attacker might combine individual vulnerabilities to seriously compromise a network. For larger networks, the number of possible vulnerability combinations to consider can be overwhelming. In this chapter, we describe a mature system that implements TVA. This tool considers combinations of modeled attacker exploits on a network and then discovers attack paths (sequences of exploits) leading to specific network targets. The discovered attack paths allow an assessment of the true vulnerability of critical network resources. TVA automates the type of labor-intensive analysis usually performed by penetration-testing experts. It encourages inexpensive "what-if" analyses, in which candidate network configurations are tested for overall impact on network security. It also allows for the computation of network-hardening options that protect given critical resources while requiring minimal network changes.

To meet network availability requirements, there must usually remain some residual vulnerability after all protective measures have been applied. In such cases, we must then rely on the detect/react phases of security. While we cannot predict the origin and timing of attacks, TVA can reduce their impact by providing knowledge of the possible attack paths through the network. For example, TVA attack graphs can be used to correlate and aggregate network attack events, across platforms as well as across the network. TVA attack graphs can also provide the necessary context for optimal reaction to attacks.

13.2 Topological Analysis of Network Vulnerability

Because of the interdependencies of exploits across the network, a topological approach is necessary for full understanding of attack vulnerability. The traditional approach of considering network components in isolation and reporting vulnerabilities independent of one another is clearly insufficient. TVA models vulnerabilities and combines them in ways that real network attackers might do. The result is the discovery of all attack paths through a network.

Figure 13.2 shows the overall flow of processing in TVA. There are three inward flows of information: a model of the network configuration, a knowledge base of modeled attacker exploits, and a desired attack simulation scenario. From these, TVA then simulates incremental attacks through the network, thereby discovering all possible attack paths (organized as a graph) to the given critical network resources. Various innovative visualization capabilities support interactive analysis of resulting attack graphs, while keeping visual complexity manageable. TVA can also use the attack graphs to compute optimal network protection measures.

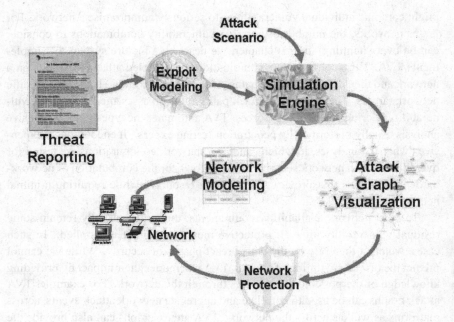

Fig. 13.1   Topological Vulnerability Analysis (TVA). Using network configuration and modeled attacker exploits, multi-step attacks are simulated based on the given attack scenario. The resulting attack graph is analyzed through interactive visualization, or used to formulate optimal network protection.

To model the various elements of the network and network attack events, our TVA system automatically processes the output of various network scanning and logging tools. It can combine scans from various network locations, building a complete map of connectivity to vulnerable services throughout the network, and can map actual intrusion events to elements of the resulting attack graph.

Figure 13.2 shows the inputs to our TVA system (current and proposed). The inputs occur in two phases. In the pre-attack *protect* phase, network scan tools provide information about network configuration and known vulnerabilities. For this, we can map vulnerability scanner output directly to corresponding vulnerable services on network machines. Our system is currently integrated with the Nessus[2] (open-source) and Retina[3] vulnerability scanners, and integration with the FoundScan[4] vulnerability scanner is under development. Or we can map the output of asset discovery tools (detected software on a machine) to the known vulnerabilities for each software package. For this, our system is integrated with Symantec Discovery,[5] which we map to known vulnerabilities through

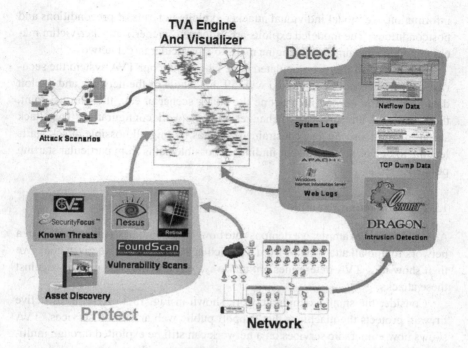

Fig. 13.2 Inputs to TVA system. During protect phase, pre-attack scans are used to build input models of the network. During the detect phase, actual attack events are mapped to the predicted attack graph.

integration with Symantec DeepSight[6] (a direct feed of the Bugtraq[7] vulnerability data). Cross-referencing data, including MITRE's Common Vulnerabilities and Exposures (CVE),[8] are used to correlate vulnerabilities across various sources.

In the *detect* phase, the TVA system maps detected attack events to their corresponding elements of the predicted attack graph. This provides the context for correlating events, predicting the next possible attack steps, and responding in the best way. It also helps remove clutter by prioritizing those predicted exploits that are correlated with recent real-time data. Our TVA system is currently integrated with the Snort intrusion detection system.[9] Integration with other intrusion detection systems (e.g., Dragon) is also possible, as well as with other sources of real-time data, such as web server logs (e.g., Apache and Microsoft IIS), operating system logs, and network traffic data (e.g., Netflow and TCP Dump).

To keep our TVA input exploit model current, we monitor emerging cyber threats, in the form of vulnerabilities that are discovered for particular software and the ways in which attackers can exploit these vulnerabilities. From this threat information, we model individual attacker exploits in terms of preconditions and postconditions. The modeled exploits are in terms of generic attacker/victim machines, which the simulation engine maps to a particular target network.

Because of all this pre-populated data, when using our TVA system the security analyst need not be burdened with all the details of the network and exploit data. All that is needed is to define the attack scenario, e.g., the starting point, the attack goal, and any what-if changes to the network configuration. The attack scenario could also be less constrained, such as finding all possible attack starts leading to one or more goals, or finding all possible paths from particular starting points.

13.3 A Motivating Example

As a motivating example, we demonstrate how TVA combines vulnerabilities in a network to find all attack paths from a particular starting point to a given goal. We then show how TVA determines optimal ways of hardening the network against these attacks.

Consider the small example network shown in Fig. 13.3. Here, a restrictive firewall protects the machines that support public web and email services. TVA shows how vulnerable services on a network can still be exploited through multi-step attacks, when the attacker cannot access them directly.

The firewall implements the following policy to restrict access to the network from the outside:

- Incoming web traffic is permitted only to the web server, which is running Microsoft IIS.
- Incoming email traffic is permitted to the mail server.
- Incoming FTP is blocked because the mail server is running wu_ftpd, which has a history of vulnerabilities.
- All other incoming traffic is blocked.

For this example, we populate the TVA network model through Nessus scans. In particular, we scan the web server and mail server from *outside* the firewall, to obtain vulnerable connectivity from the initial attack vantage point. We also scan these two servers from *behind* the firewall, showing any subsequent vulnerable connectivity once the attacker gains entry into the network. These scan results are merged to form an overall model of the network for TVA.

The attack goal for this example is to obtain super user (root) access on the mail server, starting from outside the network. This is not directly possible because (1) there are no known vulnerabilities for the version of sendmail running

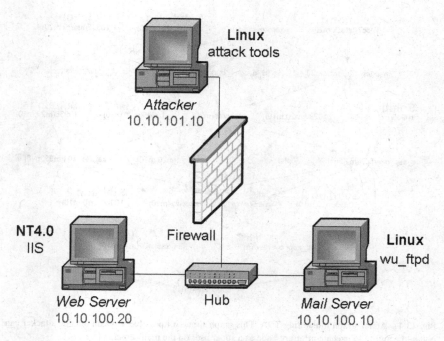

Fig. 13.3 Small example network for illustrating TVA. The firewall allows web traffic to the web server, allows email traffic to the mail server, and blocks all other incoming traffic.

on the mail server, and (2) the firewall blocks access to the vulnerable wu_ftpd service from the attack machine. TVA analyzes whether the attack goal can be realized indirectly, i.e., through a sequence of multiple exploits.

Figure 13.4 shows the resulting TVA attack graph for the example network in Fig. 13.3. Here, shaded ovals are simulated attacker exploits. For each exploit, incoming edges represent preconditions, all of which must be met for the exploit to be successful. Then for each exploit, outgoing edges represent postconditions, i.e., the conditions induced when the exploit is successful. Preconditions with the 5-digit Nessus identifiers represent connections to vulnerable network services detected by Nessus.

The initial condition *execute(attack)* represents the fact that the attacker can execute arbitrary code on his own machine. This enables three separate exploits from *attack* to the web server (machine *m20*). Each of these exploits provides the ability to execute arbitrary code on the web server. This subsequently enables four new exploits from the web server to the mail server (machine *m10*), each yielding the ability to execute arbitrary code on the mail server. Two of these exploits

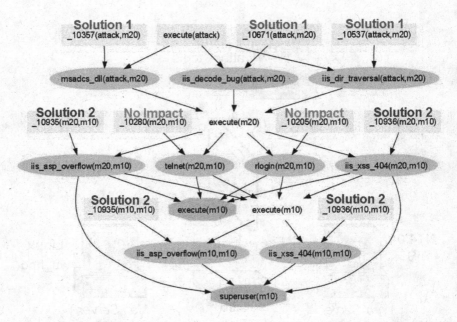

Fig. 13.4   Attack graph illustrating TVA. This graph shows all possible ways an outside attacker can obtain the ability to execute arbitrary code as a super user on the mail server.

provide access at a super user level of privilege. The other two exploits provide user-level privilege only, but two subsequent local privilege escalation exploits on the mail server provide other paths to super user.

Finding such attack paths is a unique TVA capability. Vulnerability scanning tools connected outside the firewall report only the IIS vulnerabilities on the web server. Such scans from inside the firewall would report the vulnerable wu_ftpd service, but TVA is required to build an attack path from the outside through the web server to the mail server. While easy enough for an experienced penetration tester on such a small network, it becomes unmanageable for networks where voluminous outputs must be analyzed for large numbers of machines.

TVA can not only find attack graphs, but can also use these graphs for finding optimal solutions for hardening the network. In particular, though TVA we can find combinations of network-hardening measures that prevent a given attack scenario, while requiring a minimal number of changes to the network configuration. Figure 13.4 illustrates this. For this network, one such solution is to remediate (e.g., patches or firewall blocking) the three vulnerabilities from *attack* to *m20*. Hardening these three vulnerabilities is necessary and sufficient for preventing the

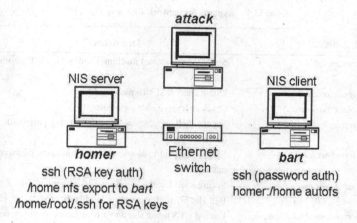

Fig. 13.5 Network illustrating TVA minimal-cost network hardening. Complicated interdependencies among exploits are to be resolved to optimal hardening measures.

attack goal. The other solution is to harden the two vulnerabilities on *m10* that enable the four exploits yielding super user access. Interestingly, TVA shows that hardening the other two vulnerabilities on *m10* (yielding user-level access only) has no impact on blocking access to the goal, i.e., hardening them is neither necessary nor sufficient.

The next section describes the TVA process for optimal network hardening in more detail.

13.4 Minimal-Cost Network Hardening

Attack graphs reveal threats by predicting combinations of attacker exploits that compromise given critical resources. But alone, they do not directly provide a solution to remove the threat. Finding such solutions manually can be tedious and error prone, especially for larger and less secure networks.

TVA automates the task of hardening a network against multistep attacks. Unlike previous approaches whose solutions are in terms of attacker exploits,[10–12] our solutions are in terms of network configuration elements. These solutions are therefore more enforceable, because the configuration elements can be independently hardened, whereas exploits are usually consequences of other exploits and cannot be disabled without removing the root causes. Also, our solutions are optimal in the sense that they incur minimal cost in terms of changes to the network.

Consider the network in Fig. 13.5, which we model using multiple layers of the TCP/IP stack. This example shows how complicated interdependencies among

Table 13.1   Exploits for network shown in Fig. 13.5.

Exploit	Description
arp_spoof	Spoof (impersonate) machine identity via ARP poison attack
ypcat_passwd	Dump encrypted NIS password file
crack_passwd	Crack encrypted user password(s)
scp_upload_pw	Secure shell copy, upload direction, using password authentication
scp_download_pw	Secure shell copy, download direction, using password authentication
ssh_login_pw	Secure shell login using password authentication
rh62_glibc_bof	Red Hat 6.2 buffer overflow in glibc library
create_nfs_home_ssh_pk_su	Exploit NFS home share to create secure shell key pair used for super user authentication
ssh_login_pk_su	Secure shell login using public key authentication

exploits can be resolved to an optimal set of hardening measures. It also demonstrates how purely exploit-based hardening approaches are insufficient for network hardening, i.e., that solutions in terms of network configuration elements are needed.

In Fig. 13.5, an Ethernet switch provides connectivity at the link layer. At the transport layer, unused services have been removed, secure shell replaces FTP, telnet and other cleartext password-based services, and there is tcpwrapper protection on RPC services. Application-layer trust relationships further restrict NFS and NIS domain access. The exploits and network configuration elements (exploit conditions) for this example are described in Table 13.1 and Table 13.2, respectively.

Figure 13.6 shows the attack graph for the network in Fig. 13.5 modeled via the exploits and network conditions in Table 13.1 and Table 13.2. Using our previously described algorithm for minimal-cost hardening,[13,14] we traverse the attack graph to construct a logical expression for the attack goal g (execute code as super user on machine *homer*) in terms of the initial network conditions:

$$g = (\alpha\beta\chi + \alpha\beta\chi\delta\epsilon).(\phi\gamma).(\alpha\beta\chi).\eta$$
$$= \alpha\beta\chi\phi\gamma\eta$$

The attack graph has been reduced to an expression that leads to simple choices for network hardening. Note that two initial conditions in the graph do

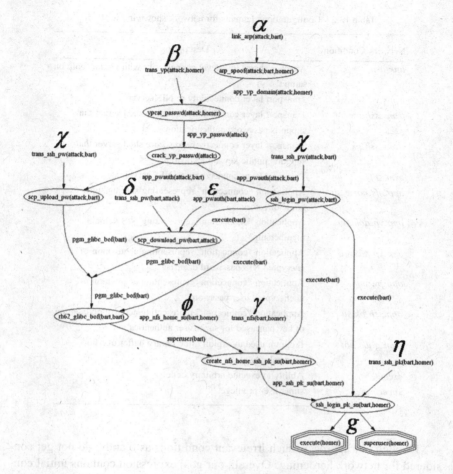

Fig. 13.6 Attack graph illustrating TVA minimal-cost network hardening. A logical expression is formed for the attack goal g in terms of initial network conditions.

not appear in the expression for goal *g*:

$$(i) \quad \delta \equiv trans\_ssh\_pw(bart, attack), \ and$$

$$(ii) \quad \varepsilon \equiv app\_pwauth(bart, attack).$$

These drop out in this fashion:

$$rh62\_glibc\_bof(bart, bart) = \alpha\beta\chi + \alpha\beta\chi\delta\varepsilon$$
$$= \alpha\beta\chi(1 + \delta\varepsilon)$$
$$= \alpha\beta\chi$$

Table 13.2   Configuration elements for network shown in Fig. 13.5.

Network Condition	Description
link_arp	Attacker shares link-level connectivity with victim (both on same LAN)
trans_yp	Transport layer connectivity to NIS server
trans_ssh_pw	Transport layer connectivity to secure shell server that supports password authentication
trans_ssh_pk	Transport layer connectivity to secure shell server that supports public key authentication
trans_nfs	Transport layer connectivity to NFS server
app_nfs_home_su	Application "connection" representing sharing super user's home directory
app_yp_domain	Application "connection" representing NIS domain membership
app_yp_passwd	Application "connection" representing acquisition of encrypted NIS password database
app_pwauth	Application "connection" representing acquisition of unencrypted user password
app_ssh_pk_su	Application "connection" representing acquisition/creation of key pair used for super user authentication
pgm_glibc_bof	Program used to exploit glibc library buffer overflow vulnerability
execute	Ability to execute arbitrary code
superuser	super user privilege

Through our approach, such irrelevant conditions as δ and ε do not get considered for network hardening. Overall, our goal expression contains initial conditions that are both necessary and sufficient for network-hardening decisions.

This kind of sufficiency is not present in previous approaches to network hardening via exploit set minimization. These approaches search for minimal sets of exploits, in which every exploit is needed in reaching the goal. In this example, there are two such minimal exploit sets:

- All exploits except *scp_upload_pw(attack,bart)*, and
- All exploits except *scp_download_pw(bart, attack)*.

For network hardening using these minimal exploit sets, we must assume that all exploits in the union of the minimal exploit sets must be stopped. In this example, we would therefore conclude that *scp_download_pw(bart,attack)* must be stopped, even though stopping it has no effect on the attacker reaching the goal.

Also, hardening initial condition *trans_ssh_pw(attack,bart)* simultaneously stops two exploits, i.e., *scp_upload(attack,bart)* and *ssh_login_pw(attack, bart)*. This would not be apparent by considering minimal exploit sets only. In other words, a single initial condition could control many exploits. In general, relationships among initial conditions and exploits can be many-to-many and complex. To solve the network-hardening problem, analysis must be at the level of network elements rather than exploits.

Our TVA network-hardening solutions not only prevent attacks against given critical resources, but also allow choices with minimal cost in network changes. In Fig. 13.6, the expression $g = \alpha\beta\chi\phi\gamma\eta$ implies that hardening any one of these will protect the goal:

(1) *link_arp(attack,bart)*,
(2) *trans_yp(attack,homer)*,
(3) *trans_ssh_pw(attack,bart)*,
(4) *app_nfs_home_su(bart,homer)*,
(5) *trans_nfs(bart,homer)*, or
(6) *trans_ssh_pk(bart,homer)*.

Implementing Solutions 2, 5, or 6 would require shutting down critical network services. Solution 1 requires hard-coding IP/MAC address relationships. Solution 4 requires removing the super user home directory file share. Solution 3 requires using public-key authentication rather than password authentication. Among all these options, Solution 3 is the best (lowest-cost) choice.

13.5 Attack Graph Visualization

One of the greatest challenges in TVA is managing the complexity of the resulting attack graphs, particularly for larger and less secure networks. Visualization is a natural choice for conveying complex graph relationships to the user. Still, attack graphs in their raw form often yield overly cluttered views that are difficult to understand, as in Fig. 13.7. Therefore, in developing our TVA system, we have devoted considerable effort in managing attack graph visual complexity.

Our TVA attack graphs scale quadratically rather than exponentially,[15] so that graphs such as Fig. 13.7 can be computed in a fraction of a second. However, when shown in their full complexity, such graphs are too complicated for easy comprehension.

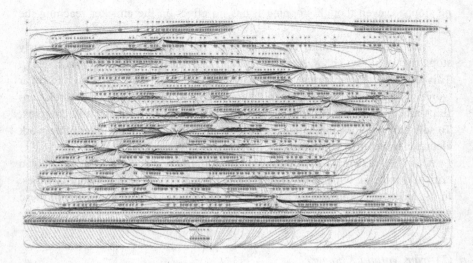

Fig. 13.7   Attack graph visual complexity. In their raw form, TVA attack graphs can overwhelm an analyst.

To manage visual complexity of attack graphs, our TVA system employs so-phisticated methods of recursive graph clustering.[16] This is illustrated in Fig. 13.8. Here, Fig. 13.8(a) is the original attack graph in its full complexity. Figure 13.8(b) shows the same attack graph, this time aggregated to the level of machines and the sets of exploits between each pair of them. In Figure 13.8(c), this is further aggre-gated to sets of machines with unlimited connectivity to one another's vulnerabil-ities (e.g., subnets). In Figure 13.8(d), subnets are collapsed to single vertices, as are the exploits between them. Thus each level of aggregation provides a progres-sively summarized (less complicated) view of the attack graph.

In our TVA system, analysts can start with high-level overviews, and drill down through successive levels of detail as desired. The system begins with the graph automatically aggregated based on known network attributes. The analyst can also interactively aggregate graph elements as desired. Graph visualization and interaction is done through our custom Visio-style user interface.

In this way, arbitrarily large and complex attack graphs can be explored through manageable, meaningful interactive displays. Figure 13.9 shows such interactive attack graph visualization, showing how the analyst can show arbitrary levels of detail across the graph all within a single view. In this example, several hundred host machines are included in the attack graph.

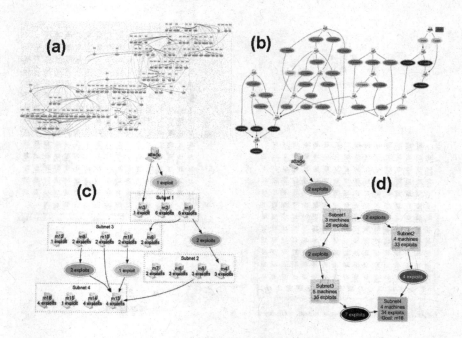

Fig. 13.8 Recursively clustered attack graphs. At each level of clustering, the attack graph view becomes progressively summarized and simplified.

13.6 Real-Time Event Correlation

Once actual attacks are detected, the TVA attack graph provides the necessary context for correlating intrusion detection events and predicting next possible attack steps. In other words, we embed incoming intrusion alarms in the TVA attack graph, which is based on known vulnerabilities across the network. While multi-step intrusion alarm correlation has been proposed in the past,[17] it lacks predictive power without the context provided by our vulnerability-based attack graphs. Further, using our pre-computed attack graphs, we can correlate alarms faster than typical intrusion detection systems can generate them.[18–20]

From the TVA attack graph predicting all possible attacks, incoming intrusion alarms are assigned to their corresponding predicted exploits. We then visualize the joint predicted/observed attack graph, as shown in Fig. 13.10. Here, red ovals are detection events placed in the predicted attack graph. In this example, one red event immediately follows another in the graph, thus correlating these as a possible two-step attack. In this way, isolated detection events can be quickly assessed as possible multi-step attacks. This approach also reduces false alarms,

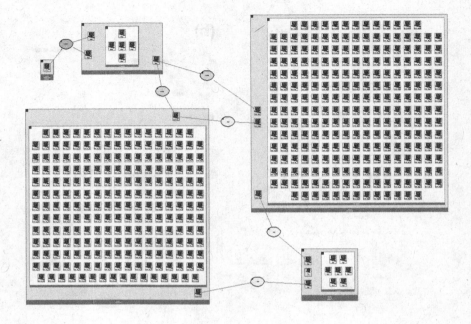

Fig. 13.9 Interactive TVA attack graph visualization. Arbitrarily complex attack graphs can be explored through interactive displays, with mixed levels of detail within a single graph view.

i.e., the graph contains only those attacks that the network is actually vulnerable against.

With TVA attack graphs, we can also provide recommendations in response to detected attacks. For example, in Fig. 13.10 the orange ovals are predicted exploits that immediately follow intrusion alarms in the attack graph. Since these are the next possible exploits an attacker could take (based on known vulnerabilities), stopping them is a high-priority for containing the attack. The blue ovals in this figure are predicted exploits that are further away from the detected attacks, and are therefore less time-critical. Without the predictive power of our vulnerability-based TVA attack graphs, we could perform alarm correlation (red ovals) only.

Our attack response recommendations are optimal in the sense that they address the exact next-step vulnerabilities in the network — no more and no less. For example, rather than blocking traffic from an entire subnet (an overly cautious and disruptive response based on limited information), our responses could give precise blocking rules down to a single vulnerable host port.

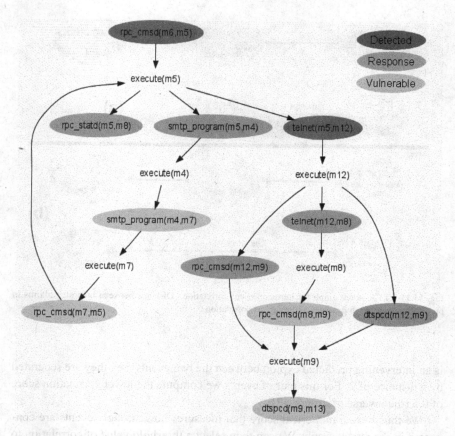

Fig. 13.10 Graph of predicted and actual attacks. Attack events detected in real time are embedded in TVA graph of predicted attacks, providing context for event correlation and attack response.

With TVA attack graphs, we can also predict missed events (false negatives) when correlating detection events into multi-step attacks. In this way, we account for uncertainty in the accuracy of our intrusion detection systems. For example, we can use attack graph distances as measures of causal correlation between detection events.[18] This is illustrated in Fig. 13.11. Here, real-time intrusion events are assigned to their corresponding predicted exploits. Correlation scores are then computed as the inverse of event distance in the graph (higher correlation for shorter distances).

In the case of Fig. 13.11(a), the two intrusion events are directly connected in the graph (unity distance), giving the maximum possible correlation score (unity) between the events. In Fig. 13.11(b), rather than being directly connected, there

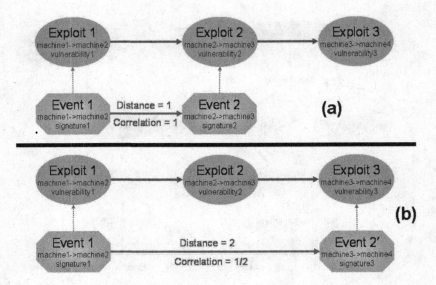

Fig. 13.11   TVA attack graph for intrusion event correlation. Distance between incoming alarms in predicted attack graph provides a measure of correlation.

is an intervening predicted exploit between the two events, i.e., they are separated by a distance of 2. For this pair of events we compute the lower correlation score of 0.5 (the inverse of distance 2).

We thus have a numerical score that measures how strongly events are connected in the attack graph. We can then select a threshold value of correlation to form multi-step attack attacks from isolated alarms. That is, event pairs that are sufficiently well correlated can be combined into a single multi-step attack. This is an extension of the idea shown in Fig. 13.10, now taking into account missed detections.

We can further refine this analysis by including recent event history when computing correlations. The idea is that occasional missed detections should be scored higher than isolated events that happen to occur nearby. For example, a missed detection (distance of 2) within a series of unity-distance events should be scored higher than a pair of distance-2 events among unrelated (large-distance) ones. As shown in Fig. 13.12, we can apply local signal averaging operations to enhance the contrast between regions of higher and lower correlation. In this way, we detect multi-step attacks with greater confidence, to address the uncertainty of our intrusion detection systems.

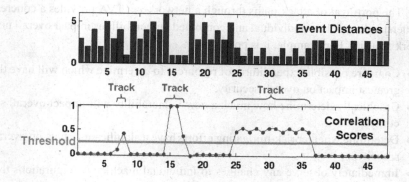

Fig. 13.12 Signal averaging to improve event correlation scores. Averaging attack-graph distances for recent events provides robust event correlation in the face of intrusion detection uncertainties.

13.7 Conclusions and Outlook

To protect our critical networks, we must understand not only individual system vulnerabilities, but also their interdependencies. The TVA approach places vulnerabilities and their protective measures within the context of overall network security by modeling their interdependencies via attack graphs. The analysis of attack graphs provides alternative sets of protective measures that guarantee safety of critical systems, ranked by cost, e.g., maximum service availability and/or minimum number of required protective measures. Through this unique new capability, administrators are able to determine the best sets of protective measures that should be applied in their environment.

Our TVA system monitors the state of network assets, maintains models of network vulnerabilities and residual risk, and combines these to produce models that convey the impact of individual and combined vulnerabilities on overall security posture. The central product of this system is a graph-based model showing all the ways an attacker can penetrate a network, built from models of network vulnerabilities and attacker exploits.

TVA is not a mere cross-referencing of security data — it is a framework for general-purpose modeling, analysis, and visualization of network penetration. Our TVA system provides a unique new capability, transforming raw security data into a roadmap that lets one proactively prepare for attacks, manage vulnerability risks, and have real-time situational awareness. It supports both offensive (e.g., penetration testing) and defensive (e.g., network hardening) applications, across all phases (protect, detect, react) of the information security lifecycle.

The portrayal of attack paths through a network via TVA provides a concrete understanding of how individual and combined vulnerabilities impact overall network security. For example, it is possible to

- Compare possible expenditures of resources to determine which will have the greatest impact on overall security,
- Graphically determine how much a new vulnerability will impact·overall security,
- Determine whether risk-mitigating efforts have a significant impact on overall security, or
- Immediately observe any changes to individual machine configurations that increase the overall risk to the enterprise.

Our TVA system transforms raw security data into a model of all possible attack paths into a network. In providing this new capability, we have met key technical challenges, including the design of appropriate models, efficient model population, effective visualizations and decision support tools, and the development of scalable mathematical representations and algorithms. Our system addresses all these challenges, and delivers a product that offers truly unique capabilities among security tools.

Acknowledgments

This material is based upon work supported by Homeland Security Advanced Research Projects Agency under the contract FA8750-05-C-0212 administered by the Air Force Research Laboratory/Rome; by Air Force Research Laboratory/Rome under the contract FA8750-06-C-0246; by Army Research Office under the grant W911NF-05-1-0374; by Federal Aviation Administration under the contract DTFAWA-04-P-00278/0001; and by the National Science Foundation under grants CT-0627493, IIS-0242237, and IIS-0430402. Any opinions, findings, and conclusions or recommendations expressed in this material are those of the authors and do not necessarily reflect the views of the sponsoring organizations.

References

1. S. Jajodia, S. Noel, and B. O'Berry (2005) Topological Analysis of Network Attack Vulnerability, in *Managing Cyber Threats: Issues, Approaches and Challenges*, V. Kumar, J. Srivastava, A. Lazarevic (eds.), Kluwer Academic Publisher.
2. R. Deraison, *Nessus*, http://www.nessus.org, last retrieved April 2007.
3. 3eEye Digital Security, *Retina Network Security Scanner*, last retrieved April 2007.

4. Foundstone, *FoundScan Frequently Asked Questions,* `http://www.foundstone.com/pdf/foundscan\_general\_faq.pdf,` last retrieved April 2007.
5. Symantec Corporation, *Symantec Discovery Overview,* `http://www.symantec.com/enterprise/products/overview.jsp?pcid=1025\&pvid=923\_1,` last retrieved April 2007.
6. Symantec Corporation, *Symantec DeepSight Threat Management System,* `http://tms.symantec.com/Default.aspx,` last retrieved April 2007.
7. Security Focus, *Bugtraq Vulnerabilities,* `http://www.securityfocus.com/vulnerabilities,` last retrieved April 2007.
8. MITRE Corporation, *CVE — Common Vulnerabilities and Exposures,* `http://cve.mitre.org/,` last retrieved April 2007.
9. Sourcefire, *Snort — The De Facto Standard for Intrusion Detection/Prevention,* `http://www.snort.org/,` last retrieved April 2007.
10. O. Sheyner, J. Haines, S. Jha, R. Lippmann, J. Wing, "Automated Generation and Analysis of Attack Graphs," in *Proceedings of the IEEE Symposium on Security and Privacy,* Oakland, CA.
11. S. Jha, O. Sheyner, J. Wing, Two Formal Analyses of Attack Graphs, in *Proceedings of the 15th Computer Security Foundation Workshop,* Nova Scotia, Canada.
12. P. Ammann, D. Wijesekera, S. Kaushik, Scalable, Graph-Based Network Vulnerability Analysis, in *Proceedings of the 9th ACM Conference on Computer and Communications Security,* Washington, DC.
13. S. Noel, S. Jajodia, B. O'Berry, M. Jacobs, Efficient Minimum-Cost Network Hardening via Exploit Dependency Graphs, in *Proceedings of the 19th Annual Computer Security Applications Conference,* Las Vegas, Nevada.
14. L. Wang, S. Noel, S. Jajodia, Minimum-Cost Network Hardening Using Attack Graphs, *Computer Communications,* 29(18), pp. 3812-3824.
15. S. Noel, S. Jajodia, Managing Attack Graph Complexity through Visual Hierarchical Aggregation, in *Proceedings of the ACM CCS Workshop on Visualization and Data Mining for Computer Security* Fairfax, Virginia.
16. S. Noel, M. Jacobs, P. Kalapa. S. Jajodia, Multiple Coordinated Views for Network Attack Graphs, in *Proceedings of the 2nd International Workshop on Visualization for Computer Security,* Minneapolis, Minnesota.
17. P. Ning, Y. Cui, D. Reeves, Constructing Attack Scenarios through Correlation of Intrusion Alerts, in *Proceedings of the 9th ACM Conference on Computer and Communications Security,* Washington, DC.
18. S. Noel, E. Robertson, S. Jajodia, Correlating Intrusion Events and Building Attack Scenarios through Attack Graph Distances, in *Proceedings of the 20th Annual Computer Security Applications Conference,* Tucson, Arizona.
19. L. Wang, A. Liu, S. Jajodia, An Efficient and Unified Approach to Correlating, Hypothesizing, and Predicting Network Intrusion Alerts, in *Proceedings of the 10th European Symposium on Research in Computer Security,* Milan, Italy.
20. L. Wang, A. Liu, S. Jajodia, Using Attack Graphs for Correlating, Hypothesizing, and Predicting Network Intrusion Alerts, *Computer Communications,* 29(15).

Chapter 14

New Malicious Code Detection Using Variable Length n-Grams

Subrat Kumar Dash*, D. Krishna Sandeep Reddy† and Arun K Pujari‡

Artificial Intelligence Lab, University of Hyderabad,
Hyderabad - 500 046, India

Most of the commercial antivirus software fail to detect unknown and new malicious code. In order to handle this problem generic virus detection is a viable option. Generic virus detector needs features that are common to viruses. Recently Kolter et al.[18] propose an efficient generic virus detector using n-grams as features. The fixed length n-grams used there suffer from the drawback that they cannot capture meaningful long sequences. In this paper we propose a new method of variable-length n-grams extraction based on the concept of episodes and demonstrate that they outperform fixed length n-grams in malicious code detection. The proposed algorithm requires only two scans over the whole data set whereas most of the classical algorithms require scans proportional to the maximum length of n-grams.

Contents

*subrat.dash@gmail.com
†krishnasandeep.reddy@yahoo.co.in
‡akpcs@uohyd.ernet.in

14.1 Introduction

Malicious code is any code added, changed, or removed from a software system to intentionally cause harm or subvert the system's intended function.[21] Any computer system is vulnerable to malicious code whether or not it is attached to other systems. Examples of malicious code include viruses, Trojan horses, worms, back doors, spyware, Java attack applets, dangerous ActiveX and attack scripts. Combining two or more of these categories can lead to fatal attack tools. Recent *Gartner* report[14] ranked viruses and worms as top security threats and hence detecting malicious code has become one of the prime research interests in the field of information security. In this paper, we concentrate on new malicious code detection especially computer viruses. These are self-replicating software entities that attach themselves to other programs. There are two approaches that are poles apart in virus detection. One approach is too generic which includes Activity monitors and Integrity management systems. The other approach, *Signature based virus detection*, is too specific and also popular. Almost all the commercial antivirus products rely on this approach. In this, a database of virus signatures is maintained and a program is detected as a virus program based on the presence of these virus signatures. There are two main disadvantages in this technique.[21]

- Unknown and new viruses will easily escape the detection by simple defenses, like code obfuscation, as their signatures are not present in the database.[6]
- This technique is not scalable. As the number of known viruses increases, the size of the signature database increases and also the time of checking a file for virus signatures increases.

Generation of virus signature is a cumbersome process and is prone to generating false positives on benign programs. Unlike signature based detection, generic virus detector uses features that are common to most of the viruses and that characterize only the virus programs. The assumption here is that viruses have certain typical characteristics in common and these characteristics are not present in benign programs. For instance, most of the virus writers use virus generating toolkits,[28] for example PS-MPC (Phalcon-Skism Mass-Produced Code Generator), to write and compile their code. It is believed that more than 15,000 variants of viruses were generated using this kit alone and there are more than 130 such kits available. The viruses generated using a toolkit have certain features that are specific to the toolkit, the compiler and also the programming environment.[3] Fred Cohen, in his seminal paper,[7] proved that there is no algorithm that can detect the set of all possible computer viruses (returning "true" if and only if its input is an object infected with some computer virus).

It is clear from the foregoing discussion that a generic virus detector is though desirable is very difficult to accomplish in its true form. A practical approach is to develop a machine learning based detector where the detection process learns the generic characteristics from a set of known examples but avoids identifying any specific signatures. Recently there have been several such attempts for detection of malicious code.

Kephart et al.[17] propose the use of Neural Networks to detect boot sector malicious binaries. Using a Neural Network classifier with all bytes from the boot sector malicious executables as input, it is shown that unknown boot sector malicious executables can be successfully identified with low false positive rate. Later, Arnold et al.[2] apply the same techniques to Win32 binaries. Motivated by the success of data mining techniques in host based and network based intrusion detection system, Schultz et al.[26] propose several data mining techniques to detect different types of malicious executables. In a companion paper[27] the authors develop a UNIX mail filter that detects malicious Windows executables based on the above work. Byte sequences are used as the set of features as machine codes are most informative to represent executables. The RIPPER algorithm of rule discovery and Naive Bayes classifiers are used for classification in this study. Abou-Assaleh et al.[3] observe that *n*-grams extracted from byte sequences can be used as effective features. They use Common *n*-gram (CNG) as the features and propose a k-NN classification for detecting computer virus. Kolter et al.[18] independently realise that *n*-grams can possibly be used as a set of features. However, as the set of all *n*-grams is very large, it is proposed to use few of them selected based on their information gain. Kolter et al.[18] also investigate several classification techniques, which are implemented in WEKA[30] and boosted J48 algorithm reportedly gave good results. In a recent paper,[22] Michael Cai et al. compared seven different feature selection measures and four different classification algorithms in identifying malicious executables. Surprisingly, they did not include information gain, which is very popular, in their comparision. Static analysis is also attempted,[6,19] where analysis of program is done without executing it. Dynamic analysis which combines testing and debugging to detect malicious activities by running a program includes wrappers,[4] sandboxing[1] etc. Behavior blocker is a method used in Bloodhound technology (Symantec) and ScriptTrap technique (Trend Inc.).

It is also interesting to note that many of these techniques use byte *n*-grams as the basic features to detect malicious codes. Byte *n*-grams are overlapping continuous substrings, collected in a sliding-window fashion where the windows of fixed size of *n* slides one byte at a time. N-grams have the ability to capture implicit features of the input that are difficult to detect explicitly. Byte *n*-grams can be viewed as features in the present context when an executable program is viewed as a sequence of bytes. N-grams have been successfully used for a long time in a wide variety of problems and domains, including information retrieval, language identification, automatic text categorization, computational immunology, authorship attribution etc. In many domains, techniques based on n-grams gave very good results. For instance, in natural language processing, n-grams can be used to distinguish between documents in different languages in multi-lingual collections and to gage topical similarity between documents in the same language. Some of the good features of n-grams are simplicity, robustness and efficiency. On the other hand, the problem which can appear in using n-grams is exponential explosion. It is clear that many of the algorithms with n-grams are computationally too expensive even for $n = 5$ or $n = 6$.

Importance of byte n-gram in detecting computer virus has been realised more than once in the computer virology research. In 1994, a byte n-gram-based method is used for automatic extraction of virus signatures.[17] The major difficulty in considering byte n-grams as a feature is that the set of all byte n-grams obtained from the set of byte strings of virus as well as of the benign programs is very large. There are several feature selection techniques proposed[31] and the important ones are *document frequency* and *information gain*. Recently, it is shown[25] that class-wise document frequency can be a better feature selection technique.

The main disadvantage of fixed-length n-grams is that they cannot capture meaningful n-gram sequences[10] of different lengths. Though variable length n-grams are previously used in intrusion detection[9,10,16,20] and text categorization,[5,13] no attempts have been made to use them in malicious code detection. Recently, we have demonstrated[9] that episodes as variable length n-grams can be used for IDS. In this paper we propose a very elegant and novel method of detection of malicious codes by extracting the common features of virus programs as variable length n-grams, based on the concept of episodes.[8] The proposed detection technique works as follows.

We use an efficient method of extracting variable length n-grams (*episodes*) and select class-wise relevant n-grams. We take m relevant n-grams from each of the two classes, virus and benign, to get M features which are used in vector space model representation. We use supervised classification techniques such as boosted J48 and demonstrate that variable length n-gram method of detection is better than fixed length approach.

The rest of the paper is organized as follows. In Section 2, we propose an episode discovery algorithm that is used for a set of sequences. The episode discovery algorithm given by Dash et al.[9] is only for one sequence at a time. But in the present case, it is necessary to find episodes for a set of sequences. In Section 3 we describe the concept of relevant episodes. Section 4 and 5 describe the experimental setup and results of the proposed method and comparative study with earlier methods. Conclusion follows in Section 6.

14.2 Episode Discovery Algorithm

Episodes are meaningful subsequences of a continuous sequence. Any method of finding episodes in a large sequence is essentially finding the break points in the sequence and hence can be viewed as sequence segmentation method.[15] There are many proposals of segmenting sequence in general and specifically segmenting time series in the fields of telecommunications, speech processing, signal processing, text processing etc. The objective of sequence segmentation can be different in different contexts.[12] In the present context, we are interested in segmenting categorical sequence into meaningful subsequences. Cohen et al.[8] proposed an efficient method of sequence segmentation based on voting experts approach. We adopt this algorithm for the present context and extend the same to handle multiple sequences.

The main idea of Cohen et al.[8] algorithm is that it has two experts who vote their inference for break points. The frequency expert measures the likelihood of a subsequence to be an episode based on its frequency of occurrence. The higher the frequency of subsequences, the lower is the chance that it contains a break point. Similarly the entropy expert measures the entropy at a point. If an element associates (precedes) itself with several other elements then it has higher probability of being a break point compared to an element which precedes only few elements. This phenomenon is captured by computing the entropy of the association. The original algorithm combines the frequency and entropy scores at each location of the sequence and identifies the possible location of break points. The subsequence between two consecutive break points is an episode. In order to efficiently handle the computation of entropy and frequency along different locations in a sequence, it is proposed[8] to represent this information in a trie structure.

In order to build the trie for a sequence, a user specified parameter d, the depth of the trie is needed. The sequence is read one symbol at a time and its preceding $d - 1$ symbols are also taken. Each of the d subsequences are inserted into the trie with frequency value set to 1 if, the subsequence is not yet present in the trie. Else, the frequency of the subsequence is incremented by 1. Dash et al.[9] have given the complete algorithm for construction of trie.

Definition 14.1 Entropy of a node ($e(x)$) refers to the entropy of the sequence from the root node to the concerned node (x). Let $f(x)$ be the frequency of the node x and $x_1, x_2, \ldots, x_\ell$ be its child nodes. The probability of the subsequence represented at node x_1, denoted as $p(x_1)$, is given by

$$P(x_1) = \frac{f(x_1)}{f(x)}$$

The entropy of x is given by

$$e(x) = -\sum_{i=1}^{\ell} p(x_i) \log p(x_i)$$

It can be noted that the entropy for the leaf nodes is zero. Now, each of the nodes in the trie has two parameters, frequency (f) and entropy (e). We standardize these parameters[8] for each of the level in the trie taking means ($\overline{f}$, $\overline{e}$) and standard deviations ($\sigma_{\overline{f}}$, $\sigma_{\overline{e}}$). Both the parameters contribute equally in finding the break points by assigning scores to the probable positions.

In the present context, we extend the algorithm to determine episodes from a set of sequences. The simple extension would mean that we construct one trie for each sequence. But we propose to store the information of all sequences in a single trie structure. Thus we can capture the frequency and entropy of any element over all the sequences together. The method of construction of trie for multiple sequences is illustrated in Example 1.

Example 1: Let us consider the following set of four sequences.
$S_1 = (\text{bf } 0e \text{ } 3a \text{ } bf \text{ } d8 \text{ } 3a \text{ } 3a \text{ } bf)$
$S_2 = (3a \text{ } bf \text{ } bf \text{ } 0e \text{ } 3a \text{ } 3a \text{ } bf \text{ } bf \text{ } d8 \text{ } 3a)$
$S_3 = (\text{bf } d8 \text{ } 3a \text{ } bf \text{ } 0e \text{ } 3a \text{ } bf \text{ } 0e \text{ } 3a \text{ } 0e \text{ } 3a)$
$S_4 = (0e \text{ } 3a \text{ } bf \text{ } d8 \text{ } 3a \text{ } 3a \text{ } bf \text{ } 3a \text{ } bf \text{ } 0e \text{ } 3a)$

The trie structure for S_1 can be obtained by using the trie construction algorithm[9] with $d = 4$ as shown in Figure 14.1. From the trie structure we note that (bf) appears 3 times in S_1, (3a bf) appears twice and (3a 3a) appears once. The structure captures the frequency of n-grams of different lengths (at most $d - 1$). We embed S_2 on this structure to get the trie representing S_1 and S_2 (Figure 14.2). The trie with S_1, S_2 and S_3 is shown in Figure 14.3. The final trie structure after considering all the sequences is shown in Figure 14.4. The algorithm for finding episodes from each of the sequences, using the above obtained combined trie structure (T_k), is given in Figure 14.5.

For each of the sequence S_t we take a sliding window of length $k(= d - 1)$. Let $x_1, x_2, \ldots, x_k$ be the elements falling in the window at one instance. For each of the k possible break points in the window, we examine the frequency and entropy as follows.

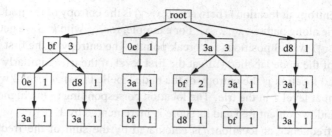

Fig. 14.1 Trie for S_1 in Example 1 with depth=4.

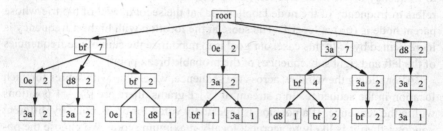

Fig. 14.2 Trie after inserting S_2 in Fig.14.1.

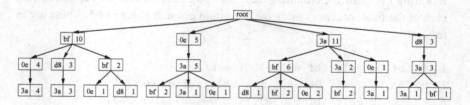

Fig. 14.3 Trie after inserting S_3 in Fig. 14.2.

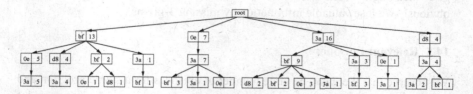

Fig. 14.4 Trie after inserting S_3 in Fig. 14.3.

The entropy at location i (between x_i, x_{i+1}) is the entropy of the node x_i at level i of the trie along path $x_1, x_2, ..., x_i$. For example, for S_1, with $k=3$ we get a window (bf|0e|3a|) with 3 positions for break points. The entropy at the first location is entropy of the node labelled (bf) at the first level of the trie. Similarly entropy at the second location is the entropy of the node labelled (0e) at level 2 with (bf) as the parent at level 1 in the trie. The location corresponding to the highest entropy in the window is identified and its score is incremented by 1.

The frequency at location i is calculated by the sum of the frequencies of subsequences $(x_1...x_i)$ and $(x_{i+1}...x_k)$. For example, the frequency at the first location of the window (bf|0e|3a|) is $f(\text{bf}) + f(\text{0e 3a})$, where $f(\text{bf})$ refers to frequency of the node labelled (bf) at the first level of the trie and $f(\text{0e 3a})$ refers to frequency of the node labelled (3a) at the second level of the trie whose parent node is (0e) at level 1. The score at the location with highest frequency is incremented by 1. In this case, our goal is to maximize the sum of the frequencies of the left and right subsequences of the probable break point.

After sliding the window across the sequence, we end up with scores for each location in the sequence. In a stream of $|S_t|$ 1-grams, there are $|S_t| - 1$ positions within the sequence. If a position is repeatedly voted for break point by different windows then it is likely to accrue a locally-maximum score. We choose the positions with local maximum of the score as break points of the episode.

Example 1 (Contd.): Continuing Example 1, we can find the list of episodes from each of the four sequences using the algorithm given in Figure 14.5. These are as follows.

$E_1 = ((\text{bf 0e 3a}), (\text{bf d8 3a}), (\text{3a bf}))$
$E_2 = ((\text{3a bf}), (\text{bf 0e 3a}), (\text{3a bf}), (\text{bf d8 3a}))$
$E_3 = ((\text{bf d8 3a}), (\text{bf 0e 3a}), (\text{bf 0e 3a}), (\text{0e 3a}))$
$E_4 = ((\text{0e 3a}), (\text{bf d8 3a}), (\text{3a bf}), (\text{3a bf 0e 3a}))$

Observing the above episodes, it is clear that episode discovery algorithm gives meaningful sequences. If we consider fixed length n-grams, say n = 2, then obviously we lose valuable information by missing 3-grams.

14.3 Relevant Episodes

We observe that there can be large number of episodes for a single program and when we consider all the programs in the training set the set of distinct episodes becomes very large. We introduce here two novel concepts- *relevant episodes* for a class and a new feature selection measure, namely *class-wise episode frequency*.

Algorithm: to find episodes for a set of sequences S
Input: T_d (trie of depth d) and S

    do for each of the node $x \in T_d$
        calculate entropy $e(x)$
    enddo
    do for each level L_i of T_d
        find mean frequency $(\overline{f})$ and mean entropy $(\overline{e})$
        find standard deviations ($\sigma_{\overline{f}}$ and $\sigma_{\overline{e}}$) taking $\overline{f}$ and $\overline{e}$ respectively
        $f(x) = \frac{f(x) - (\overline{f})}{\sigma_{\overline{f}}}$ for $x \in L_i$
        $e(x) = \frac{e(x) - (\overline{e})}{\sigma_{\overline{e}}}$ for $x \in L_i$
    enddo
    do for each sequence $S_t(s_1, s_2, \ldots, s_{|S_t|}) \in S$
        Episodes $E_t = \phi$
        initialize $score(i) = 0$, for $1 \leqslant i \leqslant |S_t|$
        do for $i = 1$ to $(|S_t| - k + 1)$
            take a window of length k starting at position i in S
            do for each of the k possible boundary positions in the window
                find $Max_j\{f(s_i, \ldots, s_{i+j}) + f(s_{i+j+1}, \ldots, s_{i+k-1})\}, 0 \leqslant j < k$
                increment $score(i+j)$
                find $Max_j\{e(s_i, \ldots, s_{i+j})\}, 0 \leqslant j < k$
                increment $score(i+j)$
            enddo
        enddo
        $start = 1$
        do for $i = 2$ to $(|S_t| - 1)$
            if $(score(i) > score(i-1))$ and $(score(i) > score(i+1))$
                $end = i$
                add $(s_{start}, \ldots, s_{end})$ to E_t
                $start = end$
            endif
        enddo
        add $(s_{start}, \ldots, s_{|S_t|})$ to E_t
    enddo

Fig. 14.5 Episode discovery algorithm for a set of sequences using a combined trie.

The main aim of feature selection is to identify a smaller set of features that are representative of the virus and benign classes.

Let the set of virus programs be V and the set of benign programs be B.

Definition 14.2 The class-wise episode frequency of an episode with respect to a class V (or, B) is the number of times it occurs in the class V (or, B).

While the term frequency is a kind of global measure, the class-wise episode frequency is a local measure with respect to a class. The main advantage of class-wise episode frequency is that we can analyze each class independently. This saves memory requirement as we handle only episodes of one class at a time.

For each executable program t in a class T of programs, let E_t be the set of all episodes. The set of all episodes for T, $E(T)$ is $\cup_{t \in T} E_t$. We assume that the elements of $E(T)$ are arranged in the non-increasing order of class-wise episode frequency.

Definition 14.3 We define $E^k(T)$ as the *relevant episodes* for T which is a subset of $E(T)$ containing only first k elements.

Thus the relevant episodes for classes V and B are $E^k(V)$ and $E^k(B)$, respectively. We get the set of relevant episodes for the whole training data as $E^k(V) \bigcup E^k(B)$.

With the set of relevant episodes for the data set, we build the vector space model which is a concept derived from information retrieval. An executable program is represented as a vector of $t_1, t_2, \ldots, t_M$, where $t_i (1 \leqslant i \leqslant M)$ is a binary (0-1) value denoting the occurrence of the ith relevant episode. The value 1 represents the occurrence of an episode and its absence is represented by 0. Thus each unique relevant episode corresponds to a dimension. Our training set consists of a set of labeled vectors – the vector representation of the set of programs together with the respective class label (virus or benign).

As we observed, the detection problem reduces essentially to supervised classification problem. Several algorithms exist[23] for supervised classification like support vector machine, decision tree, neural networks etc. We use the metaclassifier Ada Boost with J48 as base classifier available in WEKA.[30] The reason for choosing this particular classifier is that the authors of previous work[18] in this area claimed that they got the best results for this classifier.

14.4 Experimental Setup

No standard data set is available for the detection of malicious executables unlike intrusion detection. Data sets (i.e. viruses) collected from the website VX Heavens[29] were used previously.[3,18] The benign executables were collected from their respective laboratories. We collected 250 viruses from VX Heavens[29] and 250 benign executables from our lab. For viruses, we used only the loader programs; we did not use the infected programs in our analysis. At present, we only concentrate on viruses.

Each executable in the dataset is converted to hexadecimal codes in an ASCII format. Before training, the set of programs for each class V and B are used to build the trie and from the tries, the set of episodes for each individual executables are extracted. The top m episodes in order of class-wise frequency are selected from each class. These are combined to get a set of relevant episodes of cardinality M with duplicates removed. This result in a vector space model of the dataset of size 500 rows $\times$ M columns. For unbiased evaluation of our approach, we used stratified ten-fold cross-validation. In this cross-validation method, we partition the data into ten disjoint sets of equal size and select one partition as testing set and the remaining nine are combined to form the training set. This process is repeated 10 times.

14.4.1 *Classifier*

The classification is done with the metaclassifier AdaBoost M1 with J48 as base classifier. AdaBoost, short for Adaptive Boosting, is a meta-algorithm that and can be used in conjunction with many other learning algorithms to improve their performance. AdaBoost is adaptive in the sense that subsequent classifiers built are tweaked in favor of those instances misclassified by previous classifiers. It is less susceptible to the overfitting problem than most learning algorithms. These algorithms are implemented in WEKA,[30] which is a collection of machine learning algorithms for solving data mining problems implemented in Java and open sourced under the GPL. For the classifier we use the default values given by WEKA to evaluate our approach.

We compared our method with that of the method proposed by Kolter et al.[18] for two reasons. First, this method is so far the best method of generic virus detection using n-grams. The other reason is that this method takes fixed length n-grams as features and information gain as the measure for feature selection. We use the same dataset for both the methods for comparison. In the work by Kolter et al.,[18] the experiments are done on two data sets, one with 476 malicious and 561 benign executables and the other set contains 1971 malicious and 1651 benign executables. Information gain is used as feature selection measure. After extracting all n-grams along with the information gain value, the n-grams are sorted in decreasing order of information gain and top k n-grams are taken and vector space model is formed. Out of several classifiers, it is claimed that the best results are obtained for boosted J48 algorithm. In order to compare our work, we implemented their technique of using information gain as feature selection measure and J48 algorithm as classifier on our data. We give the results as ROC curves.

14.4.2 *ROC Curves*

In Data Mining, ROC (Receiver Operating Characteristics) curves are used to compare classification capability of different algorithms. An ROC curve is a technique for visualizing, organizing and selecting classifiers based on their performance.[11] It is a two-dimensional depiction of classifier performance. To compare classifiers we may need to reduce ROC performance to a single scalar value representing expected performance. A common method is to calculate the area under the ROC curve. The area under ROC curve specifies the probability that, when we draw one positive and one negative example at random, the decision function assigns a higher value to the positive than to the negative example. The more the area under the ROC curve of an algorithm, the more robust and better it is in classification. Infact, just visual inspection of ROC graphs is enough to compare the performance of classifiers.

14.5 Experimental Results and Discussions

We experimented with n as 2, 3 and 4, whereas the profile length, M, is taken as 100 and 500. We give the ROC curves in Figures 14.6-14.9.

Figure 14.6 shows the ROC curves of the proposed method and the fixed-length n-gram approach (with $n = 2$) with profile length as 100. It is evident from the graph that the proposed method gives a 100% detection rate with a false positive rate of 0.11 whereas the fixed-length method attains it with a false positive rate of 0.216. Even for fixed-length of $n=3$ and 4, proposed method gives consistantly better results in comparison to fixed-length n-gram approach as can be seen from figures 14.7 and 14.8.

For profile length of 500, as given in Figure 14.9, the accuracy of our method is more pronounced than the fixed-length approach. The proposed method achieved a detection rate of 70.8% with zero false positive, and 100% detection with 2.6% false positive rate. For the same profile length fixed-length n-gram method, for $n=4$, shows the lowest detection rate of 15% with 1.5% false positive rate and attains 100% detection rate with 20% false positive rate.

Taking area under ROC curve as performance criteria, from the visual inspection of the ROC curves it is clear that our proposed method outperforms the method proposed by Kolter et al.[18] in all cases. Based on the experimen-

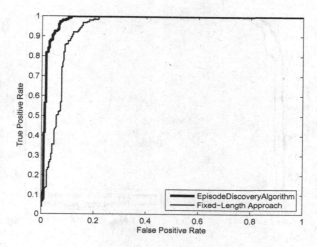

Fig. 14.6   Fixed vs Variable length *n*-grams: Profile length = 100, *n* = 2 for fixed length *n*-grams.

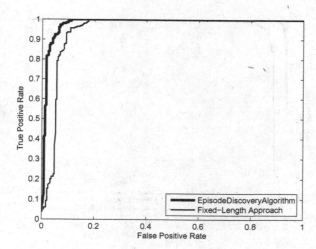

Fig. 14.7   Fixed vs Variable length *n*-grams: Profile length = 100, *n* = 3 for fixed length *n*-grams.

tation we infer the following. The variable-length *n*-grams approach is better than fixed length *n*-grams approach as the fixed-length *n*-grams do not capture the long meaningful sequences while keeping the size and number of *n*-grams manageable.

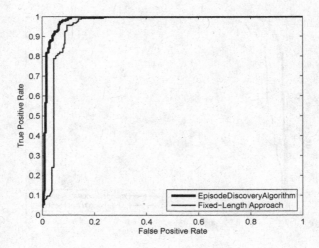

Fig. 14.8   Fixed vs Variable length *n*-grams: Profile length = 100, *n* = 4 for fixed length *n*-grams.

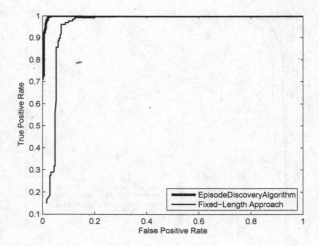

Fig. 14.9   Fixed vs Variable length *n*-grams: Profile length = 500, *n* = 4 for fixed length *n*-grams.

Moreover it is also observed that possibly class-wise episode frequency is a better measure for feature selection than the information gain. This is also demonstrated in detail in the context of fixed length *n*-grams in a recent work by Reddy et al.[25]

When it comes to advanced computer viruses like polymorphic viruses, static analysis methods do not work. To tackle these viruses, static analysis methods should be combined with dynamic analysis methods for efficient detection. For example, polymorphic virus consists of three components – decryption routine, mutation engine and virus body. Since the mutation engine and the virus body are encrypted and decryption routine is different in each replication, it is not possible to directly apply any static analysis method (including our method) to detect this virus. Instead we can use a dynamic analysis technique (like sandboxing) that trick a polymorphic virus into decrypting and revealing itself.[24] On this decrypted virus we can use the static analysis method. Here we assume that a polymorphic virus must decrypt before it can execute normally.

14.6 Conclusion

The main objective of the present work is to establish that proper feature extraction and selection technique can help in efficiently detecting virus programs. We showed here that episodes as variable length n-grams is better than usual fixed length n-grams. We also showed that selecting frequent episodes in terms of class-wise relevance is a better feature selection technique. We demonstrate that a supervised classification by the proposed feature selection method gives better accuracy and less false positives in comparison to earlier proposed methods.

In n-gram approach attaching semantic meaning for the relevant n-grams (episodes) is not yet explored by us. Our future work will be to develop a semantic aware method and include different kinds of malicious and benign executables in our training data.

Acknowledgment

This research is supported by Ministry of Communications and IT, Govt of India under the grant no. 12(22)/04-IRSD dated: 04.02.2004.

References

1. K. G. Anagnostakis, S. Sidiroglou, P. Akritidis, K. Xinidis, E. Markatos, and A. D. Keromytis, Detecting targeted attacks using shadow honeypots, in *Proceedings of the 14th USENIX Security Symposium* (2005).
2. W. Arnold, and G. Tesauro, Automatically generated Win32 heuristic virus detection. In *Proceedings of the 2000 International Virus Bulletin Conference* (2000).
3. T. A. Assaleh, N. Cercone, V. Keselj, and R. Sweidan, Detection of new malicious

code using N-grams signatures, in *Proceedings of the Second Annual Conference on Privacy, Security and Trust* (2004) 193–196.

4. R. Balzer, N. Goldman, Mediating Connectors, in *Proceedings of the 19th IEEE International Conference on Distributed Computing Systems Workshop*, Austin, TX (1999), pp. 73–77.

5. W. Cavnar, J. Trenkle, N-gram based text categorization, in *Proceedings of SDAIR-94, 3rd Annual Symposium on Document Analysis and Information Retrieval* (1994), pp. 161–175.

6. M. Christodorescu, S. Jha, Static analysis of executables to detect malicious patterns, in *Proceedings of the 12th USENIX Security Symp.*, Washington, DC, August (2003), pp. 169–186.

7. F. Cohen, Computational aspects of computer viruses, *Journal of Computers and Security* **8** (1989) 325–344.

8. P. Cohen, B. Heeringa, and N. M. Adams, An unsupervised algorithm for segmenting categorical timeseries into episodes, in *Proceedings of the ESF Exploratory Workshop on Pattern Detection and Discovery*, London, UK, September (2002), pp. 49-62.

9. S. K. Dash, K. S. Reddy, and A. K. Pujari, Episode Based Masquerade Detection, in *Proceedings of 1st International Conference on Information Systems Security*, Kolkata, December (2005), LNCS Vol. 3803, pp. 251–262.

10. H. Debar, M. Dacier, M. Nassehi, and A. Wespi, Fixed vs. variable-length patterns for detecting suspicious process behavior, *Journal of Computer Security* **8**(2/3) (2000).

11. T. Fawcett, An introduction to ROC analysis, *Pattern Recognition Letters* **27**(8) (2006) 861–874.

12. L. Firoiu, Segmenting Time Series with a Hybrid Neural Networks – Hidden Markov Model, `citeseer.ist.psu.edu/firoiu02segmenting.html` (2002).

13. J. Furnkranz, A study using *n*-gram features for text categorization, Technical Report OEFAI-TR-9830, Austrian Research Institute for Artificial Intelligence (1998).

14. Gartner Inc.: `www.gartner.com/press_releases/asset_129199_11.html` (2005).

15. A. Gionis, and H. Mannila, Segmentation algorithms for time series and sequence data, *SIAM International Conference on Data Mining*, Newport Beach, CA (2005).

16. G. Jiang, H. Chen, C. Ungureanu, and K. Yoshihira, Multi-resolution abnormal trace detection using varied-length N-grams and automata, in *Proceedings of the Second International Conference on Autonomic Computing* (2005), pp. 111–122.

17. J. O. Kephart, G. B. Sorkin, W. C. Arnold, D. M. Chess, G. J. Tesauro, and S. R. White, Biologically inspired defenses against computer viruses, in *Proceedings of IJCAI '95*, Montreal, August (1995), pp. 985–996.

18. J. K. Kolter, and M. A. Maloof, Learning to detect malicious executables in the wild, in *Proceedings of the Tenth ACM SIGKDD International Conference on Knowledge Discovery and Data Mining* (2004), pp. 470–478.

19. R. W. Lo, K. N. Levitt, and R. A. Olsson, MCF: A malicious code filter, *Computers & Society* **14**(6) (1995) 541–566.

20. C. Marceau, Characterizing the behavior of a program using multiple-length N-grams, in *Proceedings of the 2000 Workshop on New Security Paradigms* (2001), pp. 101–110.

21. G. McGraw, and G. Morrisett, Attacking Malicious Code: A Report to the Infosec

Research Council, IEEE Software, September/October (2000).

22. D. Micheal Cai, Maya Gokhale and J. Theiler, Comparison of feature selection and classification algorithms in identifying malicious executables, *Journal of Computational Statistics & Data Analysis* **51** (2007) 3156–3172.
23. T. Mitchell, *Machine Learning*, (McGraw-Hill, New York, 1997).
24. C. Nachenberg, Understanding and managing polymorphic viruses, *The Symantec Enterprise Papers*, Vol XXX.
25. D. K. S. Reddy, and A. K. Pujari, *N*-gram Analysis for New Computer Virus Detection, *Journal of Computer Virology* **2** (2006) 231–239.
26. M. G. Schultz, E. Eskin, E. Zadok, and S. J. Stolfo, Data mining methods for detection of new malicious executables, in *Proceedings of IEEE Symposium on Security and Privacy* (2001), pp. 38–49.
27. M. G. Schultz, E. Eskin, E. Zadok, M. Bhattacharyya, and S. J. Stolfo. MEF: Malicious Email Filter, A UNIX mail filter that detects malicious windows executables, in *Proceedings of USENIX Annual Technical Conference* (2001).
28. P. Szor, *The Art of Computer Virus Research and Defense* (Addison Wesley, 2005).
29. VX Heavens: `http://vx.netlux.org`.
30. I. Witten, and E. Frank, Data mining: Practical machine learning tools and techniques with Java implementations (Morgan Kaufmann, San Francisco, CA, 2000).
31. Y. Yang, and J. O. Pedersen, A comparative study on feature selection in text categorization, in *Proceedings of 14th International Conference on Machine Learning* (1997), pp. 412–420.

Chapter 15

Overview of State-of-the-Art in Digital Image Forensics

H. T. Sencar and N. Memon

Department of Computer and Information Science,
Polytechnic University,
Brooklyn, NY 11201, USA

Digital images can now be easily created, altered, and manipulated with no obvious traces of having been subjected to any of these operations. There are currently no established methodologies to verify the authenticity and integrity of digital images in an automatic manner. Digital image forensics is an emerging research field with important implications for ensuring the credibility of digital images. In an attempt to assist these efforts, this chapter surveys the recent developments in the field of digital image forensics. Proposed techniques in the literature are categorized into three primary areas based on their focus: image source identification, discrimination of synthetic images, and image forgery detection. The main idea of the proposed approaches in each category is described in detail, and reported results are discussed to evaluate the potential of the methods.

Contents

15.1 Introduction

In the analog world, an image (a photograph) has generally been accepted as a "proof of occurrence" of the depicted event. In today's digital age, the creation and manipulation of digital images is made simple by low-cost hardware and software tools that are easily and widely available. As a result, we are rapidly reaching a situation where one can no longer take the authenticity and integrity of digital images for granted. This trend undermines the credibility of digital images presented as evidence in a court of law, as news items, as part of a medical record or as financial documents since it may no longer be possible to distinguish whether a given digital image is the original or a (maliciously) modified version or even a depiction of a real-life occurrences and objects.

This is especially true when it comes to legal photographic evidence. The Federal Rules of Evidence are shaped and drafted to deal with conventional (analog) photography. Digital photography, on the other hand, is fundamentally different from conventional photography in the way it is created, stored, and edited. Federal Rules do not currently set forth requirements for the admissibility of digital images, and, therefore, traditional notions of relevancy and authentication currently govern. Moreover, the problem becomes much more complicated (with possibly far more severe consequences) when the digital image is synthetically generated to convey the depiction of a non-existent scene or object as the existing safeguards are not well suited to verify the integrity and authenticity of such visual evidence. The struck down of a 1996 child pornography law that prohibited the possession and distribution of synthetically generated images by United States Supreme Court, in April 2002, is very important in this context.[1] This ruling brought with it an immediate need for tools and techniques that can reliably discriminate natural images from the synthetic ones in order to be able to prosecute abusers. Another pressing issue concerning digital imagery is the ease with which processing tools and computer graphics algorithms can be used to modify images. The increasing appearance of digitally altered forgeries in mainstream media and on the internet is an indication of the serious vulnerability that cast doubt on integrity of all digital images. In,[2] some well known examples of digital tampering can be found.

To address these immediate problems, digital image forensics research aims at uncovering underlying facts about an image. For example digital image forensics techniques look for authoritative answers to questions such as:

- Is this image an "original" image or was it created by cut and paste operations from different images?
- Does this image truly represent the original scene or was it digitally tampered

to deceive the viewer?
- What is the processing history of the image?
- What parts of the image has undergone processing and up to what extent?
- Was the image acquired by a source manufactured by vendor X or vendor Y?
- Did this image originate from source X as claimed?

The above questions are just a few examples of issues faced routinely by investigation and law enforcement agencies. However, there is a lack of techniques and methodologies that could determine the origin and potential authenticity of a digital image. Although digital watermarks have been proposed as a tool to provide authenticity to images, it is a fact that the overwhelming majority of images that are captured today do not contain a digital watermark. And this situation is likely to continue for the foreseeable future. Hence in the absence of widespread adoption of digital watermarks, we believe it is imperative to develop techniques that can help us make statements about the origin, veracity and nature of digital images.

The past few years have seen a growth of research on image forensics. The work has focused mainly on three types of problems:

(1) Image source identification to determine through what data acquisition device a given image is generated, *e.g.,* digital-camera or scanner. This entails associating the image with a class of sources that have common characteristics (*i.e.,* device model) and matching the image to an individual source device.
(2) Discrimination of synthetic images from real images to identify computer generated images which does not depict a real-life occurrence.
(3) Image forgery detection to determine whether a given image has undergone any form of modification or processing after it was initially captured.

To address these problems several techniques have been proposed. In this chapter, we will give an overview of state-of-the-art digital image forensics techniques. The outline of the chapter is as follows. In Section 15.2, we review various image source identification techniques. This is followed by an overview of techniques for differentiating synthetic images in Section 15.3. Image forgery (tamper) detection techniques are described in Section 15.4. Finally, our conclusions and open problems will be given in Section 15.5.

15.2 Image Source Identification

Image source identification research investigates the design of techniques to identify the characteristics of digital data acquisition device (*e.g.,* digital camera, cam-
corder, and scanner) used in generation of an image. These techniques are expected to achieve two major outcomes. The first is the class (model) properties
of the source, and the second is the individual source properties. Essentially, the
two outcomes refer two different operational settings. In determining the class
properties, typically, a single image is available for evaluation and the source information is extracted through analyzing the image. In obtaining individual source
properties, however, both an image and the potential source device or a number
of images known to be acquired by the source is available for evaluation, and the
analysis determines if the characteristics of the image in question matches to those
of the source.

The success of image source identification techniques depend on the assumption that all images acquired by an image acquisition device will exhibit certain
characteristics that are intrinsic to the acquisition devices because of their (proprietary) image formation pipeline and the unique hardware components they deploy
regardless of the content of the image. (It should be noted that such devices generally encode the device related information, like model, type, date and time, and
compression details, in the image header, *e.g.,* EXIF header. However, since this
information can be easily modified or removed, it cannot be used for forensics
purposes.) Due to prevalence of digital camera images, research has primarily focused on source digital camera identification and scanner identification research
is just starting.

15.2.1 *Image Formation in Digital Cameras and Scanners*

The design of image source identification techniques requires an understanding of
the physics and operation of these devices. The general structure and sequence of
stages of image formation pipeline remains similar for almost all digital cameras
and scanners, although much of the details are kept as proprietary information of
each manufacturer. Below, we will describe the basic structure for a digital camera
and scanner pipeline.

Digital Camera Pipeline: Consumer level digital cameras consist of a lens
system, sampling filters, color filter array, imaging sensor, and a digital image
processor.[3] The lens system is essentially composed of a lens and the mechanisms to control exposure, focusing, and image stabilization to collect and control
the light from the scene. After the light enters the camera through the lens, it goes

through a combination of filters that includes at least the infra-red and anti-aliasing filters to ensure maximum visible quality. The light is then focused onto imaging sensor, an array of rows of columns of light-sensing elements called pixels. Digital cameras deploy charge-coupled device (CCD) or complimentary metal-oxide semiconductor (CMOS) type of imaging sensors. Each light sensing element of sensor array integrates the incident light over the whole spectrum and obtains an electric signal representation of the scenery. Since each imaging sensor element is essentially monochromatic, capturing color images requires separate sensors for each color component. However, due to cost considerations, in most digital cameras, only a single sensor is used along with a color filter array (CFA). The CFA arranges pixels in a pattern so that each element has a different spectral filter. Hence, each element only senses one band of wavelength, and the raw image collected from the imaging sensor is a mosaic of different colors and varying intensity values. The CFA patterns are most generally comprised of red-green-blue (RGB) and cyan-magenta-yellow (CMY) color components. The measured color values are passed to a digital image processor which performs a number of operations to produce a visually pleasing image. As each sub-partition of pixels only provide information about a number of color component values, the missing color values for each pixel need to be obtained through demosaicing operation. This is followed by other forms of processing like white point correction, image sharpening, aperture correction, gamma correction and compression. Although the operations and stages explained here are standard stages in a digital camera pipeline, the exact processing detail in each stage varies from one manufacturer to the other, and even in different camera models manufactured by the same company.

Scanner Pipeline: Conventional consumer scanners are composed of a glass pane, a bright light source (often xenon or cold cathode fluorescent) which illuminates the pane from underneath, and a moving scan head that includes lenses, mirrors, a set of filters, and the imaging sensor, whether CCD, CMOS or contact image sensors (CIS).[4] (Drum scanners which have been typically used for high-end applications use photomultiplier tubes.) To obtain color scans, typically, three rows (arrays) of sensors with red, green, and blue filters are utilized. During scanning, the imaging sensor and light source move across the pane (linear motion). The light strikes the image, reflects, and is then reflected by a series of mirrors to the scanner lens. The light passes through the lens and is focused onto imaging sensors to be later digitized. The resolution of a scanner depends on both the number of elements of the imaging sensor (horizontal resolution) and the step size of the scan head motor (vertical resolution). The hardware resolution of the scanner can be reduced down by either down-sampling or less commonly through activating only some elements of the CCD array.

15.2.2 *Source Model Identification*

The work in this field has been primarily focused on digital cameras. The features that are used to differentiate camera-models are derived based on the differences in processing techniques and the component technologies. For example, the optical distortions due to a type of lens, the size of the imaging sensor, the choice of CFA and the corresponding demosaicing algorithm, and color processing algorithms can be detected and quantitatively characterized by analysis of the image. The deficiency of this methodology, in general, is that many models and brands use components by a few manufacturers, and processing steps/algorithms remain same or very similar among different models of a brand. Hence, reliable identification of a source camera-model depends on characterization of various model dependent features as briefly explained below.

15.2.2.1 *Image Features*

Inspired by the success of universal steganalysis techniques, Kharrazi et al.[5] proposed a similar approach to identify source camera-model. In essence, a select number of features designed to detect post-processing are incorporated with new features to fingerprint camera-models. The 34 features include color features (*e.g.,* deviations from gray world assumption, inter-band correlations, gamma factor estimates), image quality metrics, and wavelet coefficient statistics. These features are then used to construct multi-class classifiers. The results obtained on moderate to low compressed images taken by 4 different camera-models yielded an identification accuracy of 97%. When experiments are repeated on five cameras where three of them are of the same brand, the accuracy is measured to be 88%. Tsai et al.[6] later repeated this study using a different set of cameras and reported similar results. In their work,[7] Celiktutan et al. took a similar approach to differentiate between cell-phone camera-models by deploying binary similarity measures as features.[8] In this case, the identification accuracy among nine cell-phone models (of four different brands) is determined as 83%. There are two main concerns regarding this type of approaches. First is that as they provide an overall decision, it is not clear as to what specific feature enables identification which is very important in forensic investigations and in expert witness testimonies. Second concern is the scalability of performance with the increasing number of digital cameras in the presence of hundreds of digital cameras. Hence, in general, this approach is more suitable as a pre-processing technique to cluster images taken by cameras with similar components and processing algorithms.

15.2.2.2 *CFA and Demosaicing Artifacts*

The choice of CFA and the specifics of the demosacing algorithm are some of the most pronounced differences among different digital camera-models. In digital cameras with single imaging sensors, the use of demosacing algorithms is crucial for correct rendering of high spatial frequency image details, and it uniquely impacts the edge and color quality of an image. Essentially, demosaicing is a form of interpolation which in effect introduces a specific type of inter-dependency (correlations) between color values of image pixels. The specific form of these dependencies can be extracted from the images to fingerprint different demosaicing algorithms and to determine the source camera-model of an image. In,[9] Popescu et al. demonstrated that expectation/maximization (EM) algorithm can be used to estimate the (linear interpolation) filter coefficients by re-interpolating digital camera images (after down-sampling to remove existing traces of interpolation) with eight different CFA interpolation algorithms. The average accuracy in pair-wise differentiation over all pairs of interpolation algorithms is obtained as 97%. To fingerprint demosaicing algorithms used in different digital camera-models Bayram et al.[10,11] deployed EM algorithm, assuming a linear model for interpolation within a 5x5 window, and analyzed patterns of periodicity in second order derivates of rows and columns of pixels in moderately smooth and very smooth image parts, respectively. The estimated filter coefficients and the periodicity features are used as features in construction of classifiers to detect source camera-model. The accuracy in identifying the source of an image among four and five camera-models is measured as 86% and 78%, respectively, using images captured under automatic settings and at highest compression quality levels.

Alternatively, Long et al.[12] considered analyzing the modeling error due to the linear interpolation model and identifying demosaicing algorithm based on the characteristics of this error, rather than using the estimated interpolation filter coefficients. (*I.e.,* the difference between the actual pixel values in the image and their reconstructed versions as a weighted sum of 13 neighboring pixels.) They realized this by computing the autocorrelation of the error over all image. Then, the (13x13) autocorrelation matrices obtained from many images are combined together and subjected to principal component analysis to determine the most important components which are then used as features in building a classifier. They reported that an accuracy of more than 95% can be achieved in identifying the source of an image among four camera-models and a class of synthetic images and studied the change in performance under compression, noise addition, gamma correction and median filtering types of processing.

Later, Swaminathan et al.[13] enhanced this approach by first assuming a CFA pattern, thereby discriminating between the interpolated and un-interpolated pixel locations and values—an advantage over EM algorithm, and estimating the interpolation filter coefficients corresponding to that pattern (assuming a linear model within a 7x7 window) for each of three activity regions, *e.g.,* smooth, horizontal gradient, and vertical gradient. Then, the un-interpolated color values are interpolated with respect to the assumed CFA pattern with the obtained filter and the error with the resulting newly interpolated image and the actual image is computed. The CFA pattern of an image is determined by searching over all valid CFA patterns to minimize the resulting error, and the demosaicing algorithms are differentiated through the use of classifiers built based on estimated filter coefficients. The corresponding identification accuracy is determined by applying the method to images taken by 16 camera-models under different compression levels and it is reported to be 84%.

15.2.2.3 *Lens Distortions*

In their work,[14] Choi et al. proposed the utilization of lens radial distortion, which deforms the whole image by causing straight lines in object space to be rendered as curved lines. Radial distortion is due to the change in the image magnification with increasing distance from the optical axis, and it is more explicit in digital cameras equipped with spherical surfaced lenses. Therefore, manufacturers try to compensate for this by adjusting various parameters during image formation which yields unique artifacts. To quantify these distortions, the paper extends a first-order radial symmetric distortion model, which expresses undistorted radius (from optical axis) as an infinite series of distorted radius, to second order. These parameters are computed assuming a straight line model by first identifying line segments which are supposed to be straight in the scene and computing the error between the actual line segments and their ideal straight forms. Later, these parameters are used as features to build classifiers in a framework similar to.[5] The measurements obtained from images captured with no manual zooming and flash and at best compression level by three digital camera-models resulted with an identification accuracy of approximately 91%. These features are also incorporated with those of earlier proposed ones[5] and similar overall identification accuracy is reported.

15.2.3 *Individual Source Identification*

The ability to match an image to its source requires identifying unique characteristics of the source acquisition device. These characteristics may be in the form of hardware and component imperfections, defects, or faults which might arise due to inhomogeneity in the manufacturing process, manufacturing tolerances, environmental effects, and operating conditions. For example, the aberrations produced by a lens, noise in an imaging sensor, dust specks on a lens will introduce unique but mostly imperceptible artifacts in images which can later be extracted to identify the source of the image. The main challenge in this research direction is that reliable measurement of these minute differences from a single image is very difficult and they can be easily eclipsed by the image content itself. Another challenge is that these artifacts tend to vary in time and depend on operating conditions. Therefore, they may not always yield positive identification. Following approaches are proposed to utilize such characteristics in image source identification.

15.2.3.1 *Imaging Sensor Imperfections*

This class of approaches to source matching aims at identifying and extracting systematic errors due to imaging sensor, which reveal themselves on all images acquired by the sensor in a way independent of the scene content. These errors include sensor's pixel defects and pattern noise which has two major components, namely, fixed pattern noise and photo response non-uniformity noise. The initial work in the field has been done by Kurusowa et al.[15] in which fixed pattern noise caused by dark currents in (video camera) imaging sensors is detected. Dark current noise refers to differences in pixels when the sensor is not exposed to light and it essentially behaves as an additive noise. Therefore, it can be easily compensated within the camera by first capturing a dark frame and subtracting it from the actual readings from the scene, thereby hindering the applicability of the approach. Geradts et al.[16] proposed matching the traces of defective pixels, *e.g.,* hot pixels, cold/dead pixels, pixel traps, cluster defects, for determining the source camera. Their experiments on 12 cameras showed the uniqueness of the defect pattern and also demonstrated the variability of the pattern with operating conditions. However, ultimately, such defects also cannot be reliably used in source identification as most cameras deploy mechanism to detect such defects and compensate them through post-processing.

The most promising and reliable approach in this field is proposed by Lukas et al.[17] to detect the pixel non-uniformity noise, which is the dominant component of the photo-response non-uniformity pattern noise arising due to different sensitivity of pixels to light. The main distinction of this approach as compared to earlier ones is that the correction of this noise component requires an operation called flat-fielding which in essence requires division of the sensor readings by a pattern extracted from a uniformly lit scene before any non-linear operation is performed. Since obtaining a uniform sensor illumination in camera is not trivial, most digital cameras do not flat-field the resulting images. The key idea of the method is to denoise the image by wavelet based denoising algorithm so that the resulting noise residue contains the needed noise components. However, since the underlying image model used in denoising is an idealistic one the residue signal also contains contributions from the actual image signal. Hence to eliminate the random component of the noise, denoising is applied to a set of images (captured by the same camera) and the corresponding noise residues are averaged to obtain the reference pattern of a given digital camera. Later, to determine whether a given image is captured by a digital camera, the noise pattern extracted from the individual image is correlated with the reference pattern of the digital camera. A decision is made by comparing the measured correlation statistic to a pre-determined decision threshold. The results obtained from (high quality) images taken by 9 cameras yielded 100% identification accuracy.

To determine the false-positive and true-detection performance of the scheme proposed in[17] under a more realistic setting Sutcu et al.[18] performed experiments on an image dataset with roughly 50K randomly selected images and observed that some of the tested cameras yield false-positive rates much higher than the expected values. To compensate for false-positives the authors proposed coupling the approach of[17] with source-model identification methodology. In this case, during the extraction of the pattern the demosaicing characteristics of the source camera-model are also determined as described in.[11] When a decision is to be made in matching an image to a potential source camera, it is also required that the class properties of the camera extracted from the individual image is also in agreement with those of the source camera. It is shown that this approach is very effective in reducing the false-positive rate with a marginal reduction in the true-detection rate. In,[19] Fridrich et al. proposed enhancements to the noise extraction scheme by deploying pre-processing techniques to reduce the contributions of image noise and to gain robustness against compression.

Khanna et al.[20,21] extended sensor noise extraction methodology to also include scanned images and to enable source scanner identification. The main difference between the imaging sensors deployed in digital camera and (flatbed) scanners is that in the former sensor is a two-dimensional array, whereas in the latter it is a one-dimensional linear array, and a scan is generated by translating the sensor over the image. As a result the noise pattern extracted from a scanned image is expected to repeat itself over all rows. Therefore, a row reference noise pattern can be obtained from a single scanned image by averaging the extracted noise (via denoising) over all rows. In [20], the authors showed that this difference in the dimension of the array can be used to distinguish between digital camera and scanner images. In realizing this, classifiers are built based on (seven) statistics computed from averaged row and column reference patterns extracted from both scanned images at hardware resolution (*e.g.*, no down-sampling) and digital camera images. In experiments, various training scenarios are considered and an average accuracy of more than 95% is achieved in discriminating digital camera images from scanned images. The methodology is also applied to source scanner identification problem with the inclusion of new features in classifier design.[21] When identifying the source scanner of an image among four scanners an average classification accuracy of 96% is achieved and when the images are compressed with JPEG quality factor 90 an accuracy of 85% is obtained.

Gou et al.[22] proposed another approach to fingerprint the scanning noise associated with different models of (flatbed) scanners. The method characterizes scanning noise by three sets of features. The first set of features are obtained by denoising the scanned images and obtaining first and second order moments of the log-absolute transformed version of the noise residue. The second set of features are obtained as the mean, variance and error due to fitting normal distributions to high frequency sub-band coefficients of one-level wavelet decomposed version of the (normalized) scanned image. The third set consists of features extracted from the first two moments of prediction error applied to smooth regions. The most distinctive of the resulting 60 features are used to construct classifiers, which yielded an identification accuracy of 90% among seven scanner models with relatively smaller size of datasets (27 uncompressed images per model). The distinguishability of the features are also compared to wavelet coefficient statistics[23] and image quality metrics[24] and shown to be better. In the context of scanner identification, one issue that needs to be further studied is the variability of scanner noise among individual scanners and determining the corresponding false-alarm rates in identifying the source scanner.

15.2.3.2 *Sensor Dust Characteristics*

Dirik et al.[25] proposed another method for source camera identification based on sensor dust characteristics of single digital single-lens reflex (DSLR) cameras which are becoming increasingly popular because of their interchangeable lenses. Essentially, the sensor dust problem emerges when the lens is removed and the sensor area is opened to the hazards of dust and moisture which are attracted to the imaging sensor due to electrostatic fields, causing a unique dust pattern before the surface of the sensor. Sensor dust problem is persistent and most generally the patterns are not visually very significant. Therefore, traces of dust specks can be used for two purposes: to differentiate images taken by cheaper consumer level cameras and DSLR cameras and to associate an image with a particular DSLR camera. However, it should be noted that the lack of a match between dust patterns does not indicate anything since the dust specks might have been cleaned. Devising an empirical dust model characterized by intensity loss and roundness properties; the authors proposed a technique to detect noise specks on images through match filtering and contour analysis. This information is used in generation of a camera dust reference pattern which is later checked in individual images. In the experiments, ten images obtained from three DSLR cameras are used in generating a reference pattern which is then tested on a mixed set of 80 images (20 taken with the same camera and 60 with other cameras) yielding an average accuracy of 92% in matching the source with no false-positives.

15.3 Identification of Synthetic Images

A great deal of progress has been made in both fields of computer vision and computer graphics and these two fields have now begun to converge very rapidly. Consequently, more realistic synthetic imagery became achievable. Today, generative algorithms are able to produce realistic models of natural phenomena, *e.g.*, waves, mountains, sky, plants, objects with geometric structure, stimulate the behavior of light, *e.g.*, ray tracing and subsurface scattering methods, and take into consideration sensitivities of human perceptual system. Moreover, the sophistication of these algorithms parallels the increasing computation power. These advances in a way defeat the whole purpose of imagery and put the credibility of digital imagery at stake. Therefore, distinguishing photo-realistic computer generated (PRCG) images from real (natural) images is a very challenging and immediate problem. Several approaches have been proposed to address this problem. Essentially, all proposed approaches are based on machine learning methods, which express the relations between features extracted from a sample set of PRCG and real images in the form of classifiers. These classifiers are later used to differentiate between the two types of images. Hence, the main difference between the proposed approaches lie in the features they use in constructing the classifiers. Another concern with this class of methods is the image sets used during training and test phases as the true performance of the method will depend on how well their characteristics represent the overall class of images they belong to.

The first approach to differentiating natural (photographic) images from PRCG images was proposed by Lyu et al.[26] based primarily on a model of natural images. In this technique, the features are designed to capture the statistical regularities of natural images in terms of statistics of three-level discrete wavelet transform coefficients. The features include first order statistics (*e.g.,* mean, variance, skewness, and kurtosis) of both sub-band coefficients at each orientation and scale and of the errors in a linear predictor of coefficient magnitude (of all spatial, scale, orientation, and color neighbors) to capture higher order statistical correlations across space, scales and orientations, resulting with 72 features in each color band. The experiments are done on 40K real and 6K PRCG images of which 32K+4.8K images were used for training the classifiers and the rest for the testing which yielded an identification accuracy of 67% at 1% false-alarm rate.

In their work,[27] Ng et al. proposed another promising approach based on identifying the distinctive (geometry-based) characteristics of PRCG images, as compared to natural images. Their technique takes into account the differences in surface and object models and differences in the acquisition process between the PRCG and real images. The selection of their features is motivated by the observations that generation of PRCG images, mostly due to issues of computational complexity, is based on polygonal surface models and simplified light transport models, and does not exhibit acquisition characteristics of hardware device *e.g.,* cameras and scanners. The 192 features used in the design of the classifier are extracted by analyzing local patch statistics, local fractal dimension, and (normalized) differential geometry quantities, *e.g.,* surface gradient, quadratic geometry, and Beltrami flow. The authors used 800 PRCG images and 1.6K real images to test their features and obtained an average identification accuracy of 83% in comparison to an accuracy of 80% by Lyu et al.'s features.[26] It is also shown that when classifiers are trained to identify the CG images that are captured by digital cameras (*i.e.,* recapturing attack), a similar performance can be achieved by both feature sets.

Another wavelet transform based method was proposed by Wang et al.[28] where features are obtained from characteristic functions of wavelet-coefficient histograms. The features are obtained by first applying three-level wavelet decomposition at each color channel and further decomposing the diagonal sub-band into four second-level sub-bands, yielding a total of 48 sub-bands, and then by obtaining the normalized histograms in each sub-band. The DFT transform of the normalized histograms are filtered by three filters (two high-pass filters and a band-pass filter) to determine their energy at different frequency component ranges. Hence, a total of 144 features are obtained. The classifier trained on half of the 4.5K natural and 3.8K PRCG images yielded detection and false-positive results comparable to those of.[26] However, it is reported the classifier did not perform uniformly (much higher false-alarm rate) on the dataset used by in.[27]

Motivated by the fact that majority of the real images are captured by digital cameras, Dehnie et al.[29] presented an approach that aims at discriminating synthetic images from digital camera images based on the lack of artifacts due to acquisition process by focusing on the imaging sensor's pattern noise. Although each digital camera has a unique noise pattern,[17] since the underlying sensor technology remains similar, it is very likely that pattern noise introduced by different digital cameras may have common statistical properties. On the other hand, to avoid lack of real-life details, such as textures and lighting, generation of PRCG requires methods that add noise to simulate such phenomena in a physically consistent manner, *e.g.*, ray tracing algorithms. Similarly, it is very likely that the noise introduced by these methods to have certain statistical properties. To test the discriminative ability of the approach, a 600 PRCG images and more than 600 digital camera images have been denoised and the statistics of the resulting noise residues are analyzed. It is shown that the first-order statistics, like skewness and kurtosis, for the two noise components are distinct and the two types of image can be discriminated with an average accuracy of 75%.

Later, Dirik et al.[30] extended this approach to also include demosaicing artifacts[11] by proposing new features to detect the use of Bayer color filter array during demosaicing and and to detect traces of chromatic aberration. These features are later incorporated with the features of[26] and tested on 1.8 K PRCG and digital camera images half of which were used for training. Test results obtained on high quality images show that the classifier designed based on only four demosaicing features perform as good as wavelet transform coefficient statistics based features alone.[26] The results obtained from both high quality and medium level compressed images show that On the other hand, the proposed single feature based on traces chromatic aberration is shown to perform slightly worse but with less sensitivity to compression in highh to medium compression levels. The results for combined features show that the proposed five features can further improve the performance of the existing methods.

15.4 Image Forgery Detection

Due to the ease with which digital images can be altered and manipulated using widely available software tools, forgery detection is a primary goal in image forensics. An image can be tampered in many ways and at varying degrees, like compositing, re-touching, enhancing, with various intents. Although, many of the tampering operations generate images with no visual artifacts, they will, nevertheless affect the inherent statistics of the image. Furthermore, the process of image manipulation very often involves a sequence of processing steps to produce visually consistent images. Typically, a forged image (or parts of it) would have undergone some common image processing operations like affine transformations (*e.g.*, scaling, rotation, shearing), compensation for color and brightness variations, and suppression of details (*e.g.*, filtering, compression, noise addition). As a result, it is very likely that tampered image statistics will also exhibit variations due to such operations. In what follows, we briefly review various techniques proposed to determine whether the image has undergone any form of modification or processing after it has been captured.

15.4.1 *Variations in Image Features*

These approaches designate a set of features that are sensitive to image tampering and determine the ground truth for these features by analysis of original (unaltered) and tampered images. These values are stored as reference values and later tampering in an image is decided based on deviation of the measured features from the ground truth. These approaches most generally rely on classifiers in making decisions. For example, to exploit the similarity between the steganalysis and image manipulation detection, Avcibas et al.[31] proposed an approach similar to[24] by utilizing image quality metrics to probe different quality aspects of images, which could be impacted during tampering. In,[24] image quality metrics are used in cooperation with classifiers to differentiate between original and altered images based on measures obtained between a supposedly modified image and its estimated original (obtained through denoising) in terms of pixel and block level differences, edge distortions, and spectral phase distortions. To ensure that the features respond only to induced distortions due to tampering and not be confused by the variations in the image content, in[31] metrics are also measured with respect to a fixed set of images. Results obtained on 200 images by subjecting them to various image processing operations at a global scale yielded an average accuracy of 80%. When the same classifiers are given 60 skillfully tampered images, the detection accuracy is obtained to be 74%.

Based on the observation that non-linear processing of a signal very often introduced higher-order correlations, Ng. et al.[32] studied the effects image splicing on magnitude and phase characteristics of bicoherence spectrum (*i.e.*, normalized bispectrum which is the Fourier transform of the third order moment of a signal). The authors modeled the discontinuity introduced at the splicing point as a perturbation of a smooth signal with a bipolar signal and showed that bipolar signals contribute to changes in bicoherence spectrum of a signal. When tested the magnitude and phase features provided a classification accuracy of 62% which can be attributed to strong higher order correlations exhibited by natural images. Later,[33] the authors augmented the existing bicoherence features with newer ones that take into consideration the sensitivity of bicoherence to edge pixel density and the variation in bicoherece features between the spliced and un-spliced parts of the image. With the inclusion of the new features the accuracy in splicing detection is reported to increase to 72%. Bayram et al.[34,35] compiled three fundamental sets of features that have been successfully used in universal steganalysis and rigorously tested their sensitivity in detecting various common image processing operations by constructing classifiers to identify images that have undergone such processing. The tested features include image quality metrics,[24] wavelet coefficient statistics,[23] binary similarity measures,[8] the joint feature set which combines all the three sets, and the core feature set which is a reduced version of joint feature set. Different types of classifiers built from these features are tested under various image manipulations, like scaling up/down, rotation, contrast enhancement, brightness adjustment, blurring/sharpening and combinations, with varying parameters. Results on 100 locally tampered images, obtained from Internet, show that joint feature set performs best with an identification accuracy of around 90%.

15.4.2 *Image Feature Inconsistencies*

This class of techniques tries to detect image tampering based on inconsistent variations of selected features across the image. These variations may be in the form of abrupt deviations from the image norm or unexpected similarities over the image. One of the earliest methods in this class exploits the presence of double JPEG compression artifacts. Recompression of an (already compressed) image at a different quality factor distorts the smoothness of DCT coefficient histograms and creates identifiable patterns in DCT coefficient histograms. When the second quantization step size is smaller, some bins in the resulting histogram will be empty (zero valued) yielding a periodic peaks-and-valleys pattern. On the other hand, if the second quantization step is larger than the first one, all histogram values will be present but due to uneven splitting and merging of bins, histogram will show periodic peak patterns.

This phenomenon has been observed and studied in[36] and[37] to determine the initial compression parameters and to detect double compressed images. Essentially, the most common form of image tampering involves splicing of images which are very likely to be compressed at different quality factors. Therefore, the spliced parts in the recompressed image will have different double compression characteristics as compared to other parts. He et al.[38] developed a workable algorithm for automatically locating the tampered parts. In the method, the coefficient histogram of each DCT channel is analyzed for double compression effects and to assign probabilities to each (8x8) DCT block of its being a doctored block. The probabilities for each block are later fused together to obtain normality map of blocks, and tampering is decided based on presence and location of clusters on this map. Experiments performed on a small number of tampered images demonstrate the success of the algorithm. Further experiments are needed to determine how the method performs under various types of image tampering.

Popescu et al.[39] proposed a method for detecting resized (parts of) images which might potentially indicate image tampering. The principle of their method is based on the fact that up-sampling (interpolation) operation introduces periodic inter-coefficient correlations (*i.e.*, all interpolated coefficients depend on their neighbors in the same manner) and re-sampling at arbitrary rates requires a combination of up-sampling and down-sampling operations to achieve the intended rate. Hence, the presence of correlation between pixels can be used to determine which parts of images underwent resizing. To extract the specific form of correlations, the authors assumed probabilistic models for the prediction errors of both interpolated and un-interpolated coefficients. The estimation of distribution parameters and grouping of coefficients are performed simultaneously by EM algorithm. Results obtained on high quality JPEG images by subjecting images to global transformations such as scaling, rotations and gamma correction yielded detection accuracy close to 100% in most cases. However, the accuracy of detecting locally tampered regions have to be further tested.

Johnson et al.[40] considered the use of lighting direction inconsistencies across an image to detect image tampering, as it is often difficult to ensure (physically) consistent lighting effects. The crux of the method lies in a technique that estimates the light source direction from a single image. Assuming a point light source infinitely far away, a surface that reflects light isotropically and has a constant reflectance, and the angle between the surface normal and the light direction is less than 90 degrees, the image intensity is expressed a function of surface normal, light source direction, and constants (*i.e.*, reflectance and ambient light terms). Surfaces of known geometry in the image (*e.g.*, plane, sphere, cylinder, etc.) in the image are partitioned into many patches and by solving the formulation for all patches and combining the results to obtain the light direction. Formulation is also applied to local light sources and multiple light sources by combining them into a single virtual light source. The applicability of the method is demonstrated on a smaller set of images.

Image tampering very often involves local sharpness/blurriness adjustments. Hence, the blurriness characteristics in the tampered parts are expected to differ in non-tampered parts. In,[41] Sutcu et al. proposed the use of regularity properties of wavelet transform coefficients to estimate sharpness/blurriness of edges to detect variations and to localize tampering. The decay of wavelet transform coefficients across scales has been employed for edge detection and quality estimation purposes previously. The proposed method first employs an edge detection algorithm to determine edge locations which is then followed by a multi-scale wavelet decomposition of the image. Edge locations are located by analyzing the edge image and corresponding maximum amplitude values of wavelet sub-band signals are determined. Then, a linear curve is fitted to the log of these maximum amplitude values and the goodness of the fit is used an indicator of sharpness/blurriness value. The potential of the method in detecting variations in sharpness/blurriness is demonstrated on both globally blurred images and tampered images with local adjustments.

Another common form of forgery is content repetition which involves copying and pasting part(s) of an image over other parts of the same image to disguise some (contextual) details in the image. Although this type of tampering can be easily detected by exhaustive search and analysis of correlation properties of the image (autocorrelation function) due to introduced correlation by content repetition, these methods are not computationally practical and do not perform well when the copied pasted parts are smaller in region. To address this problem Fridrich et al.[42] proposed a better performing (faster and accurate) method. The method obtains DCT coefficients from a window that is slid over the whole image in an overlapping manner and quantizes them. The resulting coefficients are arranged and inserted into a row matrix. The rows (of quantized DCT coefficients) are then sorted in a lexicographical order and through row-wise comparisons similar blocks are determined. The main computational cost of this algorithm is due to sorting which requires significantly less time steps as compared to brute force search, *e.g.,* an $O(n\log n)$ algorithm. To further improve the robustness of this method to possible variations Popesctu et al.[43] used an alternative representation of blocks based on principal component analysis to identify the similar blocks in the image. Similar to,[41] the coefficients in each block are vectorized and inserted in a matrix and the corresponding covariance matrix is computed. By finding the eigenvectors of the covariance matrix, a new linear basis for each image block is obtained and a new representation is obtained by projecting each image block onto selected basis vectors with higher eigenvalues to reduce dimensionality. Then, the representation of each block is lexicographically sorted and compared to determine the similar blocks. The robustness of the method in detecting tampered parts is demonstrated under ranging JPEG compression qualities and additive noise levels.

15.4.3 *Inconsistencies Concerning Acquisition Process*

As discussed earlier, image acquisition process introduces certain distinguishing characteristics in each acquired image which can be used for source identification.

Since these characteristics will be fairly uniform over the whole image, their consistency across the image can also be used for detecting and localizing tampering. Hence, this group of techniques is extensions of source identification techniques with some minor differences. For example, Swaminathan et al.[44] used inconsistencies in color filter array interpolation to detect tampered parts of an image based on their approach in.[13] After estimating the CFA pattern and the interpolation filter, the demosaiced image is reconstructed and compared to the image itself. Modeling the linear part of the post-processing as a tampering filter, its coefficients are obtained by deconvolution. These coefficients are then used in design of a classifier to detect tampering by comparing the obtained filter coefficients with a reference pattern obtained from direct camera output (*i.e.*, unaltered images). Results obtained by subjecting test images to spatial averaging, rotation, compression and resampling is reported to yield average detection accuracy of more than 90%.

Similarly, based on,[17] Lukas et al. proposed to detect and localize tampering by analyzing the inconsistencies in the sensor pattern noise extracted from an image.[45] The noise patterns obtained from various regions are correlated with the corresponding regions in the camera's reference pattern and a decision is made based on comparison of correlation results of region of interest (potentially tampered region) with those of other regions. Along the same line Popescu et al.[46] proposed to detect the presence of CFA interpolation, as described in,[9] in overlapping blocks of an image to detect tampering. Experiments were performed on a limited number of digital camera images to identify traces of CFA interpolation in each block with no tampering.

Johnson et al.[47] proposed a new approach by inspecting inconsistencies in lateral chromatic aberration as a sign of tampering. Lateral aberration is due to inability of the lens to perfectly focus light of all wavelengths onto imaging sensor, causing a misalignment between color channels that worsens with the distance from the optical center. The method treats the misalignment between color channels as an expansion (or contraction) of a color channel with respect to one another and tries to estimate the model parameters (*e.g.*, center and aberration constant) to attain alignment. The estimation of these model parameters is framed as an image registration problem and a mutual entropy metric is used to find the exact aberration constant which gives the highest mutual entropy between color channels. To detect tampering, image is partitioned into blocks and the aberration estimated in each block is compared to global estimate. Any block that deviates significantly from the global estimate is deemed to be tampered. The threshold deviation is determined experimentally under varying compression qualities; however, further experiments are needed to generalize the results and determine the dependency on image content.

Alternatively, Lin et al.[48] proposed a method to recover the response function of the camera by analyzing the edges in different patches of the image and verifying their consistency. Camera response function defines the relation between radiance values from the scene and measured brightness values in each color channel and due to this non-linear response a linear variation of the pixel irradiance at the edges will be distorted. The main idea of the method is to utilize this phenomenon by computing the inverse response function and to determine its conformance to known properties of response functions (which should be monotonically increasing with at most one inflexion point and similar to each other in each color channel). The normality of the estimated functions, from each patch, is decided by comparing them to a database of known camera response functions. For this, classifiers are designed by extracting features from the computed and available response functions and tested on a few example images to demonstrate the feasibility of the idea. Although the success of the method requires images to be of high contrast so that the color range in each patch is wide enough, this assumption can be relaxed by applying the method to source camera-model identification problem.

15.5 Conclusions and Outlook

There is a growing need for digital image forensics techniques, and many techniques have been proposed to address various aspects of digital image forensics problem. Although many of these techniques are very promising and innovative, they all have limitations and none of them by itself offers a definitive solution. Ultimately, these techniques have to be incorporated together to obtain reliable decisions. However, there are still two major challenges to be met by image forensics research.

- *Performance Evaluation and Benchmarking.* Essentially the foremost concern that arises with respect to forensic use of proposed techniques is the achievable performance in terms of false-alarm and true-detection/identification rates and clear understanding of the factors that affect the performance. From this point of view, many of the proposed techniques can be more accurately defined as proof of concept experiments. To further refine these methods, performance merits have to be defined more clearly and proper test and evaluation datasets have to be designed and shared.

- *Robustness Issues.* The most challenging issue that image forensics research faces is the robustness to various common and malicious image processing operations. Proposed methods are not designed and tested rigorously to perform under the most difficult conditions, and, moreover, most techniques can be easily circumvented by a novice manipulator. Since the information utilized by the image forensics techniques is mostly in imperceptible detail, it can be easily removed. It is a matter of time for such tools to be available for public use. Techniques have to be designed and evaluated with this caveat in mind.

Overcoming these challenges requires the development of several novel methodologies and thorough evaluation of their limitations under more general and practical settings. This can be achieved in collaboration with forensics experts and through their continuous feedback on the developed methods. The research effort in the field is progressing well in these directions.

References

1. The US Supreme Court Ruling in Ashcroft v. Free Speech Coalition, No. 00-795.
2. B. Goldfarb, Digital Deception.
   (Online, `http://larrysface.com/deception.shtml`).
3. J. Adams, K. Parulski, and K. Spaulding, Color Processing in Digital Cameras, *IEEE Micro*, vol. 18, no. 6, pp. 20–31 (1998).
4. How Scanners Work. (Online, `http://www.extremetech.com/article2/0,1697,1157540,00.asp`).
5. M. Kharrazi, H. T. Sencar, and N. Memon, Blind Source Camera Identification, *Proc. of IEEE ICIP* (2004).
6. M.-J. Tsai and G.-H. Wu, Using Image Features to Identify Camera Sources, *Proc. of IEEE ICASSP* (2006).
7. O. Celiktutan, I. Avcibas, B. Sankur and N. Memon, Source Cell-Phone Identification, *Proc. of ADCOM* (2005).
8. I. Avcibas, M. Kharrazi, N. Memon and B. Sankur, Image Steganalysis with Binary Similarity Measures, *EURASIP Journal on Applied Signal Processing*, vol. 17, pp. 2749–2757 (2005).
9. A. Popescu, Statistical Tools for Digital Image Forensics, Ph.D. Dissertation, Department of Computer Science, Darthmouth College (2005).
10. S. Bayram, H. T. Sencar and N. Memon, Source Camera Identification Based on CFA Interpolation, *Proc. of IEEE ICIP* (2005).
11. S. Bayram, H. T. Sencar and N. Memon, Improvements on Source Camera-Model Identification Based on CFA Interpolation, *Proc. of WG 11.9 Int. Conf. on Digital Forensics* (2006).
12. Y. Long and Y. Huang, Image Based Source Camera Identification Using Demosaicing, *Proc. of IEEE MMSP* (2006).
13. A. Swaminathan, M. Wu and K. J. Ray Liu, Non-Intrusive Forensics Analysis of Visual Sensors Using Output Images, *Proc. of IEEE ICIP* (2006).

14. K. S. Choi, E. Y. Lam and K. K. Y. Wong, Source Camera Identification Using Footprints from Lens Aberration, *Proc. of SPIE* (2006).
15. K. Kurosawa, K. Kuroki and N. Saitoh, CCD Fingerprint Method, *Proc. of IEEE ICIP* (1999).
16. Z. J. Geradts, J. Bijhold, M. Kieft, K. Kurusawa, K. Kuroki and N. Saitoh, Methods for Identification of Images Acquired with Digital Cameras, *Proc. of SPIE*, vol. 4232 (2001).
17. J. Lukas, J. Fridrich and M. Goljan, Digital Camera Identification from Sensor Pattern Noise, *IEEE Trans. Inf. Forensics and Security*, vol. 1, no. 2, pp. 205–214 (2006).
18. Y. Sutcu, S. Bayram, H. T. Sencar and N. Memon, Improvements on Sensor Noise Based Source Camera Identification, *Proc. of IEEE ICME* (2007).
19. M. Chen, J. Fridrich and M. Goljan, Digital Imaging Sensor Identification (Further Study), *Proc. of SPIE* (2007).
20. N. Khanna, A. K. Mikkilineni, G. T.-C. Chiu, J. P. Allebach and E. J. Delp, Forensic Classification of Imaging Sensor Types, *Proc. of SPIE* (2007).
21. N. Khanna, A. K. Mikkilineni, G. T.-C. Chiu, J. P. Allebach and E. J. Delp, Scanner Identification Using Sensor Pattern Noise, *Proc. of SPIE* (2007).
22. H. Gou, A. Swaminathan and M. Wu, Robust Scanner Identification Based on Noise Features, *Proc. of SPIE* (2007).
23. S. Lyu and H. Farid, Steganalysis Using Higher-Order Image Statistics, *IEEE Trans. Image Forensics and Security*, vol. 1, no. 1, pp. 111–119 (2006).
24. I. Avcibas, B. Sankur and N. Memon, Steganalysis of Watermarking and Steganography Techniques Using Image Quality Metrics, *IEEE Trans. Image Processing*, vol. 12. no. 2, pp. 221–229 (2003).
25. E. Dirik, H. T. Sencar and N. Memon, Source Camera Identification Based on Sensor Dust Characteristics, *Proc. of IEEE SAFE* (2007).
26. S. Lyu and H. Farid, How Realistic is Photorealistic?, *IEEE Trans. On Signal Processing*, vol. 53, no. 2, pp. 845–850 (2005).
27. T.-T Ng, S.-F. Chang, J. Hsu, L. Xie, M.-P. Tsui, Physics-Motivated Features for Distinguishing Photographic Images and Computer Graphics, *ACM Multimedia* (2005).
28. Y. Wang and P. Moulin, On Discrimination between Photorealistic and Photographic Images [corrected version], *Proc. of IEEE ICASSP* (2006).
29. S. Dehnie, H. T. Sencar and N. Memon, Identification of Computer Generated and Digital Camera Images for Digital Image Forensics, *Proc. of IEEE ICIP* (2006).
30. E. Dirik, S. Bayram, H. T. Sencar and N. Memon, New Features to Identify Computer Generated Images, *Proc of IEEE ICIP* (2007).
31. I. Avcibas, S. Bayram, N. Memon, B. Sankur and M. Ramkumar, A Classifier Design for Detecting Image Manipulations, *Proc. of IEEE ICIP* (2004).
32. T. Ng, S.-F. Chang and Q. Sun, Blind Detection of Photomontage Using Higher Order Statistics, *Proc. of ISCAS* (2004).
33. T. Ng and S.-F. Chang, A Model for Image Splicing, *Proc. of ICIP* (2004).
34. S. Bayram, I. Avcibas, B. Sankur and N. Memon, Image Manipulation Detection with Binary Similarity Measures, *Proc. of EUSIPCO* (2005).
35. S. Bayram, I. Avcibas, B. Sankur and N. Memon, Image Manipulation Detection, *Journal of Electronic Imaging*, vol. 15, no. 4 (2006).

36. J. Lukas and J. Fridrich, Estimation of Primary Quantization Matrix in Double Compressed JPEG Images, *Proc. of DFRWS* (2003).
37. A. C. Popescu and H. Farid, Statistical Tools for Digital Forensics, *Proc. of IHW* (2006).
38. J. He, Z. Lin, L. Wang and X. Tang, Detecting Doctored JPEG Images via DCT Coefficient Analysis, *Proc. of ECCV* (2006).
39. A. C. Popescu and H. Farid, Exposing Digital Forgeries by Detecting Traces of Re-Sampling, *IEEE Trans. Signal Processing*, vol. 53, no. 2. pp. 758–767 (2005).
40. M. K. Johnson and H. Farid, Exposing Digital Forgeries by Detecting Inconsistencies in Lighting, *Proc. of ACM Multimedia Security Workshop* (2005).
41. Y. Sutcu, B. Coskun, H. T. Sencar and N. Memon, Tamper Detection Based on Regularity of Wavelet Transform Coefficients, *Proc. of IEEE ICIP* (2007).
42. J. Fridrich, D. Soukal and J. Lukas, Detection of Copy-Move Forgery in Digital Images, *Proc. of DFRWS* (2003).
43. A. C. Popescu and H. Farid, Exposing Digital Forgeries by Detecting Duplicated Image Regions, Technical Report, TR2004-515, Dartmouth College, Computer Science.
44. A. Swaminathan, M. Wu and K. J. R. Liu, Image Tampering Identification Using Blind Deconvolution, *Proc. of IEEE ICIP*, (2006).
45. J. Lukas, J. Fridrich and M. Goljan, Detecting Digital Image Forgeries Using Sensor Pattern Noise, *Proc. of SPIE* (2006).
46. A. C. Popescu and H. Farid, Exposing Digital Forgeries in Color Filter Array Interpolated Images, *IEEE Trans. Signal Processing*, vol. 53, no. 10, pp. 3948–3959 (2005).
47. M. K. Johnson and H. Farid, Exposing Digital Forgeries through Chromatic Aberration, *Proc. of ACM Multimedia Security Workshop* (2006).
48. Z. Lin, R. Wang, X. Tang and H.-Y. Shum, Detecting Doctored Images Using Camera Response Normality and Consistency Analysis, *Proc. of CVPR* (2005).

Chapter 16

Privacy Preserving Web-Based Email

Kevin R. B. Butler, William Enck, Patrick Traynor,
Jennifer Plasterr and Patrick D. McDaniel

Systems and Information Infrastructure Security Laboratory
The Pennsylvania State University
University Park, PA 16802 USA
{butler,enck,traynor,plasterr,mcdaniel}@cse.psu.edu

Recent web-based applications offer users free service in exchange for access
to personal communication, such as on-line email services and instant messaging. The inspection and retention of user communication is generally intended
to enable targeted marketing. However, unless specifically stated otherwise by
the collecting service's privacy policy, such records have an indefinite lifetime
and may be later used or sold without restriction. In this paper, we show that
it is possible to protect a user's privacy from these risks by exploiting mutually oblivious, competing communication channels. We create virtual channels
over online services (e.g., Google's Gmail, Microsoft's Hotmail) through which
messages and cryptographic keys are delivered. The message recipient uses a
shared secret to identify the shares and ultimately recover the original plaintext.
In so doing, we create a wired "spread-spectrum" mechanism for protecting the
privacy of web-based communication. We discuss the design and implementation of our open-source Java applet, Aquinas, and consider ways that the myriad
of communication channels present on the Internet can be exploited to preserve
privacy.

Contents

16.1 Introduction

Internet users hemorrhage personal information. Almost every interaction on the web is scanned (directly or indirectly) by some party other than those directly involved in the transaction. Tracking cookies, web bugs, and other tools are used by advertisers to follow users as they move from site to site across the Internet.[1] Less scrupulous groups rely upon spyware to surreptitiously acquire personal information. Such information can be warehoused, collated with other sources, and stored indefinitely.

Recently, however, a more active means of collecting personal information has become common: users expose their personal communications to service providers in exchange for free online applications such as email and instant messaging. As promoted, access to this information allows online providers to personalize the user experience by offering targeted advertisements.[2] The revenue generated by connecting users and vendors has historically fueled much of the growth of the Internet, and is the major source of revenue for many websites. Hence, user profiling is an often positive and possibly necessary element of online life.

However, the communications provided by users of these new services such as free email can be used to develop profiles that extend far beyond simply online habits. By allowing these services to scan the contents of every message that passes through their system, they provide commercial interests with insight into their daily and sometimes highly personal lives. The contents of such communications are further susceptible to interception and examination by repressive regimes.[3–5] Such practices are becoming the norm in web-based applications. Unfortunately, the legal devices for protecting user privacy against abuse or sale of this information are few, and those that do exist are often ineffective.[6]

We assert that users need not sacrifice their right to privacy in exchange for any service. Just as customers of the postal service have come to expect that their messages will only be read by the intended recipient, so too should users of web-based services be guaranteed privacy in their communications. We demonstrate that strong confidentiality is attainable for all Internet users, regardless of the privacy policy of these online services.

In this paper, we introduce *Aquinas*, an open source tool designed to provide email privacy while maintaining plausible deniability against the existence of the unobservable (covert) communication. The Aquinas client provides privacy using a hybrid scheme; we employ cryptography to secure communication, steganography to hide the existence and substance of ciphertext, and multipath delivery to ensure compromised accounts or intercepted messages provide little information to an adversary. All email messages are initially encrypted and protected with a *message authentication code* (MAC) to ensure confidentiality and integrity. The key and ciphertext are then carefully divided into *shares*. The shares are embedded in emails using steganographic tools and sent to the recipient via multiple email accounts established at competing services such as Yahoo! Mail, Gmail, and Hotmail. When the recipient receives the ciphertext and key shares, Aquinas reconstructs the key and ciphertext. The ciphertext is decrypted and the contents validated to obtain the plaintext message.

Aquinas is an open-source Java applet. While the mechanisms and distributed nature of content delivery make the current iteration of Aquinas highly robust against multi-party collusion and third-party scanning, it is our intention to allow anyone to contribute additional algorithms and functionality to the codebase. This diversity of operation means that ultimately, the ability of any entity to detect or prevent private communications through web-based email services will be severely curtailed.

Through its use of multiple channels for message delivery, Aquinas's design mimics wireless "spread-spectrum" protocols, which use a pseudo-random pattern of radio channels in order to prevent eavesdropping and jamming. Even with the observation of some subset of channels, an adversary gains no usable information about the true nature of a message's contents. In Aquinas, an adversary needs to intercept email on all used mail accounts to gain *any* information about the user communication. Because no web service can feasibly intercept all communication, user profiling is not possible.

Aquinas differs significantly from existing email privacy tools such as PGP.[7] Existing tools seek to secure the missives between known users typically using highly secure keys, i.e., public keys. Conversely, Aquinas seeks to enable mobile and lightweight communication; users need not have any physical data beyond a single password in their head. Moreover, Aquinas seeks to secure communication in environments where integration with existing tools is not available, such as with free email accounts. That is not to say that Aquinas provides a superset of features of these tools. Specifically, Aquinas does not provide all the guarantees that other systems may, e.g., non-repudiation. However, Aquinas is robust to compromise due to the generation of new keys for each message. We believe that this forward-security in combination with portability make this mechanism a highly attractive means of addressing privacy.

The remainder of this paper is organized as follows: Section 16.2 gives an overview of our approach to solving these issues; Section 16.4 discusses additional issues facing the use of privacy preserving software; Section 16.3 examines the specifics of the our implementation; Section 16.5 examines the related work in this field; Section 16.6 offers concluding thoughts and future directions for this work.

16.2 Design

We first define the goals of Aquinas and consider the threats and adversaries we seek to protect against. The latter parts of this section describe the protections in Aquinas and the mechanisms for their implementation.

16.2.1 *Goals*

The high-level design goals of Aquinas include:

Confidentiality: No adversary should be able to obtain information about the existence or content of email communication.

Integrity: The integrity of all communication must also be preserved, i.e., any modification of the message should be detectable by the recipient.

Ease of use: Aquinas should not require that the user understand or directly use any sophisticated concepts such as cryptography or steganography. Additionally, the tool should provide a user experience consistent with traditional email applications.

The systemic requirements of Aquinas are somewhat more mundane. We do not want to place a requirement on the user for having to install software beyond a simple web browser, or to provide complex data, e.g., maintain keyrings. The implications of this are that all security-relevant data needed to receive email from a single user should be derivable from a password. The second implication is that the tool should be able to execute on arbitrary platforms.

In addition, we want to maximize the flexibility of the services that can be used; to that end, we wish to be able to easily integrate Aquinas with any communication service available on the Internet. Finally, we require the tool to be extensible in order to accommodate future functionality.

16.2.2 *Threat Analysis*

Users of web-based email services are subject to a variety of threats against their privacy and security. Below, we consider possible adversaries and the motivation and attacks they may employ.

Threats may arise from corporate adversaries. For the application providers that run web-based email services, there is a strong interest in profiling their users for revenue generation. Information about users can be sold to marketing agencies or directly to other companies interested in advertising products to their target demographics. The information gleaned about a user through profiling email can be arbitrarily detailed; through sufficiently optimized data-mining techniques, even users reticent to reveal personal information may unwittingly divulge many more personal details than they realize. If information is sent without any form of obfuscation, it is trivial for the adversary to intercept communications; any party between the user and the application provider will also have unfettered access to this information.

There are environments where protections such as message confidentiality may not be allowed: the email provider may disallow encrypted or unrecognizable content, or the network used for information transmission may have similar restrictions. Even when hidden channels are used, vulnerabilities may still be manifested. As information flows to and from an email account, the account will be subject to *channel decay* over time: an adversary collecting copies of the transferred information will be able to use the amassed data to more easily mount an attack against the channel. In addition, the probability of an adversary learning of a channel's existence will increase with time.

An additional adversary with a similar reward model to the application provider can be the webmail user's ISP. Defending against these attacks presents a tangential set of challenges. We consider adversarial ISPs in greater detail in section 16.2.5.

While the goals of adversarial companies are largely financially-based, political adversaries may represent a greater threat to some users. Repressive political states have shown little compunction about using Internet activity logs to target and persecute dissidents.[3-5] These adversaries can be significantly more determined to discover information about their target than businesses, and have full access to all records and logs of activity. We can consider the political adversary to have all of the same tools at their disposal as the corporate adversary, plus the ability to compel multiple application providers to turn over all information they possess, or force those companies into collusion. This could create very serious consequences for a dissident attempting to keep their communications hidden from a regime.

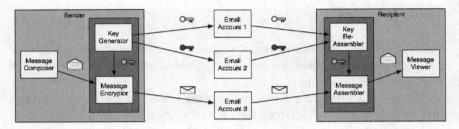

Fig. 16.1   A sample message and key delivery flow. The sender encrypts the plaintext and then
embeds it into carefully select covertext using steganography. The message containing the hidden
content is then sent as shares to one or more accounts owned by the recipient. Each of the key shares
used to create the encryption key are then sent to different destination email accounts. The recipient's
client checks all of the accounts, reassembles the key and ciphertext from the shares, and recovers the
plaintext. The separate emails from different SMTP servers to prevent reassembly by adversaries.

16.2.3 *Email Protection*

Figure 16.1 provides an overview of how email messages are protected by
Aquinas. After a message is composed, the email is encrypted and steganographic
techniques are applied to conceal the nature of the information being sent. We use
symmetric cryptography as the encryption mechanism, in contrast to alternative
email schemes, which use public-key cryptography. Because public-key systems
require the use of a trusted third party for endorsing user identities, selecting parameters for key encryption, and proving credentials—a non-trivial problem that
has not been entirely solved in a satisfactory manner[8]—as well as a full associated
infrastructure, we found that this architecture would not fit within the goals of our
system. While use of symmetric cryptography necessitates initial establishment of
a shared secret (typically in an out-of-band fashion), we felt this was an adequate
tradeoff.

Symmetric cryptography requires both the sender and recipient to agree on a
key. Obviously, we do not want to send the key in the same email as the ciphertext.
A simple solution is to send the key and ciphertext is separate emails, but if both
are sent through the same mail service, the adversary still has access to both. The
solution is to split both the key and ciphertext into multiple shares and send each
part through multiple mail services.

The encryption process is straightforward. The sender begins by creating some
number of keys. These keys are combined via XOR (herein noted as $\oplus$) to create
the encryption key[a]. Using some symmetric cryptographic algorithm, e.g. AES,
the message ciphertext is created. However, encryption alone is not sufficient

---

[a]The encryption key cannot be determined unless *all* of the key shares are known.

protection for the email, as a service provider could easily detect that an encrypted message was sent. A sender may wish to plausibly deny that sensitive information has been transmitted, and the presence of ciphertext in a message alludes to the transmission of unknown information. To make the emails appear innocuous, the message and key shares are passed through a steganographic filter (e.g., SNOW[9]), obscuring the email with *covertext* that provides no insight as to the real message contents.

Once the message has been encrypted and protected with a MAC, it is steganographically obscured with covertext. The resulting message is sent in an email to one of the recipient's accounts. The key shares are also hidden through steganographic techniques, and these messages are sent to different accounts. At this point, the message and key shares are distributed among multiple, independently administered email servers, and the message contents, mass collusion notwithstanding, are secured from unauthorized observers.

The recipient begins the decryption process by downloading both the message and key shares. From the recipient's point of view, the key to decrypt the message is the recipient email accounts. Once downloaded, the recipient applies the steganographic filter to eliminate the covertext and retrieve the ciphertext and key shares. The key shares are combined with $\oplus$ to create the decryption key, and the ciphertext is decrypted.

16.2.4 *Design Detail*

Our key distribution approach is an example of *multipath delivery*. This method leverages the distributed nature of Internet services to create and multiplex orthogonal channels in the form of multiple email accounts. An analogous means of communications, known as *spread spectrum*, has been used for more than fifty years. Given some range of radio spectrum with x discernible frequencies, messages are transmitted using some pseudorandom sequence of frequencies known only to sender and receiver. An adversary attempting to eavesdrop on communications has a probability $(1/x)^p$ of overhearing the entire message over p time periods. As x and p increase, the ability of an attacker to successfully intercept communications quickly approaches zero. The application of such a technique to the Web makes interception by an adversary an even more daunting task. While the radio spectrum arguably has a limited number of frequencies, the number of channels in which data can be injected into and across the Internet are arguably infinite. We demonstrate the use of Aquinas with key shares carried across multiple email addresses; however, with little additional extension, we can store key shares and messages in web log comments, chat rooms, newsgroups, and a variety of other locations. If we consider each of these particular channels equivalent to a different frequency in the spread-spectrum analogy, then we see the vast number of virtual frequencies afforded to us.

Each of these email accounts used to send the shares should be located at domains operated by different providers. This method of key delivery is robust to collusion for a number of reasons. Competition will deter collusion: any information about a user that a provider is able to garner or derive that is not known to the provider's competitors generates a competitive advantage. Because providers are competing for revenue from advertisers, having unique insights into customer profiles will be rewarded by allowing more targeted marketing to those users, making advertising more lucrative and profitable. Hence, providers desire to keep this information as private as possible, and colluding with other providers would necessitate providing information on the user. This creates a *competitive disincentive* for the provider to engage in collusion. Additionally, even if an adversary is to discover that a message is hidden within an email, they must still recover all n key shares along with the message in order to decrypt it, making this system robust to the compromise of up to $n - 1$ key shares.

The recipient, using the Aquinas client, checks her disparate *message* and *key* email accounts for shares. Aquinas downloads all of the messages and then searches through the headers for a flag identifying the keys for a specific message. Demultiplexing via the $\oplus$ operation is performed on all n key shares, providing the recipient with key K. The actual data contained within the email is then uncovered and decrypted using K. The real message from the sender is then displayed for the recipient.

The communication process is no more difficult from a user's standpoint than using a traditional mail program. Specifically, a user must enter the multiple outgoing (SMTP) and incoming (POP3) email servers that are to be used to deliver messages. With the *address book* feature in Aquinas, allowing storage of multiple users per email address, this information only needs to be entered once.

16.2.5 *Adversarial ISPs*

Many users rely on a single service provider to transit their information to the greater Internet. The consequence, however, is that this ISP has access to all of the information sent through its network. By implication, this means that all of the messages sent by the Aquinas user will pass through their home provider who can collect data, even though the destinations of these messages may be disparate email services providers.

Key management does not help in this case because all n channels are implicitly revealed. However, the user has recourse through use of the SSL protocol. SSL provides end-to-end data protection between the user and the email provider, making information unreadable to an ISP attempting to passively eavesdrop on messages. Aquinas supports the use of SSL in order to thwart the ISP threat. With SSL, however, there is some information leakage; the adversary can learn the destination of the packets (but not the destination of the email) by examining the IP header. Thus, while the content of the messages will be unknowable, the fact that information is being transferred to an email provider will be leaked. By observing this information, the ISP could learn all of the providers used and instantiate collusion with them. To hide evidence of the destination, the user could make use of proxies, such as anonymous remailers and other anonymous routing services.[10,11] Additionally, to lower the probability of an adversary detecting the existence of a channel formed by the email account, the user can periodically abandon their accounts and set up new ones for communication.

An alternative solution to the ISP threat exists that does not require the use of SSL between a user and their email provider. Security can be implemented through *chaffing and winnowing*[12] with email accounts. By including email accounts not used during the email communication, the adversarial ISP will have to choose the correct subset of accounts that correspond to a message. A brute-force approach based on combinatorics rapidly becomes infeasible for the adversary. For example, if the user transmits a message with 40 shares, but only 20 of those are used to construct the message, the adversary will be required to search through the $\binom{40}{20}$, or nearly 138 billion, combinations.

16.2.6 *Key Negotiation and Management*

Bootstrapping communication between users requires a mechanism outside of Aquinas to be used. Out-of-band key communication through methods such as speaking over the phone or meeting in person is possible; alternately, a mechanism such as PGP could be used for the initial setup. While the user would have to be on a trusted machine that has PGP installed to perform this transaction, once the initial key setup was complete, the user can then communicate using any terminal with the recipient.

We propose that a directory of users be stored in a publicly accessible repository. Each set of email addresses associated with a user can be stored within this space. The addresses can be public because it is their particular combination used for an email transmission that is the secret. Part of the initial communication between two users can include transmission of a shared secret between the two parties. This can be very simple, such as the phrase "secretpassword"[b]. A permutation sequence can then be calculated by using this secret as a key. For example, AES-128 has a keyspace of 2^{128} entries. Encoding the secret as a value (e.g., converting "secretpassword" in its decimal representation) allows us to use it as a key. If there are 40 email addresses associated with a user, the keyspace can be binned into 40 intervals, and the generated number will fall into one of these bins, generating one of the email addresses that will comprise the key share. The resulting value is then encrypted with the key and another interval is selected based on the new output. This process is repeated until there are 20 unique addresses selected. By negotiating a new secret (for example, through email communication), a new combination of addresses used as key shares can be selected. The following matrix illustrates the series of transformations that generates the values to be binned:

$$
\begin{bmatrix}
k_0 = h(\text{"secret password"}) \\
k_1 = E(k_0, k_0) \\
k_2 = E(k_1, k_1) \\
\vdots \\
k_{20} = E(k_{19}, k_{18})
\end{bmatrix}
$$

Note that email is not the only method by which keys and ciphertext may be delivered. The open functionality inherent to the Internet allows any means of sending data to become a covert channel for communication. A combination of keys placed in weblog referrer logs, instant messages, BitTorrent[13] and other P2P file sharing systems, streaming audio and video, newsgroup postings, and any number of disposable or community email accounts can be used to keep the contents of any message secret. This method of key and content distribution creates a wired "spread-spectrum" effect, effectively using servers across the Internet like unique "frequencies". This technique thereby obfuscates the ability to determine that communication has occurred at all. Because of the sheer vastness of the web, the ability to prohibit privacy on this medium is *virtually impossible*.

16.3 Implementation

Aquinas is principally designed to support a simple and user-friendly interface. In order to retain the convenience of web-based email, Aquinas is required to be accessible via the Internet. Ideally, this portability should be machine independent to allow use by the widest possible community. For these reasons, we developed

---

[b] Shared secrets should be picked carefully to avoid dictionary attacks.

Aquinas using Java. Our goals, however, were not merely to allow use on their primary home or work machines (although this use is encouraged); rather, we wanted to ensure that users could protect their communications no matter where they were or what machine they were using, such as a terminal at an Internet cafe[c]. Accordingly, we have designed Aquinas to run as an applet. The Aquinas Java applet and source-code are freely available from:

http://siis.cse.psu.edu/aquinas.html

The fully functional applet is linked off this page and was used as a method of communication by the authors during the writing of this paper.

Mail services are handled through the javax.mail package. The current version of the software includes support for POP3-based services. While a number of domains offer IMAP connectivity, many of the major web-based email providers including Gmail do not currently include such functionality at the time of this writing. We emulate IMAP via the POP3 TOP command; the message headers are all that is downloaded until a user requests the message body itself[d].

Because the client must communicate with a number of servers other than the one on which it is hosted and creates state with the address book feature, we were forced to create Aquinas as a trusted applet. The disadvantage with this approach is that the Java Virtual Machine's sandboxing mechanisms are turned off, giving the applet access to the user's file system. We provide source code for our application for inspection and a self-signed applet, allowing a per use exception to sandbox restrictions. Note that the user must either accept the certificate or turn off sandboxing for the applet to be usable in a browser. Unfortunately, some browsers do not have this capability and thus a native operating system Java VM may be necessary.

Aquinas uses the SNOW steganographic tool,[9] a Java codebase that uses whitespace at the end of lines to hide data. All steganographic transformations are handled through a generic API. Hence, additional steganographic tools may be quickly integrated into Aquinas with little effort. This wrapper class also contains multiple interfaces to accommodate the use of MIME-type forgery. Both the key and message emails make calls to this tool.

Figure 16.2 shows a screenshot of what the Gmail scanner sees as the content of an email sent using Aquinas. The plaintext of the message, however, is displayed in Figure 16.3. We performed extensive tests with emails protected by different steganographic covertexts, to determine how they would be handled by

---

[c]Note that users must still be cognizant of their surroundings and the machines they use if Aquinas is used in an untrusted location such as a remote kiosk. We cannot and do not protect against physical attacks such as keystroke loggers on remote terminals.

[d]The POP3 command for downloading a message's header, but not its body, is TOP <message#> 0

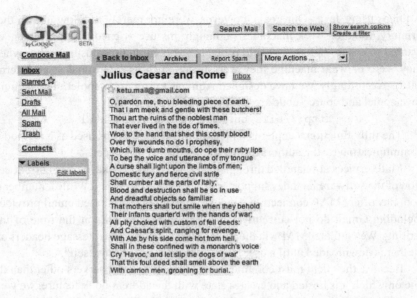

Fig. 16.2   A screenshot of the content of an email sent from Aquinas to a Gmail account.

Gmail and other providers. While Gmail sometimes showed advertisements pertaining to the content of the covertext, none of these advertisements reflected the keywords or terms found in the plaintext message. This indicates to us that the real message transmitted stayed private and was protected from profiling.

16.4 Discussion

Aquinas extends the confidential nature of email by allowing message contents to remain secret until being read by the intended recipients thereby redefining the endpoint of web-based email as the user. Its portability, imperceptibility and forward-security through unique session keys make the use of Aquinas more attractive than many more traditional schemes. We therefore consider several issues of the secure use and implementation of Aquinas in the following subsections.

16.4.1 *Preserving Privacy*

Although the mechanisms discussed in this paper can provide security against profile generation and data mining, users of these solutions must still be cognizant of other privacy issues. Specifically, in spite of the use of encryption and steganography, it is still possible for information leakage to occur. The selection of cover

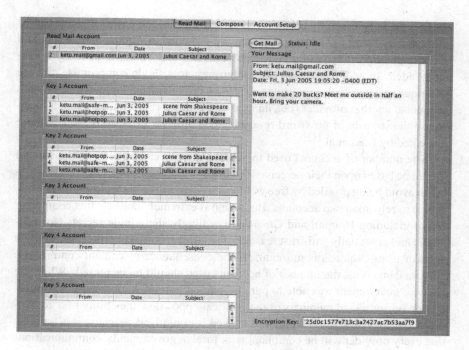

Fig. 16.3   A screenshot of the recovered plaintext of the email displayed in Figure 16.2.

text, for example, provides data that can be scanned and associated with a user. If a user were to select text from a website with radical political statements or adult material, that information may still be affiliated with the user in spite of there being no actual relationship between the two parties in the real world. To mitigate this threat, we suggest using neutral text, such as the "Terms of Service" or "Frequently Asked Questions" pages available at the websites hosting the email. By doing this, a user exposes only the fact that they use a service (which is already known to the service provider).

The sender should also be aware of the paths that key shares take. For example, if all data were to cross a particular domain either during the sending or receiving process, all of the data necessary to create the keys for decryption would be readily available. It is therefore critical that users take advantage of as many unique channels as possible to provide maximum security.

Users should take additional precautions when deciding upon names for email accounts. While identically named accounts at a number of major free email providers would be easy for people to remember, they also increase the ease with which collusion between providers can occur. The tradeoff between ease and security must be carefully considered by each user. Much of this tradeoff can be mitigated by using the address book feature provided in Aquinas. As a standard security practice, the use of unique passwords across accounts in also highly recommended. In addition to providing robustness to a single compromise, the use of unique passwords also prevents one service provider from logging in to a user's account at another provider (i.e., unapproved collusion[14]). Simple methods to increase the security of password re-use include browser extensions such as those presented by Ross et al.[15]

The number of accounts used to achieve privacy can be set by the user and should be based upon their perceived threats. For example, someone simply wanting to avoid being profiled by free web-based email providers and advertisers may decide to rely upon two accounts. Because it is extremely unlikely that competing forces including Hotmail and Gmail will willingly share trade secrets (for economic and potentially anti-trust reasons), the effort required to protect the average account using Aquinas is minimal. If the consequence of content compromise is more dangerous, the number of accounts used should be increased. While the Chinese government was able to put pressure on Yahoo! Mail to turn over information on suspected members of the political opposition, the ability of a government to achieve the same if Aquinas is used is minimized. Because it is unlikely that every provider will be compliant with foreign governments, communications can be protected from this sort of interception. One way to realistically implement a significant increase in the number of accounts would be for users to aggregate and share accounts within larger communities. In a design similar to the Crowds,[16] users could receive and forward mail on behalf of other users within their community while maintaining plausible deniability of the communication details.

Techniques leveraging the temporal spacing of messages can also help to protect against traffic analysis attacks. As mentioned in Section 16.2.5, a user can include chaffing and winnowing techniques to increase their security. For example, slowly sending shares over the course of an hour forces an adversary to consider all egress traffic during that period. A small alteration to the current version of Aquinas would allow it to continuously emit low volumes of traffic to randomly chosen websites and accounts. Shares included within this stream would be significantly more difficult to detect.

Due to the nearly infinite number of ways in which data can be injected into the Internet, the probability of an adversary selecting all of the correct repositories is incalculably small. Even in the unlikely event of an adversary having perfect knowledge of the accounts used for communication, a user can still be protected. Assuming that 40 messages are again used, but that the number of keys used is decided out of band (perhaps as part of account selection as in Section 16.2.6), an adversary is would be required to try up to $2^n - 1$, or nearly 1.1 trillion, combinations of messages. The action of selecting accounts therefore becomes equivalent to encryption by an additional, unrelated key. If the accounts are unknown, the size of this key is arguably infinite. In the worst case, the key size of the secondary in this example is 40-bits. Users uncomfortable with such a key length can increase robustness by changing the algorithm used to generate the encryption key from the key shares. If the $\oplus$ operation is replaced by an order-dependent technique (such as alternating multiplication and division of key shares according to the account selection scheme in Section 16.2.6), the adversary will instead have to try $\sum_{k=1}^{n} {}_nP_k$ permutations, as between 1 and n shares in the correct order could be required to reassemble the key. This operation has time complexity $O(n!)$. With 40 messages, more than $1.6 * 10^{48}$ permutations would be required to uncover the key. As this is much larger than the number of brute-force attempts to recover a 128-bit key, a user is sufficiently protected against even the strongest adversaries.

16.4.2 *Resiliency*

While offering robustness to the collusion of multiple service providers, the multipath key and message delivery mechanism described in this paper is not without its own limitations. For example, if an email service provider were to determine that a message contained a key, simply deleting the message would prevent the intended recipient from decrypting and reading their mail. A message mistakenly classified as spam would have similarly deleterious effects, as the user would have difficulty differentiating real messages amongst the torrent of spam messages most email users receive.

Shamir's *threshold secret sharing*[17] could be used to make Aquinas robust against share loss. This technique works by creating the key K from the combination of n key shares. K can be reconstructed as long as k key shares (where $n = 2k - 1$) are in the possession of the recipient. The advantage to this scheme is that it allows for $k - 1$ key shares to be lost (or delivered late) without affecting the ability of the recipient to decrypt and read their email. If spam filtering were to become an issue, this scheme would be more robust, as it would allow the intended recipient to still read their encrypted messages without all n keys. While this approach is secure to the compromise of up to $k - 1$ key shares, if $k < n$, messages can be decrypted with fewer keys than in the currently implemented scheme.

Robustness based upon the perceived threat of an adversary could also be incorporated as a keying mechanism. For example, a user may decide that the overhead of increasing the number of email accounts is greater than the protection offered from a keying scheme based on threshold secret sharing. One simple extension to the multipath mechanism is to increase the number of accounts to which copies of key shares are sent. A user could opt to send the same key share to multiple accounts. In so doing, fewer cooperating adversaries would be necessary to reconstruct keys. A more elegant solution would be to use a mechanism based on *error correcting codes* (ECC). By attaching tags containing a few extra bytes to the end of each key, it becomes possible to reconstruct K with only a subset of all n key shares. The size of this subset (and the attached ECC) needed to recreate K can be adjusted to suit the specific expected adversary. The threshold secret sharing, multi-share delivery and error correcting code alternatives are all under consideration for future versions of this software.

16.5 Related Work

Privacy on the Internet is not guaranteed for users in general, and can be ambiguously defined even where it exists.[18] Often, users believe that they have online privacy but really have no guarantees to that effect.[19] To mitigate these shortcomings, many privacy-preserving tools have been created and deployed, protecting numerous aspects of a user's online activities.

Methods of securing non-web-based email have been extensively studied. Solutions such as Privacy Enhanced Mail (PEM)[20] and its successor, Secure MIME (S/MIME),[21] provide confidentiality, integrity, and non-repudiation for email messages. With PEM, this is accomplished through the construction of a full certificate hierarchy within a public key infrastructure (PKI); this has proven to be unwieldy in practice. For S/MIME, cryptographically transformed messages are sent as attachments within email, with key validation performed through a PKI. Pretty Good Privacy (PGP)[7] is another system for providing confidentiality and integrity of email that does not rely on the use of a PKI. A user forms a *web of trust* by trusting certain entities she communicates with, which in turn has other trusted relationships. The transitive certification paths of trust among these relationships are used to authenticate the source of email. Confidentiality can be provided by the mailer itself, with tools such as *ssmail*, a patch for the *sendmail*[22] mail transfer agent.

The *Off-the-record Email* (OTR) system[23] works at the user level, with dynamic key management performed between the two parties using it. Additionally, OTR provides non-recoverability of email messages once they have been deleted, even if the private keys used to generate the cryptographic operations have been revealed. However, while forward secrecy is assured, plausible deniability is not: an agent monitoring traffic will observe that encrypted information is being transmitted to the recipient.

While privacy within web-based email services has been largely absent, one solution is offered by SAFe-mail.net.[24] This system supplies confidentiality and integrity through the use of a PKI that is run by SAFe-mail themselves. Because the service handles both certificates and user email, however, it has access to all of a user's information, allowing them to arbitrarily link and use this data.

Secure publication of data is another area where privacy can be crucial, in order to protect the authors of controversial documents from reprisal. The ability to publish without the fear of retribution has been tremendously important to citizens throughout history. The Federalist papers in the United States brought forth the ideals that ultimately became enshrined in the Constitution, but many of the authors published anonymously to avoid reprisal. More recently, the former Soviet-bloc countries witnessed the rise of *samizdat*, the process of anonymously publishing and distributing information banned by the government.[25] Publius[26] is a tool that facilitates secure publishing on the Internet, using threshold keying (discussed further in Section 16.4) to preserve anonymity. Other systems, including Free Haven,[27] provide anonymous storage and retrieval. Free Haven uses a secure network of devices within a community of servers to manage shares while maintaining anonymity. Documents are broken into shares in a manner similar to Publius, but shares keep track of server reliability, where less trust is afforded to servers that drop shares. In this way, Free Haven offers the censorship-resistant qualities of Publius while also providing greater server accountability. Similarly, Freenet,[28] a distributed system for storage, provides anonymous content storage and dynamic growth of the network through the addition of new nodes.

Many of these tools have been useful in keeping communications private and secure; in particular, PGP has been extensively used by human rights organizations around the world. However, in virtually all cases, the fact that communication has taken place can be divined through the presence of encrypted data, or information has been transferred through private services. To this point, there have not been any solutions that allow for encrypted and steganographically concealed communications that transmit information solely through public channels and publicly available services.

16.6 Conclusion

This work has introduced Aquinas, an open source tool for preserving the privacy of user communication carried by web-email services. Each message is initially encrypted with a random symmetric key. The resulting ciphertext and key are both divided into shares. Each share is hidden in randomly chosen cover-text using steganography and sent through an independent web email account. Clients reconstitute the ciphertext and keys from shares received via the appropriate accounts. The result is decrypted to obtain the original message. We use email accounts in an analogous manner to the multiple channels employed in spread-spectrum communications. More generally, we show that the retention of one's privacy is possible regardless of the policies imposed by the providers of these web-based services.

Future extensions to this work will incorporate a variety of new image and linguistic steganography techniques, allowing users to more fully obfuscate their communications. Additionally, we will implement features that support the distribution of ciphertext shares across multiple accounts, and will continue to improve the usability of our interface as directed by user input. Such an approach also begs extension to the panoply of channels available throughout the Internet. Our future work will not only explore these diverse channels, but also develop a formal framework for reasoning about the security provided by them.

Acknowledgements

We gratefully acknowledge Simon Byers and Dave Kormann for their assistance in formulating the problem scenario and their input. We would also like to thank Matthew Kwan, author of the SNOW steganographic tool, for graciously allowing us to use SNOW within our Aquinas client. SNOW is open-source for purposes of this project and other non-commercial applications, but not open-source in general.

References

1. W. Roger, Surfer beware: Advertiser's on your trail, DoubleClick tracks online movements, *USA Today*, p. 01.B, Jan. 26, 2000.
2. D. Peppers and M. Rogers, The One to One Future: Building Relationships One Customer at a Time, *Doubleday*, New York, 1993.
3. BBC News, Chinese man 'jailed due to Yahoo', http://news.bbc.co.uk/2/hi/asia-pacific/4695718.stm, February, 2006.
4. Reporters Without Borders, Information supplied by Yahoo! helped journalist Shi Tao get 10 years in prison, http://www.rsf.org/article.php3?id_article=14884, September, 2005.
5. Reporters Without Borders, Yahoo! implicated in third cyberdissident trial, http://www.rsf.org/article.php3?id_article=17180, April 20, 2006.
6. Electronic Frontier Foundation, http://www.eff.org.
7. P. R. Zimmermann, *The official PGP user's guide*, MIT Press, Cambridge, MA, 1995.
8. C. M. Ellison and B. Schneier, Ten Risks of PKI: What You're Not Being Told About Public-Key Infrastructure, *Computer Security Journal*, vol. 16, pp. 1–7, 1999.

9. SNOW, *The SNOW Home Page*, http://www.darkside.com.au/snow/.

10. The Anonymizer, http://www.anonymizer.com.

11. D. Goldschlag, M. Reed and P. Syverson, Onion routing for anonymous and private Internet connections, *Communications of the ACM*, Vol. 42, pp. 39–41, 1999.

12. R. L. Rivest, Chaffing and Winnowing: Confidentiality without Encryption, *RSA CryptoBytes*, Vol. 4, 1998.

13. *BitTorrent*, http://www.bittorrent.com.

14. E. Jordan and A. Becker, *Princeton officials broke into Yale online admissions decisions*, Yale Daily News, http://www.yaledailynews.com/article.asp?AID=19454, July 25, 2002.

15. B. Ross, C. Jackson, N. Miyake, D. Boneh and J. Mitchell, Stronger Password Authentication Using Browser Extensions, *Proceedings of the 14th USENIX Security Symposium*, Baltimore, MD, USA, 2005.

16. M. K. Reiter and A. D. Rubin, Crowds: anonymity for Web transactions, *ACM Transactions on Information and System Security*, Vol. 1, pp. 66–92, 1998.

17. A. Shamir, How to share a secret, *Communications of the ACM*, Vol. 22, pp. 612-613, 1979.

18. L. Palen and P. Dourish, Unpacking "privacy" for a networked world, *CHI '03: Proceedings of the SIGCHI conference on Human factors in computing systems*, pp. 129–136, 2003.

19. R. L. Mcarthur, Reasonable expectations of privacy, *Ethics and Inf. Tech.*, Vol. 3, pp. 123–128, 2001.

20. S. T. Kent, Internet Privacy Enhanced Mail, *Communications of the ACM*, Vol. 36, pp. 48–60, 1993.

21. B. Ramsdell, *S/MIME Version 3 Message Specification*, IETF, RFC, No. 2633, 1999.

22. B. Costales and E. Allman, *Sendmail(2nd ed.)*, O'Reilly & Associates, Inc., Sebastopol, CA, USA, 1997.

23. P. Henry and H. Luo, Off-the-record email system, *Proceedings of IEEE INFOCOM 2001*, pp. 869–877, 2001.

24. SAFe-mail.net, *SAFe-Mail Features*, May, 2005, http://www.safe-mail.net/help/SAFeMailFeatures.html.

25. G. Saunders, *Samizdat: Voices of the Soviet Opposition*, Pathfinder Press, Atlanta, GA, USA, 1974.

26. M. Waldman, A. D. Rubin and L. F. Cranor, Publius:' A robust, tamper-evident, censorship-resistant, web publishing system, *Proc. 9th USENIX Security Symposium*, pp. 59–72, 2000.

27. R. Dingledine, M. J. Freedman and D. Molnar, The Free Haven Project: Distributed Anonymous Storage Service, *International Workshop on Designing Privacy Enhancing Technologies*, Springer, pp. 67–95, 2001.

28. I. Clarke, O. Sandberg, B. Wiley and T. W. Hong, Freenet: a distributed anonymous information storage and retrieval system, *International Workshop on Designing Privacy Enhancing Technologies*, Springer-Verlag, pp. 46–66, 2001.

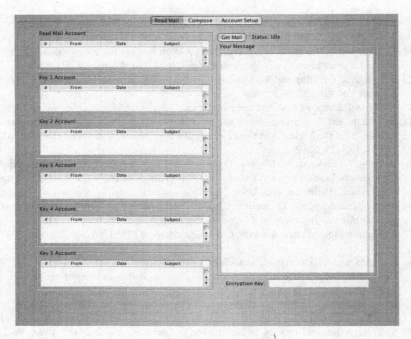

Fig. 16.4   A screenshot of the Aquinas client's window for reading email. The left-hand side of the panel displays the headers of waiting messages.

Appendix

Interface of Aquinas

The Aquinas GUI is written using the `javax.swing` libraries and is separated into three different panels. The first panel, shown in Figure 16.4, allows a user to view their email. The left-hand side of the panel contains the message and key email accounts that display the downloaded headers of awaiting messages. When a user clicks on a message, its contents are displayed in the frame on the right-hand side of the panel. As the user clicks on the key shares associated with a given message, the decoded contents of that message are displayed on the screen. Should the integrity of the message be altered while in transit, the content frame displays a message warning the user of the change.

The second panel, shown in Figure 16.5, provides users space to compose new emails. The fields in the upper-left portion of the panel the "To (Data):" field (where the email containing the hidden content is sent), the "To (Keys):" field (which specifies the comma-separated accounts to which key shares will be sent)

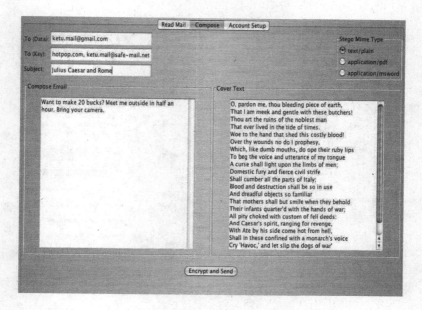

Fig. 16.5   A screenshot of the Aquinas client's window for composing new messages.

and a "Subject" field. The preferred methods of steganography are currently selectable through the radio buttons in the upper-right corner of the panel. The real message to be delivered to the sender is composed in the large window on the left-hand side of the panel. The cover text used to hide the content, is entered into the pane on the right. It should be noted that because SNOW[9] hides text within whitespace at the end of a message, no actual cover text is needed in the current version of Aquinas, i.e., the message will consist of nothing but whitespace. For the purpose of maintaining plausible deniability, however, including viable cover text in this window is still suggested. The selection of specific cover text is discussed in Section 16.4.

The third and final panel, shown in Figure 16.6, provides an interface for a user to enter the accounts through which email is delivered. In order to simplify this process, drop-down menus with account options including preset POP3 capable, web-based service (and their requisite information) are included. The actual process of registering for the accounts at multiple providers is left to the user; however, in order to facilitate this process, we list a number of these services with which free POP3 email access is provided[e]. Aquinas supports both SSL and un-

---

[e]These services include, but are by no means limited to Gmail (`www.gmail.com`), Hotmail (`www.hotmail.com`), HotPOP (`www.hotpop.com`), SAFe-mail (`www.safe-mail.net`)

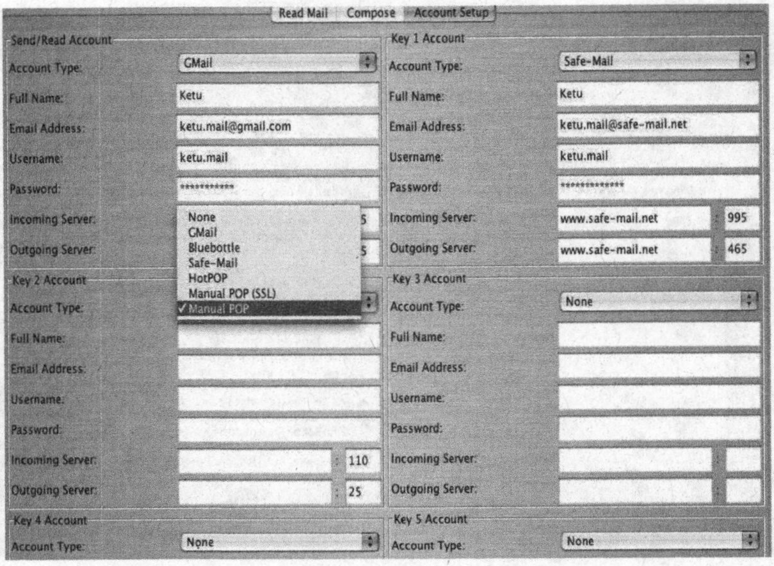

Fig. 16.6   A screenshot of the Aquinas client's window for account setup information.

encrypted connections to accounts. Additionally, Aquinas also allows users to save these settings and addresses into a password-encrypted cookie[f]. In so doing, only an initial setup is required and future use is not encumbered by having to remember multiple passwords.

The current implementation allows for up to five key accounts (and therefore five key shares) to be used to provide message confidentiality and integrity. As is discussed in detail in Section 16.4, there are security advantages to using additional accounts. Should the desire for a greater number of key accounts exist, simply modifying the MAX_KEYS constant and re-compiling the code automatically recreates the Aquinas GUI with the desired number of key accounts.

We conducted experiments to better understand the scanning mechanisms associated with each web-based email provider. Seeding messages with commercial keywords revealed that content within the "Subject" and "Body" fields of the email is harvested for creating targeted advertisements. Data in other fields including "To", "From" and miscellaneous X-Headers was not included in the scanning process. It was also discovered that the contents of attachments, regardless of the name of those attachments were not examined. Forged MIME types were

---

[f]This cookie can be stored on any web server and be pointed to so as to allow remote users the same ease of use.

similarly ignored. Plaintext messages hidden using the mechanisms in Aquinas were also tested against spam filtration. After extensive testing, all messages were delivered to the recipient address without being flagged.

At the time of writing, the mechanisms included with Aquinas are more than sufficient to preserve the privacy of both sender and receiver.